2ND EDITION

ANALYSIS

with an Introduction to Proof

Steven R. Lay
Aurora University

PRENTICE HALL, *Englewood Cliffs, New Jersey 07632*

Library of Congress Cataloging-in-Publication Data

Lay, Steven R.
Analysis with an introduction to proof / Steven R. Lay. -- 2nd ed.
p. cm.
Includes bibliographical references.
ISBN 0-13-033267-4
1. Mathematical analysis. 2. Proof theory. I. Title.
QA300.L428 1990
515--dc20 89-38271
CIP

Editorial/production supervision: Merrill Peterson
Interior design: Joan Stone
Cover design: Edsal Enterprises
Manufacturing buyer: Paula Massenaro

Printed in the United States of America
10 9 8 7 6 5

ISBN 0-13-033267-4

Prentice-Hall International (UK) Limited, *London*
Prentice-Hall of Australia Pty. Limited, *Sydney*
Prentice-Hall Canada Inc., *Toronto*
Prentice-Hall Hispanoamericana, S.A., *Mexico*
Prentice-Hall of India Private Limited, *New Delhi*
Prentice-Hall of Japan, Inc., *Tokyo*
Simon & Schuster Asia Pte. Ltd., *Singapore*
Editora Prentice-Hall do Brasil, Ltda., *Rio de Janeiro*

Analysis

Contents

†Optional section

Preface

The first edition of this book was titled *Analysis: An Introduction to Proof.* It identified the main goals as being a study of analysis in a context that pays special attention to developing the students' ability to read and write proofs. The emphasis of this second edition remains the same. The slight change in the title is intended to indicate more clearly the balance between analysis and the study of proofs, with the main weight being placed on analysis.

The substantive changes in the text itself for the second edition are threefold:

1. The material on the topology of $\mathbb{R}$ has been expanded somewhat and divided into two sections. This makes for an easier transition into these fundamental topics
2. An optional section on metric spaces has been added at the end of Chapter 3. This enables us to study topological properties in a new setting that is rich in applications and variations. In particular, it is helpful to be able to illustrate neighborhoods in two and three dimensions.
3. An optional section on continuity in metric spaces has been added at the end of Chapter 5. It depends on the earlier optional section in Chapter 3.

The overall organization of the book remains the same. The first chapter takes a careful (albeit nontechnical) look at the laws of logic and then examines how these laws are used in the structuring of mathematical arguments. The second chapter discusses the two main foundations of analysis: sets and functions. This provides an elementary setting in which to practice the techniques encountered in the previous chapter.

Chapter 3 develops the properties of the real number $\mathbb{R}$ as a complete ordered field and introduces the topological concepts of neighborhoods, open sets, closed sets, and compact sets. The remaining chapters cover the topics usually included in an analysis of functions of a real variable: sequences, continuity, differentiation, integration, and series.

Throughout the book, attention has been given to making the text readable and understandable to the student. There are numerous examples and 40 figures that illustrate the concepts being explained. To encourage student interaction with the material, nearly a hundred practice problems have been included. Generally, these problems are not very difficult, and it is intended that students should stop to work them as they read. The answers are given at the end of each section just prior to the exercises. The student should also be encouraged to read (if not attempt) most of the exercises. They are viewed as an integral part of the text and vary in difficulty from the routine to the challenging. Those exercises that are used in a later section are marked with an asterisk. Hints for many of the exercises are included at the back of the book. These hints should be used only after a serious attempt to solve an exercise has proved futile.

The text has been written in a way designed to provide flexibility in the pacing of topics. If only one term is available, the first chapter can be assigned as outside reading. Chapter 2 and the first half of Chapter 3 can be covered quickly, again with much of the reading being left to the student. By so doing, the remainder of the book can be covered adequately in a single semester. Alternatively, depending on the background and interests of the students, one can concentrate on developing the first five chapters in some detail. By placing a greater emphasis on the early material, the text can be used in a "transitional" course whose main goal is to teach mathematical reasoning and to illustrate its use in developing an abstract structure. It is also possible to skip derivatives and integrals and go directly to series, since the only results needed from these two chapters will be familiar to the student from beginning calculus. A thorough treatment of the whole book would require two quarters.

I appreciate the helpful comments that I have received from users of the first edition. In particular, I would like to thank Professors Paul McGuire, Roger B. Nelson, Thomas A. Metzger, Daniel Sweet, Brian M. Scott, Juan Gatica, and Peter Colwell. I am also grateful to Craig Borkowf, a student at the University of Chicago, for his numerous suggestions.

Steven R. Lay
Aurora, Illinois

Analysis

1

Logic and Proof

To be able to understand mathematics and mathematical arguments, it is necessary to have a solid understanding of logic and the way in which known facts can be combined to prove new facts. Although many people consider themselves to be logical thinkers, the thought patterns developed in everyday living are only suggestive of and not totally adequate for the precision required in mathematics. In this first chapter we take a careful look at the rules of logic and the way in which mathematical arguments are constructed. Section 1 presents the logical connectives that enable us to build compound statements from simpler ones. In Section 2 we discuss the role of quantifiers. In Sections 3 and 4 we analyze the structure of mathematical proofs and illustrate the various proof techniques by means of examples.

Section 1 LOGICAL CONNECTIVES

The language of mathematics consists primarily of declarative sentences. If a sentence can be classified as true or false, it is called a **statement**. The truth or falsity of a statement is known as its truth value. For a sentence to be a statement, it is not necessary that we actually know whether it is true or false, but it must clearly be the case that it is one or the other.

1.1 EXAMPLES Consider the following sentences.

(a) Two plus two equals four.
(b) Every continuous function is differentiable.
(c) $x^2 - 5x + 6 = 0$.
(d) A circle is the only convex set in the plane that has the same width in each direction.
(e) If n is an integer greater than 2, then $x^n + y^n = z^n$ has no positive integral solutions.

Sentences (a) and (b) are statements since (a) is true and (b) is false. Sentence (c), on the other hand, is true for some x and false for others. If we have a particular context in mind, then (c) will be a statement. In Section 2 we shall see how to remove this ambiguity. Sentences (d) and (e) are more difficult. You may or may not know whether they are true or false, but it is certain that each sentence must be one or the other. Thus (d) and (e) are both statements. [It turns out that (d) can be shown to be false, and the truth value of (e) has not yet been established.]

1.2 PRACTICE Which of the following sentences are statements?

(a) If x is a real number, then $x^2 \geqslant 0$.
(b) Seven is a prime number.
(c) Seven is an even number.
(d) This sentence is false.

In studying mathematical logic we shall not be concerned with the truth value of any particular simple statement. To be a statement, it must be either true or false (and not both), but it is immaterial which condition applies. What will be important is how the truth value of a compound statement is determined by the truth values of its simpler parts.

In everyday English conversation we have a variety of ways to change or combine statements. A simple statement† like

It is windy.

can be negated to form the statement

It is *not* windy.

† It may be questioned whether or not the sentence "It is windy" is a statement since the term "windy" is so vague. If we *assume* that "windy" is given a precise definition, then in a particular place at a particular time, "It is windy" will be a statement. It is customary to assume precise definitions when we use descriptive language in an example. This problem does not arise in a mathematical context because the definitions *are* precise.

The compound statement

It is windy *and* the waves are high.

is made up of two parts: "It is windy" and "The waves are high." These two parts can also be combined in other ways. For example,

It is windy *or* the waves are high.
If it is windy, *then* the waves are high.
It is windy *if and only if* the waves are high.

The italicized words above (*not*, *and*, *or*, *if ... then*, *if and only if*) are called **sentential connectives**. Their use in mathematical writing is similar to (but not identical with) their everyday usage. To remove any possible ambiguity, we shall look carefully at each and specify its precise mathematical meaning.

Let p stand for a given statement. Then $\sim p$ (read *not* p) represents the logical opposite (**negation**) of p. When p is true, then $\sim p$ is false; when p is false, then $\sim p$ is true. This can be summarized in a truth table:

p	$\sim p$
T	F
F	T

where T stands for true and F stands for false.

1.3 EXAMPLE Let p, q, and r be given as follows:

p: Today is Monday.
q: Five is an even number.
r: The set of integers is countable.

Then their negations can be written as

$\sim p$: Today is not Monday.
$\sim q$: Five is not an even number.
or
Five is an odd number.
$\sim r$: The set of integers is not countable.
or
The set of integers is uncountable.

The connective *and* is used in logic in the same way as it is in ordinary language. If p and q are statements, then the statement p *and* q (called the **conjunction** of p and q and denoted by $p \wedge q$) is true only when both p and q are true and it is false otherwise.

1.4 PRACTICE Complete the truth table for $p \wedge q$. Note that we have to use four lines in this table to include all possible combinations of truth values of p and q.

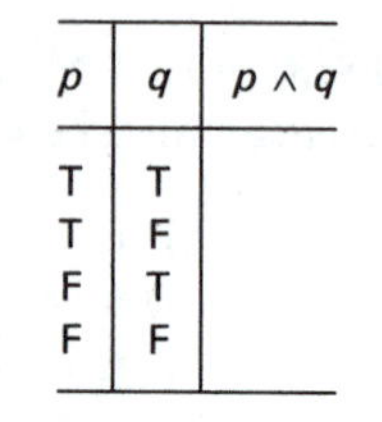

p	q	$p \wedge q$
T	T	
T	F	
F	T	
F	F	

The connective *or* is used to form a compound statement known as a **disjunction**. In common English the word *or* can have two meanings. In the sentence

We are going to paint our house yellow or green.

the intended meaning is *yellow or green but not both.* This is known as the exclusive meaning of *or.* On the other hand, in the sentence

Do you want cake or ice cream for dessert?

the intended meaning may include the possibility of having both. This inclusive meaning is the *only* way the word *or* is used in logic. Thus, if we denote the disjunction *p or q* by $p \vee q$, we have the following truth table:

p	q	$p \vee q$
T	T	T
T	F	T
F	T	T
F	F	F

A statement of the form

If p, then q.

is called an **implication** or a **conditional** statement. The if-statement p in the implication is called the **antecedent** and the then-statement q is called the **consequent**. To decide on an appropriate truth table for implication, let us consider the following sentence:

If it stops raining by Saturday, then I will go to the football game.

If a friend made a statement like this, under what circumstances could you call him a liar? Certainly, if the rain stops and he doesn't go, then he did not tell the truth. But what if the rain doesn't stop? He hasn't said what he will do then, so whether he goes or not, either is all right.

Although it might be argued that other interpretations make equally good sense, mathematicians have agreed that an implication will be called false only when the antecedent is true and the consequent if false. If we denote the implication "if p, then q" by $p \Rightarrow q$, we obtain the following table:

p	q	$p \Rightarrow q$
T	T	T
T	F	F
F	T	T
F	F	T

It is important to recognize that in mathematical writing the conditional statement can be disguised in several equivalent forms. Thus the following expressions all mean exactly the same thing:

if p, then q	q provided that p
p implies q	q whenever p
p only if q	p is a sufficient condition for q
q if p	q is a necessary condition for p

1.5 PRACTICE Identify the antecedent and the consequent in each of the following statements.

(a) If n is an integer, then $2n$ is an even number.
(b) You can work here only if you have a college degree.
(c) The car will not run whenever you are out of gas.
(d) Continuity is a necessary condition for differentiability.

The statement "p if and only if q" is the conjunction of the two implications $p \Rightarrow q$ and $q \Rightarrow p$. A statement in this form is called an **equivalence** and is denoted by $p \Leftrightarrow q$. In written form the abbreviation "iff" is frequently used instead of "if and only if." The truth table for equivalence can be obtained by analyzing the compound statement $(p \Rightarrow q) \wedge (q \Rightarrow p)$ a step at a time.

p	q	$p \Rightarrow q$	$q \Rightarrow p$	$(p \Rightarrow q) \wedge (q \Rightarrow p)$
T	T	T	T	T
T	F	F	T	F
F	T	T	F	F
F	F	T	T	T

Thus we see that $p \Leftrightarrow q$ is true precisely when p and q have the same truth values.

1.6 PRACTICE Construct a truth table for each of the following compound statements.

(a) $\sim(p \wedge q) \Leftrightarrow [(\sim p) \vee (\sim q)]$
(b) $\sim(p \vee q) \Leftrightarrow [(\sim p) \wedge [(\sim q)]$
(c) $\sim(p \Rightarrow q) \Leftrightarrow [p \wedge (\sim q)]$

In Practice 1.6 we find that each of the compound statements is true in all cases. Such a statement is called a **tautology**. We shall encounter many more tautologies in the next few sections. They are very useful in changing a statement from one form into an equivalent statement in a different (one hopes simpler) form. In 1.6(a) we see that the negation of a conjunction is the disjunction of the negations. Similarly, in 1.6(b) we learn that the negation of a disjunction is the conjunction of the negations. In 1.6(c) we find that the negation of an implication is *not* another implication, but rather is the conjunction of the antecedent and the negation of the consequent.

1.7 EXAMPLE Using Practice 1.6(a), we see that the negation of

The set S is compact and convex.

can be written as

Either the set S is not compact or it is not convex.

This example also illustrates that using equivalent forms in logic does not depend on knowing the meaning of the terms involved. It is the form of the statement that is important. Whether or not we happen to know the definition of "compact" and "convex" is of no consequence in forming the negation above.

1.8 PRACTICE Use the tautologies in Practice 1.6 to write out a negation of each statement.

(a) Seven is prime or $2 + 2 = 4$.
(b) If M is bounded, then M is compact.
(c) If roses are red and violets are blue, then I love you.

ANSWERS TO PRACTICE PROBLEMS

1.2 (a), (b), and (c)

1.4

p	q	$p \wedge q$
T	T	T
T	F	F
F	T	F
F	F	F

1.5 (a) Antecedent: n is an integer
Consequent: $2n$ is an even number
(b) Antecedent: you can work here
Consequent: you have a college degree
(c) Antecedent: you are out of gas
Consequent: the car will not run
(d) Antecedent: differentiability
Consequent: continuity

1.6 Sometimes we condense a truth table by writing the truth values under the part of a compound expression to which they apply.

(a)

p	q	$\sim(p \wedge q)$		$\Leftrightarrow$	$[(\sim p)$	$\vee$	$(\sim q)]$
T	T	F	T	T	F	F	F
T	F	T	F	T	F	T	T
F	T	T	F	T	T	T	F
F	F	T	F	T	T	T	T

(b)

p	q	$\sim(p \vee q)$		$\Leftrightarrow$	$[(\sim p)$	$\wedge$	$(\sim q)]$
T	T	F	T	T	F	F	F
T	F	F	T	T	F	F	T
F	T	F	T	T	T	F	F
F	F	T	F	T	T	T	T

(c)

p	q	$\sim(p \Rightarrow q)$		$\Leftrightarrow$	$[p$	$\wedge$	$(\sim q)]$
T	T	F	T	T	T	F	F
T	F	T	F	T	T	T	T
F	T	F	T	T	F	F	F
F	F	T	T	T	F	F	T

1.8 (a) Seven is not prime and $2 + 2 \neq 4$.
(b) M is bounded and M is not compact.
(c) Roses are red and violets are blue, but I do not love you.

EXERCISES

1.1 Write the negation of each statement.
(a) H is a normal subgroup.
(b) The set of real numbers is finite.
(c) Bob and Bill are over 6 feet tall.
(d) Seven is prime or five is even.
(e) If today is not Monday, then it is hot.
(f) If K is closed and bounded, then K is compact.

1.2 Write the negation of each statement.

(a) M is an orthogonal matrix.
(b) G is normal and H is not regular.
(c) Today is Wednesday or it is snowing.
(d) Bob and Betty are related.
(e) If it rains today, then the roof will leak.
(f) If K is compact, then K is closed and bounded.

1.3 Identify the antecedent and the consequent in each statement.

(a) I will raise the flag provided that I get there first.
(b) Normality is a sufficient condition for regularity.
(c) You can climb the mountain only if you have the nerve.
(d) If $x = 5$, then $f(x) = 2$.

1.4 Identify the antecedent and the consequent in each statement.

(a) I get sleepy in class whenever I stay up late.
(b) Two real symmetric matrices are congruent if they have the same rank and the same signature.
(c) A real sequence is Cauchy only if it is convergent.
(d) $f(x) = 5$ provided that $x > 3$.

1.5 Construct a truth table for each statement.

(a) $p \Rightarrow \sim q$
(b) $[p \wedge (p \Rightarrow q)] \Rightarrow q$
(c) $[p \Rightarrow (q \wedge \sim q)] \Leftrightarrow \sim p$

1.6 Construct a truth table for each statement.

(a) $\sim p \vee q$
(b) $p \wedge \sim p$
(c) $[\sim q \wedge (p \Rightarrow q] \Rightarrow \sim p$

1.7 Indicate whether each statement is true or false.

(a) $5 > 3$ and 4 is even.
(b) 8 is prime or $4 < 9$.
(c) 7 is even or 6 is prime.
(d) If $3 < 5$, then $7^2 = 49$.
(e) If $3 > 5$, then $7^2 < 49$.
(f) If 5 is prime, then $5^2 = 20$.
(g) If 6 is odd or 4 is even, then $4 > 5$.
(h) If $3 < 7$ implies that $5 > 9$, then 8 is prime.

1.8 Indicate whether each statement is true or false.

(a) 7 is prime and 5 is even.
(b) 7 is prime or 5 is even.
(c) $2 > 4$ or 6 is odd.
(d) If 3 is prime, then $2 + 2 = 5$.
(e) If $2 + 2 = 5$, then 3 is prime.
(f) If π is rational, then 3 is even.
(g) If $3 > 5$ and 4 is even, then $5^2 = 25$.
(h) If $7 < 5$ only if 6 is even, then 8 is odd.

1.9 Use truth tables to verify that each of the following is a tautology. Parts (a) and (b) are called *commutative laws*, parts (c) and (d) are *associative laws*, and parts (e) and (f) are *distributive laws*.

(a) $(p \wedge q) \Leftrightarrow (q \wedge p)$
(b) $(p \vee q) \Leftrightarrow (q \vee p)$
(c) $[p \wedge (q \wedge r)] \Leftrightarrow [(p \wedge q) \wedge r]$
(d) $[p \vee (q \vee r)] \Leftrightarrow [(p \vee q) \vee r]$
(e) $[p \wedge (q \vee r)] \Leftrightarrow [(p \wedge q) \vee (p \wedge r)]$
(f) $[p \vee (q \wedge r)] \Leftrightarrow [(p \vee q) \wedge (p \vee r)]$

Section 2 QUANTIFIERS

In Section 1 we found that the sentence

$$x^2 - 5x + 6 = 0$$

needed to be considered within a particular context in order to become a statement. When a sentence involves a variable such as x, it is customary to use functional notation when referring to it. Thus we write

$$p(x): x^2 - 5x + 6 = 0$$

to indicate that $p(x)$ is the sentence "$x^2 - 5x + 6 = 0$." For a specific value of x, $p(x)$ becomes a statement that is either true or false. For example, $p(2)$ is true and $p(4)$ is false.

Another way to remove the ambiguity of $p(x)$ is by using a quantifier. The sentence

$$\text{For every } x, x^2 - 5x + 6 = 0.$$

is a statement since it is false. In symbols we write

$$\forall x, p(x),$$

where the **universal quantifier** $\forall$ is read, "For every...," "For all...," "For each...," or a similar equivalent phrase. The sentence

$$\text{There exists an } x \text{ such that } x^2 - 5x + 6 = 0.$$

is also a statement, and it is true. In symbols we write

$$\exists x \ni p(x),$$

where the **existential quantifier** $\exists$ is read, "There exists...," "There is at least one...," or something equivalent. The symbol $\ni$ is just a shorthand notation for the phrase "such that."

2.1 EXAMPLE The statement

There exists a number less than 7.

can be written

$$\exists\, x \ni x < 7$$

or in the abbreviated form

$$\exists\, x < 7,$$

where it is understood that x is to represent a number. Sometimes the quantifier is not explicitly written down, as in the statement

If x is greater than 1, then x^2 is greater than 1.

The intended meaning is

$$\forall\, x, \text{ if } x > 1, \text{ then } x^2 > 1.$$

In general, if a variable is used in the antecedent of an implication without being quantified, then the universal quantifier is assumed to apply.

2.2 PRACTICE Rewrite each statement using $\exists$, $\forall$, and $\ni$, as appropriate.

(a) There exists a positive number x such that $x^2 = 5$.
(b) For every positive number M there is a positive number N such that $N < 1/M$.
(c) If $n \geqslant N$, then $|f_n(x) - f(x)| \leqslant 3$ for all x in A.

Having seen several examples of how existential and universal quantifiers are used, let us now consider how quantified statements are negated. Consider the statement

Everyone in the room is awake.

What condition must apply to the people in the room in order for the statement to be false? Must everyone be asleep? No, it is sufficient that at least one person be asleep. On the other hand, in order for the statement

Someone in the room is asleep.

to be false, it must be the case that everyone is awake. Symbolically, if

$p(x)$: x is awake,

then

$$\sim(\forall\, x, p(x)) \Leftrightarrow \exists\, x \ni \sim p(x).$$

Similarly,

$$\sim(\exists\, x \ni p(x)) \Leftrightarrow \forall\, x, \sim p(x).$$

2.3 EXAMPLES Let us look at several more quantified statements and derive their negations. Notice in part (b) that the inequality "$0 < g(y) \leqslant 1$" is a conjunction of two inequalities "$0 < g(y)$" and "$g(y) \leqslant 1$." Thus its negation is a disjunction. In a complicated statement like (c) it is helpful to work through the negation one step at a time. Fortunately, (c) is about as messy as it will get.

(a) Statement: For every x in A, $f(x) > 5$.

$$\forall\, x \text{ in } A,\ f(x) > 5.$$

Negation: $\exists\, x$ in $A \ni f(x) \leqslant 5$.

There is an x in A such that $f(x) \leqslant 5$.

(b) Statement: There exists a positive number y such that $0 < g(y) \leqslant 1$.

$$\exists\, y > 0 \ni 0 < g(y) \leqslant 1.$$

Negation: $\forall\, y > 0$, $g(y) \leqslant 0$ or $g(y) > 1$.

For every positive number y, either $g(y) \leqslant 0$ or $g(y) > 1$.

(c) Statement:

$$\forall\, \varepsilon > 0\ \exists\, N \ni \forall\, n, \text{ if } n \geqslant N, \text{ then } \forall\, x \text{ in } S, |f_n(x) - f(x)| < \varepsilon$$

Negation:

$$\exists\, \varepsilon > 0 \ni \sim[\exists\, N \ni \forall\, n, \text{ if } n \geqslant N, \text{ then } \forall\, x \text{ in } S, |f_n(x) - f(x)| < \varepsilon]$$

or

$$\exists\, \varepsilon > 0 \ni \forall\, N, \sim[\forall\, n, \text{ if } n \geqslant N, \text{ then } \forall\, x \text{ in } S, |f_n(x) - f(x)| < \varepsilon]$$

or

$$\exists\, \varepsilon > 0 \ni \forall\, N\ \exists\, n \ni \sim[\text{if } n \geqslant N, \text{ then } \forall\, x \text{ in } S, |f_n(x) - f(x)| < \varepsilon]$$

or

$$\exists\, \varepsilon > 0 \ni \forall\, N\ \exists\, n \ni n \geqslant N \text{ and } \sim[\forall\, x \text{ in } S, |f_n(x) - f(x)| < \varepsilon]$$

or

$$\exists\, \varepsilon > 0 \ni \forall\, N\ \exists\, n \ni n \geqslant N \text{ and } \exists\, x \text{ in } S \ni |f_n(x) - f(x)| \geqslant \varepsilon$$

2.4 PRACTICE Write the negation of each statement in Practice 2.2.

It is important to realize that the order in which quantifiers are used affects the truth value. For example, when talking about real numbers, the statement

$$\forall\, x\ \exists\, y \ni y > x$$

is true. That is, given any real number x there is always a real number y that is bigger than that x. But the statement

$$\exists\, y \ni \forall\, x,\ y > x$$

is false, since there is no fixed real number y that is bigger than every real number. Thus care must be taken when reading (and writing) quantified statements so that the order of the quantifiers is not inadvertently changed.

ANSWERS TO PRACTICE PROBLEMS

2.2 (a) $\exists\, x > 0 \ni x^2 = 5$.

(b) $\forall\, M > 0\ \exists\, N > 0 \ni N < \dfrac{1}{M}$.

(c) $\forall\, n$, if $n \geqslant N$, then $\forall\, x$ in A, $|f_n(x) - f(x)| \leqslant 3$.

2.4 (a) $\forall\, x > 0,\ x^2 \neq 5$.

(b) $\exists\, M > 0 \ni \forall\, N > 0,\ N \geqslant \dfrac{1}{M}$.

(c) $\exists\, n \ni n \geqslant N$ and $\exists\, x$ in $A \ni |f_n(x) - f(x)| > 3$.

EXERCISES

2.1 Write the negation of each statement.

(a) Some pencils are blue.
(b) All chairs have four legs.
(c) $\exists\, x > 1 \ni f(x) = 3$.
(d) $\forall\, x$ in A, $\exists\, y$ in $B \ni x < y < 1$.
(e) $\forall\, x\ \exists\, y \ni \forall\, z,\ x + y + z \leqslant xyz$.

2.2 Write the negation of each statement.

(a) Everyone likes Bob.
(b) All students on the basketball team are smart.
(c) $\exists\, x$ in $A \ni f(x) > y$.
(d) $\exists\, y \leqslant 2 \ni f(y) < 2$ or $g(y) \geqslant 7$.
(e) $\forall\, x > 1,\ 0 < f(x) < 4$.

2.3 Determine the truth value of each statement, assuming that x, y and z are real numbers.

(a) $\exists\, x \ni \forall\, y\ \exists\, z \ni x + y = z$.
(b) $\exists\, x \ni \forall\, y$ and $\forall\, z,\ x + y = z$.
(c) $\forall\, x$ and $\forall\, y$, $\exists\, z \ni xz = y$.
(d) $\exists\, x \ni \forall\, y$ and $\forall\, z$, $z > y$ implies that $z > x + y$.
(e) $\forall\, x$, $\exists\, y$ and $\exists\, z \ni z > y$ implies that $z > x + y$.

2.4 Determine the truth value of each statement, assuming that x, y, and z are real numbers.

(a) $\forall\, x$ and $\forall\, y$, $\exists\, z \ni x + y = z$.
(b) $\forall\, x\, \exists\, y \ni \forall\, z$, $x + y = z$.
(c) $\exists\, x \ni \forall\, y$, $\exists\, z \ni xz = y$.
(d) $\forall\, x\, \exists\, y \ni \forall\, z$, $z > y$ implies that $z > x + y$.
(e) $\forall\, x$ and $\forall\, y$, $\exists\, z \ni z > y$ implies that $z > x + y$.

Exercises 2.5 to 2.12 give certain properties of functions that we shall encounter later in the text. You are to do two things: (a) rewrite the defining condition in logical symbolism using $\forall$, $\exists$, $\ni$, and $\Rightarrow$, as appropriate; and (b) write the negation of part (a) using the same symbolism. It is not necessary that you understand precisely what each term means.

2.5 A function f is *even* iff, for every x, $f(-x) = f(x)$.

2.6 A function f is *periodic* iff there exists a $k > 0$ such that, for every x, $f(x + k) = f(x)$.

2.7 A function f is *increasing* iff for every x and for every y, if $x \leqslant y$, then $f(x) \leqslant f(y)$.

2.8 A function is *strictly decreasing* iff for every x and for every y, if $x < y$, then $f(x) > f(y)$.

2.9 A function $f: A \to B$ is *injective* iff for every x and y in A, if $f(x) = f(y)$, then $x = y$.

2.10 A function $f: A \to B$ is *surjective* iff for every y in B there exists an x in A such that $f(x) = y$.

2.11 A function $f: D \to R$ is *continuous* at $c \in D$ iff for every $\varepsilon > 0$ there is a $\delta > 0$ such that $|f(x) - f(c)| < \varepsilon$ whenever $|x - c| < \delta$ and $x \in D$.

2.12 A function f is *uniformly continuous on a set* S iff for every $\varepsilon > 0$ there exists a $\delta > 0$ such that $|f(x) - f(y)| < \varepsilon$ whenever x and y are in S and $|x - y| < \delta$.

2.13 The real number L is the *limit* of the function $f: D \to R$ at the point c iff for each $\varepsilon > 0$ there exists a $\delta > 0$ such that $|f(x) - L| < \varepsilon$ whenever $x \in D$ and $0 < |x - c| < \delta$.

2.14 Consider the following sentences:

(a) The nucleus of a carbon atom consists of protons and neutrons.
(b) Jesus Christ rose from the dead and is alive today.
(c) Every differentiable function is continuous.

Each of these sentences has been affirmed by some people at some time as being "true." Write an essay on the nature of truth, comparing and contrasting its meaning in these (and possibly other) contexts. You might also want to consider some of the following questions: To what extent is truth absolute? To what extent can truth change with time? To what extent is truth based on opinion? To what extent are people free to accept as true anything they wish?

Section 3 TECHNIQUES OF PROOF: I

In the first two sections we introduced some of the vocabulary of logic and mathematics. Our aim is to be able to read and write mathematics, and this requires more than just vocabulary. It also requires syntax. That is, we need to understand how statements are combined to form that mysterious mathematical entity known as a proof. Since this topic tends to be intimidating to many students, let us ease into it gently by first considering the two main types of logical reasoning: inductive reasoning and deductive reasoning.

3.1 EXAMPLE Consider the function $f(n) = n^2 + n + 17$. If we evaluate this function for various positive integers, we observe that we always seem to obtain a prime number. (Recall that a positive integer n is prime if $n > 1$ and its only positive divisors are 1 and n.) For example,

$$f(1) = 19$$
$$f(2) = 23$$
$$f(3) = 29$$
$$f(4) = 37$$
$$\vdots$$
$$f(8) = 89$$
$$\vdots$$
$$f(12) = 173$$
$$\vdots$$
$$f(15) = 257$$

and all these numbers (as well as the ones skipped over) are prime. On the basis of this experience we might conjecture that the function $f(n) = n^2 + n + 17$ will always produce prime numbers when n is a positive integer. Drawing a conclusion of this sort is an example of **inductive** reasoning. On the basis of looking at individual cases we make a general conclusion.

If we let $p(n)$ be the sentence "$n^2 + n + 17$ is a prime number" and we understand that n refers to a positive integer, then we can ask, is

$$\forall\, n,\ p(n)$$

a true statement? Have we proved it is true?

It is important to realize that indeed we have *not* proved that it is true. We have shown that

$$\exists\, n \ni p(n)$$

is true. In fact, we know that $p(n)$ is true for many n. But we have not proved that it is true for *all* n. How can we come up with a proof? It turns out that we cannot, since the statement "$\forall\, n,\ p(n)$" happens to be false.

How do we know that it is false? We know that it is false because we can think of an example where $n^2 + n + 17$ is not prime. (Such an example is called a **counterexample**.) One such counterexample is $n = 17$:

$$17^2 + 17 + 17 = 17 \cdot 19.$$

There are others as well. For example, when $n = 16$,

$$16^2 + 16 + 17 = 16(16 + 1) + 17 = 17^2,$$

but it only takes one counterexample to prove that "$\forall\, n,\ p(n)$" is false.

On the basis of Example 3.1 we might infer that inductive reasoning is of little value. Although it is true that the conclusions drawn from inductive reasoning have not been proved logically, they can be very useful. Indeed, this type of reasoning is the basis for most if not all scientific experimentation. It is also often the source of the conjectures that when proved become the theorems of mathematics.

3.2 PRACTICE Provide counterexamples to the following statements.

(a) All birds can fly.
(b) Every continuous function is differentiable.

3.3 EXAMPLE Consider the function $g(n, m) = n^2 + n + m$, where n and m are understood to be positive integers. In Example 3.1 we saw that $g(16, 17) = 16^2 + 16 + 17 = 17^2$. We might also observe that

$$g(1, 2) = 1^2 + 1 + 2 = 4 = 2^2$$
$$g(2, 3) = 2^2 + 2 + 3 = 9 = 3^2$$
$$\vdots$$
$$g(5, 6) = 5^2 + 5 + 6 = 36 = 6^2$$
$$\vdots$$
$$g(12, 13) = 12^2 + 12 + 13 = 169 = 13^2.$$

On the basis of these examples (using inductive reasoning) we can form the conjecture "$\forall\, n,\ q(n)$," where $q(n)$ is the statement

$$g(n, n + 1) = (n + 1)^2.$$

It turns out that our conjecture this time is true, and we can prove it. Using the familiar laws of algebra, we have

$$\begin{aligned} g(n, n+1) &= n^2 + n + (n+1) && [\text{definition of } g(n, n+1)] \\ &= n^2 + 2n + 1 && [\text{since } n + n = 2n] \\ &= (n+1)(n+1) && [\text{by factoring}] \\ &= (n+1)^2 && [\text{definition of } (n+1)^2]. \end{aligned}$$

Since our reasoning at each step does not depend on n being any specific integer, we conclude that "$\forall\, n, q(n)$" is true.

Now that we have proved the general statement "$\forall\, n, q(n)$," we can apply it to any particular case. Thus we know that

$$g(124, 125) = 125^2$$

without having to do any computation. This is an example of **deductive** reasoning: applying a general principle to a particular solution. Most of the proofs encountered in mathematics are based on this type of reasoning.

3.4 PRACTICE In what way was deductive reasoning used to prove $\forall\, n, q(n)$?

The most common type of mathematical theorem can be symbolized as $p \Rightarrow q$, where p and q may be compound statements. To assert that $p \Rightarrow q$ is a theorem is to claim that $p \Rightarrow q$ is a tautology; that is, that it is always true. From Section 1 we know that $p \Rightarrow q$ is true unless p is true and q is false. Thus, to prove that p implies q, we have to show that whenever p is true it follows that q must be true. When an implication $p \Rightarrow q$ is identified as a theorem, it is customary to refer to p as the **hypothesis** and q as the **conclusion**.

The construction of a proof of the implication $p \Rightarrow q$ can be thought of as building a bridge of logical statements to connect the hypothesis p with the conclusion q. The building blocks that go into the bridge consist of four kinds of statements: (1) definitions, (2) assumptions or axioms that are accepted as true, (3) theorems that have previously been established as true, and (4) statements that are logically implied by the earlier statements in the proof. When actually building the bridge, it may not be at all obvious which blocks to use or in what order to use them. This is where experience is helpful, together with perseverance, intuition, and sometimes a good bit of luck.

In building a bridge from the hypothesis p to the conclusion q, it is often useful to start at both ends and work toward the middle. That is, we might begin by asking, "What must I know in order to conclude that q is true?" Call this q_1. Then ask, "What must I know to conclude that q_1 is true?" Call this q_2. Continue this process as long as it is productive, thus obtaining a sequence of implications,

$$\cdots \Rightarrow q_2 \Rightarrow q_1 \Rightarrow q.$$

Then look at the hypothesis p and ask, "What can I conclude from p that will lead me toward q? Call this p_1.Then ask, "What can I conclude from p_1?" Continue this process as long as it is productive, thus obtaining

$$p \Rightarrow p_1 \Rightarrow p_2 \Rightarrow \cdots$$

We hope that at some point the part of the bridge leaving p will join with the part that arrives at q, forming a complete span:

$$p \Rightarrow p_1 \Rightarrow p_2 \Rightarrow \cdots \Rightarrow q_2 \Rightarrow q_1 \Rightarrow q.$$

3.5 EXAMPLE Let us return to the result proved in Example 3.3 to illustrate the process just described. We begin by writing the theorem in the form $p \Rightarrow q$. One way of doing this is as follows: "If $g(n, m) = n^2 + n + m$, then $g(n, n + 1) = (n + 1)^2$." Symbolically, we identify the hypothesis

$$p\colon g(n, m) = n^2 + n + m$$

and the conclusion

$$q\colon g(n, n + 1) = (n + 1)^2.$$

In asking what statement will imply q, there are many answers. One simple answer is to use the definition of a square and let

$$q_1\colon g(n, n + 1) = (n + 1)\,(n + 1).$$

By multiplying out the product $(n + 1)(n + 1)$, we obtain

$$q_2\colon g(n, n + 1) = n^2 + 2n + 1.$$

Now certainly $q_2 \Rightarrow q_1 \Rightarrow q$, but it is not clear how we might back up further. Thus we turn to the hypothesis p and ask what we can conclude. Since we wish to know something about $g(n, n + 1)$, the first step is to use the definition of g. That is, let

$$p_1\colon g(n, n + 1) = n^2 + n + (n + 1).$$

It is clear that $p_1 \Rightarrow q_2$, so the complete bridge is now formed:

$$p \Rightarrow p_1 \Rightarrow q_2 \Rightarrow q_1 \Rightarrow q.$$

This is essentially what was written in Example 3.3.

Associated with an implication $p \Rightarrow q$ there is a related implication $\sim q \Rightarrow \sim p$, called the **contrapositive**. It is easy to see using a truth table that an implication and its contrapositive are logically equivalent. Thus one way of proving an implication is to prove its contrapositive.

3.6 PRACTICE (a) Use a truth table to verify that $p \Rightarrow q$ and $\sim q \Rightarrow \sim p$ are logically equivalent.
(b) Is $p \Rightarrow q$ logically equivalent to $q \Rightarrow p$?

3.7 EXAMPLE The contrapositive of the theorem "If $7m$ is an odd number, then m is an odd number" is "If m is not an odd number, then $7m$ is not an odd number" or, equivalently, "If m is an even number, then $7m$ is an even number." (Recall that a number m is even if it can be written as $2k$ for some integer k. If a number is not even, then it is odd. It is to be understood here that we are talking about integers.) Using the contrapositive, we can construct a proof of the theorem as follows:

Hypothesis: m is an even number.

$m = 2k$ for some integer k	[definition]
$7m = 7(2k)$	[known property of multiplication]
$7m = 2(7k)$	[known property of multiplication]
$7k$ is an integer	[since k is an integer].

Conclusion: $7m$ is an even number
[since $7m$ is 2 times the integer $7k$].

This is much easier than trying to show directly that $7m$ being odd implies that m is odd.

3.8 PRACTICE Write the contrapositive of each implication in Practice 1.5.

In Practice 3.6(b) we saw that $p \Rightarrow q$ is not logically equivalent to $q \Rightarrow p$. The implication $q \Rightarrow p$ is called the **converse** of $p \Rightarrow q$. It is possible for an implication to be false, while its converse is true. Thus we cannot prove $p \Rightarrow q$ by showing $q \Rightarrow p$.

3.9 EXAMPLE The implication "If $m^2 > 0$, then $m > 0$" is false, but its converse "If $m > 0$, then $m^2 > 0$" is true.

3.10 PRACTICE Write the converse of each implication in Practice 1.5.

Another implication that is closely related to $p \Rightarrow q$ is the **inverse** $\sim p \Rightarrow \sim q$. The inverse implication is not logically equivalent to $p \Rightarrow q$, but it is logically equivalent to the converse. In fact, the inverse is the contrapositive of the converse.

3.11 PRACTICE Use a truth table to show that the inverse and the converse of $p \Rightarrow q$ are logically equivalent.

Looking at the contrapositive form of an implication is a useful tool in proving theorems. Since it is a property of the logical structure and does not depend on the subject matter, it can be used in any setting involving an implication. There are many more tautologies that can be used in the same way. Some of the more common are listed in the next example.

3.12 EXAMPLES The following tautologies are useful in constructing proofs. The first two indicate, for example, that an "if and only if" theorem $p \Leftrightarrow q$ can be proved by establishing $p \Rightarrow q$ and its converse $q \Rightarrow p$ or by showing $p \Rightarrow q$ and its inverse $\sim p \Rightarrow \sim q$. The letter c is used to represent a statement that is always false. Such a statement is called a **contradiction**. While this list of tautologies need not be memorized, it will be helpful if each tautology is studied carefully to see just what it is saying.

(a) $(p \Leftrightarrow q) \Leftrightarrow [(p \Rightarrow q) \wedge (q \Rightarrow p)]$
(b) $(p \Leftrightarrow q) \Leftrightarrow [(p \Rightarrow q) \wedge (\sim p \Rightarrow \sim q)]$
(c) $(p \Rightarrow q) \Leftrightarrow (\sim q \Rightarrow \sim p)$
(d) $p \vee \sim p$
(e) $(p \wedge \sim p) \Leftrightarrow c$
(f) $(\sim p \Rightarrow c) \Leftrightarrow p$
(g) $[(p \wedge \sim q) \Rightarrow c] \Leftrightarrow (p \Rightarrow q)$
(h) $[p \wedge (p \Rightarrow q)] \Rightarrow q$
(i) $[\sim q \wedge (p \Rightarrow q)] \Rightarrow \sim p$
(j) $[\sim p \wedge (p \vee q)] \Rightarrow q$
(k) $(p \wedge q) \Rightarrow p$
(l) $[(p \Rightarrow q) \wedge (q \Rightarrow r)] \Rightarrow (p \Rightarrow r)$
(m) $[(p_1 \Rightarrow p_2) \wedge (p_2 \Rightarrow p_3) \wedge \cdots \wedge (p_{n-1} \Rightarrow p_n)] \Rightarrow (p_1 \Rightarrow p_n)$
(n) $[(p \wedge q) \Rightarrow r] \Leftrightarrow [p \Rightarrow (q \Rightarrow r)]$
(o) $[(p \Rightarrow q) \wedge (r \Rightarrow s) \wedge (p \vee r)] \Rightarrow (q \vee s)$
(p) $[p \Rightarrow (q \vee r)] \Leftrightarrow [(p \wedge \sim q) \Rightarrow r]$
(q) $[(p \Rightarrow r) \wedge (q \Rightarrow r)] \Leftrightarrow [(p \vee q) \Rightarrow r]$

ANSWERS TO PRACTICE PROBLEMS

3.2 (a) Any flightless bird, such as an ostrich. (b) The absolute value function is continuous for all real numbers, but it is not differentiable at the origin.

3.4 The general rules about factoring polynomials were applied to the specific polynomial $n^2 + n + (n + 1)$.

3.6

(a)

p	q	$(p \Rightarrow q)$	$\Leftrightarrow$	$[(\sim q) \Rightarrow (\sim p)]$		
T	T	T	T	F	T	F
T	F	F	T	T	F	F
F	T	T	T	F	T	T
F	F	T	T	T	T	T

(b) No.

p	q	$(p \Rightarrow q)$	$\Leftrightarrow$	$(q \Rightarrow p)$
T	T	T	T	T
T	F	F	F	T
F	T	T	F	F
F	F	T	T	T

3.8 (a) If $2n$ is an odd number, then n is not an integer.
(b) If you do not have a college degree, then you cannot work here.
(c) If the car runs, then you are not out of gas.
(d) If a function is not continuous, then it is not differentiable.

3.10 (a) If $2n$ is an even number, then n is an integer.
(b) If you have college degree, then you can work here.
(c) If the car does not run, then you are out of gas.
(d) If a function is continuous, then it is differentiable.

3.11

p	q	$(q \Rightarrow p)$	$\Leftrightarrow$	$[(\sim p) \Rightarrow (\sim q)]$		
T	T	T	T	F	T	F
T	F	T	T	F	T	T
F	T	F	T	T	F	F
F	F	T	T	T	T	T

EXERCISES

3.1 Write the contrapositive of each implication.
(a) If all roses are red, then all violets are blue.
(b) H is normal if H is not regular.
(c) If K is closed and bounded, then K is compact.

3.2 Write the converse of each implication in Exercise 3.1.

3.3 Write the inverse of each implication in Exercise 3.1

3.4 Provide a counterexample of each statement.
(a) For every real number x, if $x^2 > 4$ then $x > 2$.
(b) For every positive integer n, $n^2 + n + 41$ is prime.
(c) Every triangle is a right triangle.

(d) No integer greater than 100 is prime.
(e) Every prime is an odd number.
(f) For every positive integer n, $3n$ is divisible by 6.
(g) No rational number satisfies the equation $x^3 + (x-1)^2 = x^2 + 1$.
(h) No rational number satisfies the equation $x^4 + (1/x) - \sqrt{x+1} = 0$.

3.5 Let f be the function given by $f(x) = 3x - 5$. Use the contrapositive implication to prove: If $x_1 \neq x_2$, then $f(x_1) \neq f(x_2)$.

3.6 Use the contrapositive implication to prove: If n^2 is an even number, then n is an even number. (Use the fact that a number is odd iff it can be written as $2k + 1$ for some integer k.)

3.7 In each part, a list of hypotheses is given. These hypotheses are assumed to be true. Using tautologies from Example 3.12, you are to establish the desired conclusion. Indicate which tautology you are using to justify each step.

(a) Hypotheses: $r \Rightarrow \sim s$, $t \Rightarrow s$
Conclusion: $r \Rightarrow \sim t$
(b) Hypotheses: r, $\sim t$, $(r \wedge s) \Rightarrow t$
Conclusion: $\sim s$
(c) Hypotheses: $r \Rightarrow \sim s$, $\sim r \Rightarrow \sim t$, $\sim t \Rightarrow u$, $v \Rightarrow s$
Conclusion: $\sim v \vee u$

3.8 Repeat Exercise 3.7 for the following hypotheses and conclusions.

(a) Hypotheses: $\sim r$, $(\sim r \wedge s) \Rightarrow r$
Conclusion: $\sim s$
(b) Hypotheses: $\sim t$, $(r \vee s) \Rightarrow t$
Conclusion: $\sim s$
(c) Hypotheses: $r \Rightarrow \sim s$, $t \Rightarrow u$, $s \vee t$
Conclusion: $r \vee u$

Section 4 TECHNIQUES OF PROOF: II

Mathematical theorems and proofs do not occur in isolation, but always in the context of some mathematical system. For example, in Section 3 when we discussed a conjecture related to prime numbers, the natural context of that discussion was the positive integers. In Example 3.7 when talking about odd and even numbers, the context was the set of all integers. Very often a theorem will make no explicit reference to the mathematical system in which it is being proved; it must be implied from the context. Usually, this causes no difficulty, but if there is a possibility of ambiguity, the careful writer will explicitly name the system being considered.

When dealing with quantified statements, it is particularly important to know exactly what system is being considered. For example, the statement

$$\forall x, \sqrt{x^2} = x$$

is true in the context of the positive numbers but is false when considering all real numbers. Similarly,

$$\exists\, x \ni x^2 = 25 \quad \text{and} \quad x < 3$$

is false for positive numbers and true for reals. When we introduce set notation (in Chapter 2) it will become easier to be precise in indicating the context of a particular quantified statement. For now, we shall have to write it out with words.

To prove a universal statement

$$\forall\, x,\ p(x),$$

we begin by choosing an arbitrary member x from the system under consideration and then show that statement $p(x)$ is true. The only properties that we can use about x are those properties that apply to all the members of the system. For example, if the system consists of the integers, we cannot use the property that x is even, since this does not apply to all the integers.

To prove an existential statement

$$\exists\, x \ni p(x),$$

we have to prove that there is at least one member x in the system for which $p(x)$ is true. The most direct way of doing this is to construct (produce, guess, etc.) a specific x that has the required property. Unfortunately, there is no sure-fire way to always find a particular x that will work. If the hypothesis in the theorem contains a quantified statement, this can sometimes be helpful, but often it is just a matter of working on both ends of the logical bridge until you can get them to meet in the middle.

4.1 EXAMPLE To illustrate the process of writing a proof with quantifiers, consider the following

THEOREM: For every $\varepsilon > 0$ there exists a $\delta > 0$ such that

$$1 - \delta < x < 1 + \delta \quad \text{implies that} \quad 5 - \varepsilon < 2x + 3 < 5 + \varepsilon.$$

We are asked to prove that something is true for each positive number ε. Thus we begin by letting ε be an arbitrary positive number. We need to use this ε to find a positive δ with the property that

$$1 - \delta < x < 1 + \delta \quad \text{implies that} \quad 5 - \varepsilon < 2x + 3 < 5 + \varepsilon.$$

Let us begin with the consequent of the implication. We want to have

$$5 - \varepsilon < 2x + 3 < 5 + \varepsilon.$$

This will be true if

$$2 - \varepsilon < 2x < 2 + \varepsilon,$$

and this in turn will follow from

$$1 - \frac{\varepsilon}{2} < x < 1 + \frac{\varepsilon}{2}.$$

Thus we see that choosing δ to be $\varepsilon/2$ will meet the required condition. In writing down the proof in a formal manner we would simply set δ equal to $\varepsilon/2$ and then show that this particular δ will work.

Proof: Let ε be an arbitrary positive number and let $\delta = \varepsilon/2$. Then δ is also positive and whenever

$$1 - \delta < x < 1 + \delta$$

we have

$$1 - \frac{\varepsilon}{2} < x < 1 + \frac{\varepsilon}{2},$$

so that

$$2 - \varepsilon < 2x < 2 + \varepsilon$$

and

$$5 - \varepsilon < 2x + 3 < 5 + \varepsilon,$$

as required. ▮[†]

In some situations it is possible to prove an existential statement in an indirect way without actually producing any specific member of the system. One indirect method is to use the contrapositive form of the implication and another is to use a proof by contradiction.

The two basic forms of a proof by contradiction are based on tautologies (f) and (g) in Example 3.12. Tautology (f) has the form

$$(\sim p \Rightarrow c) \Leftrightarrow p.$$

If we wish to conclude a statement p, we can do so by showing that the negation of p leads to a contradiction. Tautology (g) has the form

$$[(p \wedge \sim q) \Rightarrow c] \Leftrightarrow (p \Rightarrow q).$$

If we wish to conclude that p implies q, we can do so by showing that p and not q leads to a contradiction. In either case the contradiction can involve part of the hypothesis or some other statement that is known to be true.

[†] The symbol ▮ is used to denote the end of a formal proof.

4.2 PRACTICE Use truth tables to verify that $(\sim p \Rightarrow c) \Leftrightarrow p$ and $[(p \wedge \sim q) \Rightarrow c] \Leftrightarrow (p \Rightarrow q)$ are tautologies.

4.3 EXAMPLE To illustrate an indirect proof of an existential statement, consider the following:

> **THEOREM:** Let f be a continuous function. If $\int_0^1 f(x)\,dx \neq 0$, then there exists a point x in the interval $[0, 1]$ such that $f(x) \neq 0$.

Symbolically, we have $p \Rightarrow q$, where

$$p\colon \int_0^1 f(x)\,dx \neq 0$$

$$q\colon \exists\, x \text{ in } [0, 1] \ni f(x) \neq 0.$$

The contrapositive implication, $\sim q \Rightarrow \sim p$, can be written as

$$\text{If for every } x \text{ in } [0, 1],\ f(x) = 0, \text{ then } \int_0^1 f(x)\,dx = 0.$$

This is much easier to prove. Instead of having to conclude the existence of an x in $[0, 1]$ with a particular property, we are given that every x in $[0, 1]$ has a different property. Indeed, the proof now follows directly from the definition of the integral since each of the terms in any Riemann sum will be zero. (See Chapter 7.)

4.4 EXAMPLE To illustrate a proof by contradiction, consider the following:

> **THEOREM:** Let x be a real number. If $x > 0$, then $1/x > 0$.

Symbolically, we have $p \Rightarrow q$, where

$$p\colon x > 0$$

$$q\colon \frac{1}{x} > 0.$$

Tautology (g) in Example 3.12 says that $p \Rightarrow q$ is equivalent to $(p \wedge \sim q) \Rightarrow c$. Thus we begin by supposing $x > 0$ and $1/x \leqslant 0$. Since $x > 0$, we can multiply both sides of the inequality $1/x \leqslant 0$ by x obtain

$$(x)\left(\frac{1}{x}\right) \leqslant (x)(0).$$

But $(x)(1/x) = 1$ and $(x)(0) = 0$, so we have $1 \leqslant 0$, a contradiction to the (presumably known) fact that $1 > 0$.

Another tautology in Example 3.12 that deserves special attention is statement (q):

$$[(p \Rightarrow r) \wedge (q \Rightarrow r)] \Leftrightarrow [(p \vee q) \Rightarrow r].$$

Some proofs naturally divide themselves into the consideration of two (or more) cases. For example, integers are either odd or even. Real numbers are positive, negative, or zero. It may be that different arguments are required for each case. It is tautology (q) that shows us how to combine the cases.

4.5 EXAMPLE Suppose we wish to prove that, if x is a real number, then $x \leqslant |x|$. Symbolically, we have $s \Rightarrow r$, where

$$s\colon x \text{ is a real number}$$

$$r\colon x \leqslant |x|.$$

First, we recall the definition of absolute value:

$$|x| = \begin{cases} x, & \text{if } x \geqslant 0 \\ -x, & \text{if } x < 0. \end{cases}$$

Since this definition is divided into two parts, it is natural to divide our proof into two cases. Thus statement s is replaced by the equivalent disjunction $p \vee q$, where

$$p\colon x \geqslant 0 \quad \text{and} \quad q\colon x < 0.$$

Our theorem now is to prove $(p \vee q) \Rightarrow r$, and this we do by showing that $(p \Rightarrow r) \wedge (q \Rightarrow r)$. The actual proof could be written as follows:

> Let x be an arbitrary real number. Then $x \geqslant 0$ or $x < 0$. If $x \geqslant 0$, then by definition $x = |x|$. On the other hand, if $x < 0$, then $-x > 0$, so that $x < 0 < -x = |x|$. Thus, in either case, $x \leqslant |x|$. ▮

4.6 PRACTICE In proving a theorem that relates to factoring positive integers greater than 1, what two cases might reasonably be considered?

An alternative form of proof by cases arises when the conclusion of an implication involves a disjunction. In this situation tautology (p) of Example 3.12 is often helpful:

$$[p \Rightarrow (q \vee r)] \Leftrightarrow [(p \wedge \sim q) \Rightarrow r].$$

4.7 EXAMPLE Consider the following:

> **THEOREM:** If the sum of a real number with itself is equal to its square, then the number is 0 or 2.

In symbols we have $p \Rightarrow (q \vee r)$, where

$$p\colon x + x = x^2$$

$$q\colon x = 0$$

$$r\colon x = 2.$$

To do the proof, we shall show that $(p \wedge \sim q) \Rightarrow r$.

Proof: Suppose that $x + x = x^2$ and $x \neq 0$. Then $2x = x^2$ and since $x \neq 0$, we can divide by x to obtain $2 = x$. ▮

4.8 PRACTICE Suppose that you wish to prove the statement: If B is both open and closed, then $B = \varnothing$ or $B = X$. Use tautology (p) of Example 3.12 to state *two* different equivalent statements that could be proved instead.

We have now considered the most common forms of mathematical proof, except for proofs by induction. Induction proofs will be considered later in Chapter 3 in connection with the natural numbers. But before we close this chapter on logic and proof, a few informal comments are in order.

In formulating a proof it is important that a mathematician (that includes you!) be very careful to use sound logical reasoning. This is what we have tried to help you develop in this first chapter. But when writing down a proof it is usually unnecessary—and often undesirable—to include all the logical steps and details along the way. The human mind can only absorb so much information at one time. It is necessary to skip lightly over the steps that are well understood from previous experience so that greater attention can be focused on the part that is really new. Of course, the question of what to include and what to skip is not easy and depends to a considerable extent on the intended audience. The proofs included in this text will tend to be more complete than those in more advanced books or research papers, since the reader is presumably less sophisticated. As a student, you should also practice filling in more of the details, if for no other reason than to make sure that the details really do fill in. (At least be prepared to show your instructor why your "clearly" is clear and your "it follows that" really does follow.)

Throughout the rest of the book you will have the opportunity to read and write a great many proofs. Make the most of it! When you read a proof, analyze its structure. See what tautologies, if any, have been used. Note the important role that definitions play. Often a proof will be little more than unraveling definitions and applying them to specific cases. From time to time we shall point out the method to be used in a proof to help you see the structure that we shall be following. And when you begin to write proofs yourself, do not get discouraged when your instructor returns them covered with comments and corrections. The writing of proofs is an art, and the only way to learn is by doing.

ANSWERS TO PRACTICE PROBLEMS

4.2

p	$(\sim p \Rightarrow c)$	$\Leftrightarrow$	p
T	F T F	T	T
F	T F F	T	F

p	q	$[(p \wedge \sim q) \Leftrightarrow c]$	$\Leftrightarrow$	$(p \Rightarrow q)$
T	T	T F F T F	T	T
T	F	T T T F F	T	F
F	T	F F F T F	T	T
F	F	F F T T F	T	T

4.6 The positive integers greater than 1 are either prime or composite. They are also either odd or even. Either way of separating the integers into two cases could be reasonable, depending on the context.

4.8 If B is both open and closed and $B \neq \varnothing$, then $B = X$. If B is both open and closed and $B \neq X$, then $B = \varnothing$.

EXERCISES

4.1 Prove: There exists an integer n such that $n^2 + 3n/2 = 1$. Is this integer unique?

4.2 Prove: There exists a rational number x such that $x^2 + 3x/2 = 1$. Is this rational number unique?

4.3 Prove: For every real number $x > 3$, there exists a real number $y < 0$ such that $x = 3y/(2 + y)$.

4.4 Prove: For every real number $x > 1$, there exist two distinct positive real numbers y and z such that

$$x = \frac{y^2 + 9}{6y} = \frac{z^2 + 9}{6z}.$$

4.5 Prove: If x is rational and y is not rational, then $x + y$ is not rational. (Recall that a number is rational iff it can be expressed as the quotient of two integers.)

4.6 Prove: If x is a real number, then $|x - 2| \leqslant 3$ implies that $-1 \leqslant x \leqslant 5$.

4.7 Prove: If $x^2 + x - 6 \geqslant 0$, then $x \leqslant -3$ or $x \geqslant 2$.

4.8 Prove: If $x/(x - 1) \leqslant 2$, then $x < 1$ or $x \geqslant 2$.

4.9 Prove or give a counterexample: There do not exist three consecutive even integers a, b, and c such that $a^2 + b^2 = c^2$.

4.10 Prove or give a counterexample: There do not exist three consecutive odd integers a, b, and c such that $a^2 + b^2 = c^2$.

4.11 Prove or give a counterexample: For every positive integer n, $n^2 + 3n + 8$ is even.

4.12 Prove or give a counterexample: For every positive integer n, $n^2 + 4n + 8$ is even.

4.13 Assume that the following two hypotheses are true: (1) If the basketball center is healthy or the point guard is hot, then the team will win and the fans will be happy; and (2) if the fans are happy or the coach is a millionaire, then the college will balance the budget. Derive the following conclusion: If the basketball center is healthy, then the college will balance the budget. Using letters to represent the simple statements, write out a formal proof in the format of Exercise 3.7.

2

Sets and Functions

If there is one unifying foundation common to all branches of mathematics, it is the theory of sets. We have already seen the need for set notation in describing the context in which quantified statements are understood to apply. It is not our intention to develop set theory in a formal axiomatic way, but rather to discuss informally those aspects of set theory that are relevant to the study of analysis. In Section 5 we establish the basic notation for working with sets, and in the following two sections we apply this to the development of relations and functions. In Section 8 we compare the size of sets, giving special attention to infinite sets. In Section 9 (an optional section) we outline a set of axioms that can be used to develop formal set theory, and indicate some of the problems that are involved with the development.

Section 5 BASIC SET OPERATIONS

The idea of a set or collection of things is common in our everyday experience. We speak of a football *team*, a *flock* of geese, or a finance *committee*. Each of these refers to a collection of distinguishable objects that are thought of as a whole. In spite of the familiarity of the idea, it is essentially impossible to give a definition of "set" except in terms of synonyms that are also undefined. Thus we shall be content with the informal understanding that a set is a collection of objects characterized by some defining property that allows us to think of the objects as a whole. The objects in a set are called **elements** or **members** of the set.

It is customary to use capital letters to designate sets and the symbol $\in$ to denote membership in a set. Thus we write $a \in A$ to mean that object a is a member of set A, and $a \notin B$ to mean that object a is not a member of set B.

5.1 EXAMPLE If $A = \{1, 2, 3, 4\}$, then $2 \in A$ and $5 \notin A$.

To say that a set must be characterized by some defining property is to require that it be a clear question of fact whether a particular object does or does not belong to a particular set. We do not, however, demand that the answer to the question of membership be known. Another way to say this is to require that for any element a and any set A the sentence "$a \in A$" be a statement; that is, it must be true or false, and not both.

5.2 PRACTICE Which of the following satisfy the requirements of a set?

(a) all the current U.S. Senators from Massachusetts
(b) all the prime divisors of 1987
(c) all the tall people in Canada
(d) all the prime numbers between 8 and 10

To define a particular set, we have to indicate the property that characterizes its elements. For a finite set, this can be done by listing its members. For example, if set A consists of the elements 1, 2, 3, 4, we write $A = \{1, 2, 3, 4\}$, as in Example 5.1. If set B consists of just one member b, we write $B = \{b\}$. Thus we distinguish between the *element* b and the *set* $\{b\}$ containing b as its only element.

For an infinite set we cannot list all the members, so a defining rule must be given. It is customary to set off the rule within braces, as in

$$C = \{x: x \text{ is prime}\}.$$

This is read "C is the set of all x such that x is prime," or more simply, "C is the set of all prime numbers."

When considering two sets, call them A and B, it may happen that every element of A is also an element of B. In this case we say that A is a subset of B. This concept is of such a fundamental importance that we distinguish it by our first formal definition:

5.3 DEFINITION Let A and B be sets. We say that A is a **subset** of B (or A is **contained** in B) if every element of A is an element of B, and we denote this by writing $A \subseteq B$.† (Occasionally, we may write $B \supseteq A$ instead of $A \subseteq B$.) If $A \subseteq B$ and $A \neq B$ then A is called a **proper** subset of B.

This definition tells us what we must do if we want to prove $A \subseteq B$. We must show that

$$\text{if } x \in A, \text{ then } x \in B$$

is a true statement. That is, we must show that each element of A satisfies the defining condition that characterizes set B.

5.4 DEFINITION Let A and B be sets. We say that A is **equal** to B, written $A = B$, if $A \subseteq B$ and $B \subseteq A$.

When this definition is combined with the definition of subset, we see that proving $A = B$ is equivalent to proving

$$x \in A \Rightarrow x \in B \quad \text{and} \quad x \in B \Rightarrow x \in A.$$

It is important to note that, in describing a set, the order in which the elements appear does not matter, nor does the number of times they are written. Thus the following sets are all equal:

$$\{1, 2, 3, 4\} = \{2, 4, 1, 3\} = \{1, 2, 3, 2, 4, 2\}.$$

Since the repeated 2's in the last set are unnecessary, it is preferable to omit them.

Although we cannot now give a formal definition of them, it is convenient to name the following sets, which should already be familiar to the reader.

$\mathbb{N}$ will denote the set of positive integers.

$\mathbb{Z}$ will denote the set of all integers.

$\mathbb{Q}$ will denote the set of all rational numbers.

$\mathbb{R}$ will denote the set of all real numbers.

In constructing examples of sets it is often helpful to indicate a larger set from which the elements are being chosen. We indicate this by a slight change in our set notation. For example, instead of having to write

$$\{x \colon x \in \mathbb{R} \text{ and } 0 < x < 1\},$$

† When "if" is used in a definition it is understood to have the force of "iff." That is, we are defining "$A \subseteq B$" to be the same as "every element of A is an element of B." Essentially, a definition is used to establish an abbreviation for a particular idea or concept. It would be more accurate to write "iff" between the concept and its abbreviation, but it is common practice to use simply "if."

we may abbreviate by writing

$$\{x \in \mathbb{R}: 0 < x < 1\}.$$

The latter notation is read "the set of all x in $\mathbb{R}$ such that $0 < x < 1$" or "the set of all real numbers x such that $0 < x < 1$."

There is also a standard notation that we shall use for interval subsets of the real numbers:

$$[a, b] = \{x \in \mathbb{R}: a \leqslant x \leqslant b\}, \qquad (a, b) = \{x \in \mathbb{R}: a < x < b\}$$

$$[a, b) = \{x \in \mathbb{R}: a \leqslant x < b\}, \qquad (a, b] = \{x \in \mathbb{R}: a < x \leqslant b\}.$$

The set $[a, b]$ is called a **closed interval**, the set (a, b) is called an **open interval**, and the sets $[a, b)$ and $(a, b]$ are called **half-open** (or **half-closed**) **intervals**. We shall also have occasion to refer to the unbounded intervals:

$$[a, \infty) = \{x \in \mathbb{R}: x \geqslant a\}, \qquad (a, \infty) = \{x \in \mathbb{R}: x > a\}$$

$$(-\infty, b] = \{x \in \mathbb{R}: x \leqslant b\}, \qquad (-\infty, b) = \{x \in \mathbb{R}: x < b\}.$$

For the present time no special significance should be attached to the symbols "∞" and "$-\infty$" as in $[a, \infty)$ and $(-\infty, b]$. They simply indicate that the interval contains all real numbers greater than or equal to a, or less than or equal to b, as the case may be.

5.5 EXAMPLE Let

$$A = \{1, 3\}$$

$$B = \{3, 5\}$$

$$C = \{1, 3, 5\}$$

$$D = \{x \in \mathbb{R}: x^2 - 8x + 15 = 0\}.$$

Then the following statements (among others) are all true:

$$A \subseteq C \qquad 1 \notin D$$

$$A \nsubseteq b \qquad C \nsubseteq B$$

$$5 \in B \qquad B = D$$

$$\{5\} \subseteq B \qquad B \neq C.$$

Notice that the slash (/) through a connecting symbol has the meaning of "not." Thus $A \nsubseteq B$ is read "A is not a subset of B."

5.6 PRACTICE Let

$$A = \{1, 2, 3, 4, 5\}$$

$$B = \{x: x = 2k \text{ for some } k \in \mathbb{N}\}$$

$$C = \{x \in \mathbb{N}: x < 6\}.$$

Which of the following statements are true?

(a) $\{4, 3, 2\} \subseteq A$ (b) $3 \in B$
(c) $A \subseteq C$ (d) $\{2\} \in A$
(e) $C \subseteq B$ (f) $\{2, 4, 6, 8\} \subseteq B$
(g) $C \subseteq A$ (h) $A = C$

In Practice 5.2(d) we found that the collection D of all prime numbers between 8 and 10 is a legitimate set. This is so because the statement "$x \in D$" is clearly true or false for any object x. In fact, "$x \in D$" is aways false, since there are no prime numbers between 8 and 10. Thus D is an example of an **empty set**, a set with no members. It is not difficult to show (Exercise 5.6) that there is only one empty set, and we denote it by $\varnothing$. For our first theorem we shall prove that the empty set is a subset of every set. Notice the essential role that the definitions play in the proof. At this point, we really have nothing else to use as building blocks.

5.7 THEOREM Let A be a set. Then $\varnothing \subseteq A$.

Proof: To prove that $\varnothing \subseteq A$, we must establish that the implication

$$\text{if } x \in \varnothing, \text{ then } x \in A$$

is true. Since $\varnothing$ has no members, the antecedent "$x \in \varnothing$" is false for all x. Thus, according to our definition of implies, the implication is always true. ▮

There are three basic ways to form new sets from old ones. These operations are called union, intersection, and complementation. Intuitively, union may be thought of as putting together, intersection is like cutting down, and complementation corresponds to throwing out. Their precise definitions are as follows:

5.8 DEFINITION Let A and B be sets. The **union** of A and B (denoted $A \cup B$), the **intersection** of A and B (denoted $A \cap B$), and the **complement** of B in A (denoted $A \backslash B$) are given by

$$A \cup B = \{x \colon x \in A \text{ or } x \in B\}$$

$$A \cap B = \{x \colon x \in a \text{ and } x \in B\}$$

$$A \backslash B = \{x \colon x \in A \text{ and } x \notin B\}.$$

If $A \cap B = \varnothing$, then A and B are said to be **disjoint**.

The three set operations given above correspond in a natural way to three of the basic logical connectives:

$$x \in A \cup B \quad \text{iff} \quad (x \in A \vee x \in B)$$

$$x \in A \cap B \quad \text{iff} \quad (x \in A \wedge x \in B)$$

$$x \in A \backslash B \quad \text{iff} \quad [x \in A \wedge \sim(x \in B)].$$

5.9 PRACTICE Which logical connective corresponds to the set relationship $A \subseteq B$?

The definition of complementation may seem to be unnecessarily complicated. Why didn't we just define "$\sim B$" to be $\{x\colon x \notin B\}$? The problem is that $\{x\colon x \notin B\}$ is too large. For example, suppose that $B = \{2, 4, 6, 8\}$. Then $\{x\colon x \notin B\}$ contains all of the following (and more!):

the integers 1, 3, 5, 7, 9, 11
the real numbers greater than 25
the function $f(x) = x^2 + 3$
the circle of radius 1 centered at the origin in the plane
the Empire State Building
my uncle Wilbur

It is quite reasonable that the integers 1, 3, 5, 7 should be included in "$\sim B$," and, depending on the context, we might want to include the real numbers greater than 25 as well. But it is quite unlikely that we would want to include any of the other items. Certainly, knowing that my uncle Wilbur is not a member of the set B would contribute little to any discussion of B.

As we have observed earlier, mathematical concepts and proofs always occur within the context of some mathematical system. It is customary for the elements of the system to be called the universal set. Then any set under consideration is a subset of this universal set. If the universal set in a particular discussion were the integers $\mathbb{Z}$, then the nonintegral real numbers greater than 25 would not be included in $\{x\colon x \notin B\}$. On the other hand, if the universal set is $\mathbb{R}$, then they would be included.

5.10 EXAMPLE Let $A = \{1, 2, 3, 4\}$ and $B = \{2, 4, 6\}$ be subsets of the universal set $U = \{1, 2, 3, 4, 5, 6\}$. Then $A \cup B = \{1, 2, 3, 4, 6\}$, $A \cap B = \{2, 4\}$, $A \backslash B = \{1, 3\}$, and $U \backslash B = \{1, 3, 5\}$.

5.11 PRACTICE Fill in the blanks in the proof of the following theorem.

THEOREM: Let A and B be subsets of a universal set U. Then

$$A \cap (U \backslash B) = A \backslash B.$$

Proof: According to our definition of equality of sets, we must show that

$$[A \cap (U \setminus B)] \subseteq [A \setminus B] \quad \text{and} \quad [A \setminus B] \subseteq [A \cap (U \setminus B)]$$

or, equivalently,

$$x \in A \cap (U \setminus B) \quad \text{iff} \quad x \in A \setminus B.$$

Let us begin by showing that $x \in A \cap (U \setminus B)$ implies that $x \in A \setminus B$. If $x \in A \cap (U \setminus B)$, then $x \in A$ and $x \in$ __________, by the definition of intersection. But $x \in U \setminus B$ means that $x \in U$ and __________. Since $x \in A$ and $x \notin B$, we have $x \in$ __________, as required. Thus $A \cap (U \setminus B) \subseteq A \setminus B$.

Conversely, we must show that __________ $\subseteq$ __________. If __________, then $x \in A$ and $x \notin B$. Since $A \subseteq U$, we have $x \in$ __________. Thus $x \in U$ and $x \notin B$, so __________. But then __________ and $x \in U \setminus B$, so $x \in A \cap (U \setminus B)$. Hence $A \setminus B \subseteq A \cap (U \setminus B)$. ▌

A helpful way to visualize the set operations of union, intersection, and complementation is by use of **Venn diagrams**, as in Figures 5.1 and 5.2. In Figure 5.1 the shaded area represents the union of A and B, and in Figure 5.2 the shaded area is the intersection of A and B. In each case the large rectangle represents the universal set U. While Venn diagrams (and

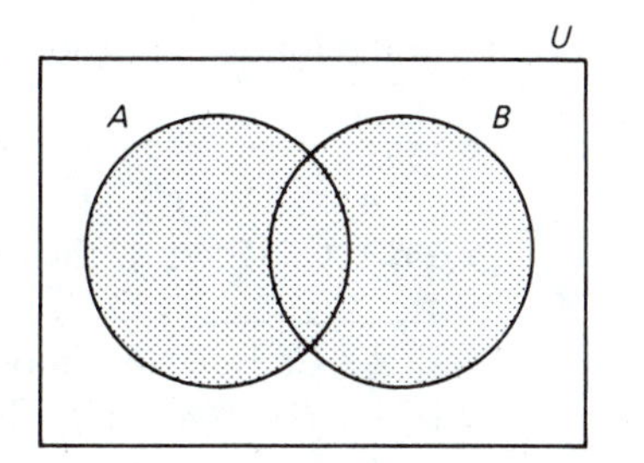

Figure 5.1 $A \cup B$

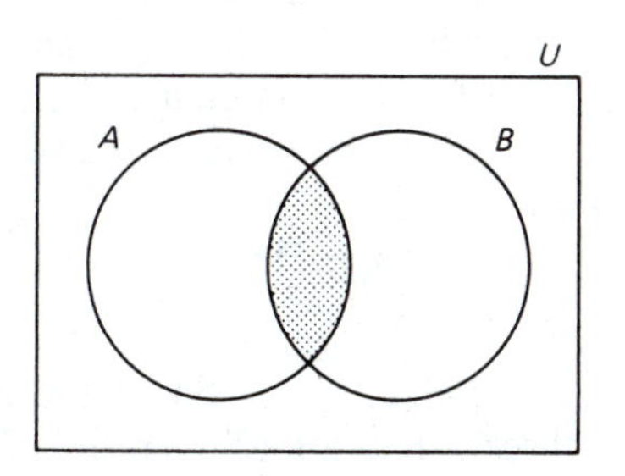

Figure 5.2 $A \cap B$

other diagrams as well) are useful in seeing the relationship between sets, and may be helpful in getting ideas for developing a proof, they should not be viewed as proofs themselves. A diagram necessarily represents only one case, and it may not be obvious whether this is a general case that always applies or whether there may be other cases as well.

5.12 PRACTICE Use a Venn diagram to illustrate $A \setminus B$.

We close this section by stating some of the important properties of unions, intersections, and complements. Two of the proofs are sketched as practice problems and the others are left for the exercises.

5.13 THEOREM Let A, B, and C be subsets of a universal set U. Then the following statements are true.

(a) $A \cup (U \setminus A) = U$
(b) $A \cap (U \setminus A) = \varnothing$
(c) $U \setminus (U \setminus A) = A$
(d) $A \cup (B \cap C) = (A \cup B) \cap (A \cup C)$
(e) $A \cap (B \cup C) = (A \cap B) \cup (A \cap C)$
(f) $A \setminus (B \cup C) = (A \setminus B) \cap (A \setminus C)$
(g) $A \setminus (B \cap C) = (A \setminus B) \cup (A \setminus C)$

5.14 PRACTICE Complete the following proof of Theorem 5.13(d).

Proof: We begin by showing that $A \cup (B \cap C) \subseteq (A \cup B) \cap (A \cup C)$. If $x \in$ __________, then either $x \in A$ or $x \in B \cap C$. If $x \in A$, then certainly $x \in A \cup B$ and $x \in A \cup C$. Thus $x \in$ __________. On the other hand, if __________, then $x \in B$ and $x \in C$. But this implies that $x \in A \cup B$ and __________, so $x \in (A \cup B) \cap (A \cup C)$. Hence $A \cup (B \cap C) \subseteq (A \cup B) \cap (A \cup C)$.

Conversely, if $y \in (A \cup B) \cap (A \cup C)$, then __________ and __________. There are two cases to consider: when $y \in A$ and when $y \notin A$. If $y \in A$, then $y \in A \cup (B \cap C)$ and this part is done. On the other hand, if __________, then since $y \in A \cup B$, we must have $y \in B$. Similarly, since $y \in A \cup C$ and $y \notin A$, we have __________. Thus __________, and this implies that $y \in A \cup (B \cap C)$. Hence $(A \cup B) \cap (A \cup C) \subseteq A \cup (B \cap C)$. ∎

Before going on to the proof of 5.13(f), let us make a couple of observations about the proof of 5.13(d). Notice how the argument divides naturally into parts, the second part being introduced by the word "conversely." This word is appropriate because the second half of the argument is indeed the converse of the first half. In the first part the point in $A \cup (B \cap C)$ was called x and in the second part the point in $(A \cup B) \cap (A \cup C)$ was called y. Why is this? The choice of a name is completely arbitrary, and in fact the same name could have been used in both parts. It is important to realize that the two parts are completely separate arguments; we start over from scratch in proving the converse and can use nothing that was derived about the point x in the first part. By using different names for the points in the two parts we emphasize this separateness. It is common practice, however, to use the same name (such as x) for the arbitrary point in both parts. When doing this we have to be careful not to confuse the properties of the points in the two parts.

We also notice that each half of the argument also has two parts or cases, the second case being introduced by the phrase "on the other hand." This type of division of the argument is necessary when dealing with unions. If $x \in S \cup T$, then $x \in S$ *or* $x \in T$. Each of the possibilities must be followed to its logical conclusion and both "bridges" must lead to the same desired result (or to a contradiction, which would show that the alternative possibility could not occur).

Finally, when proving that one set, say S, is a subset of another set, say T, it is common to begin with the phrase "If $x \in S$, then ..." It is also acceptable to begin with "Let $x \in S$" and then conclude that $x \in T$. The subtle difference between these phrases is that "Let $x \in S$" assumes that S is nonempty, so there is an x in S to choose. This might seem to be an unwarranted assumption, but really it is not. If S is the empty set, then of course $S \subseteq T$, so the only nontrivial case to prove is when S is nonempty.

5.15 PRACTICE Complete the following proof of Theorem 5.13(f).

Proof: We wish to prove that $A \backslash (B \cup C) = (A \backslash B) \cap (A \backslash C)$. To this end, let $x \in A \backslash (B \cup C)$. Then __________ and __________. Since $x \notin B \cup C$, __________ and $x \notin C$ (for if it were in either B or C then it would be in their union). Thus $x \in A$ and $x \notin B$ and $x \notin C$. Hence $x \in A \backslash B$ and $x \in A \backslash C$, which implies that __________. We conclude that $A \backslash (B \cup C) \subseteq (A \backslash B) \cap (A \backslash C)$.

Conversely, suppose that $x \in$ __________. Then $x \in A \backslash B$ and $x \in A \backslash C$. But then $x \in$ __________ and $x \notin$ __________ and $x \notin$ __________. This implies that $x \notin (B \cup C)$, so $x \in$ __________. Hence __________ $\subseteq$ __________ as desired. ▮

Up to this point we have talked about combinations of two or three sets. By repeated application of the appropriate definitions we can even consider unions and intersections of any finite collection of sets. But sometimes we want to deal with combinations of infinitely many sets, and for this we need a new notation and a more general definition.

5.16 DEFINITION If for each j in a nonempty set J there corresponds a set A_j, then

$$\mathscr{A} = \{A_j : j \in J\}$$

is called an **indexed family** of sets with J as the index set. The union of all the sets in $\mathscr{A}$ is given by

$$\cup \{A_j : j \in J\} = \{x : x \in A_j \text{ for some } j \in J\}.$$

The intersection of all the sets in $\mathscr{A}$ is given by

$$\cap \{A_j : j \in J\} = \{x : x \in A_j \text{ for all } j \in J\}.$$

Other notations for $\cup \{A_j : j \in J\}$ include

$$\bigcup_{j \in J} A_j \quad \text{and} \quad \cup \mathscr{A}.$$

If $J = \{1, 2, \ldots, n\}$, we may write

$$\bigcup_{j=1}^{n} A_j,$$

and if $J = \mathbb{N}$, the common notation is

$$\bigcup_{j=1}^{\infty} A_j.$$

Similar variations on the notation apply to intersections.

There are some situations where a family of sets has not been indexed but we still wish to take the union or intersection of all the sets. If $\mathscr{B}$ is a collection of sets, then we let

$$\bigcup_{B \in \mathscr{B}} B = \{x : \exists\, B \in \mathscr{B} \ni x \in B\}$$

and

$$\bigcap_{B \in \mathscr{B}} B = \{x : \forall\, B \in \mathscr{B}, x \in B\}.$$

5.17 EXAMPLE For each $k \in \mathbb{N}$, let $A_k = [0, 2 - 1/k]$. Then

$$\bigcup_{k=1}^{\infty} A_k = [0, 2).$$

5.18 PRACTICE For each $k \in \mathbb{N}$, let $A_k = (-1/k, 1/k)$. Find $\bigcap_{k=1}^{\infty} A_k$.

ANSWERS TO PRACTICE PROBLEMS

5.2 (a), (b) and (d) are sets. (c) is not a set unless "tall" and "in" are made precise.

5.6 (a), (c), (f), (g), and (h) are true.

5.9 $x \in A \Rightarrow x \in B$.

5.11 If $x \in A \cap (U \backslash B)$, then $x \in A$ and $x \in U \backslash B$, by the definition of intersection. But $x \in U \backslash B$ means that $x \in U$ and $x \notin B$. Since $x \in A$ and $x \notin B$, we have $x \in A \backslash B$, as required. Thus $A \cap (U \backslash B) \subseteq A \backslash B$.

Conversely, we must show that $A \backslash B \subseteq A \cap (U \backslash B)$. If $x \in A \backslash B$, then $x \in A$ and $x \notin B$. Since $A \subseteq U$, we have $x \in U$. Thus $x \in U$ and $x \notin B$, so $x \in U \backslash B$. But then $x \in A$ and $x \in U \backslash B$, so $x \in A \cap (U \backslash B)$. Hence $A \backslash B \subseteq A \cap (U \backslash B)$.

5.12

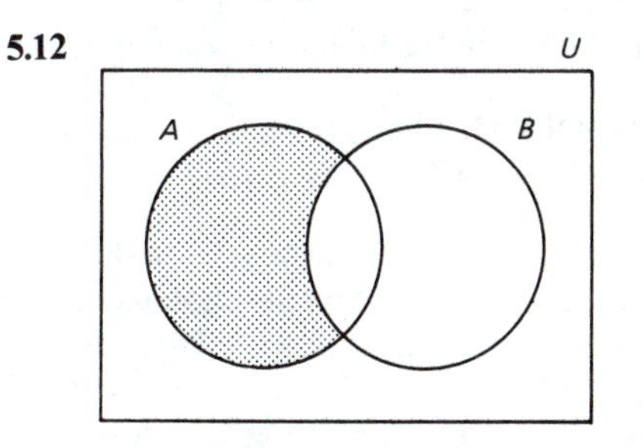

5.14 We begin by showing that $A \cup (B \cap C) \subseteq (A \cup B) \cap (A \cup C)$. If $x \in A \cup (B \cap C)$, then either $x \in A$ or $x \in B \cap C$. If $x \in A$, then certainly $x \in A \cup B$ and $x \in A \cup C$. Thus $x \in (A \cup B) \cap (A \cup C)$. On the other hand, if $x \in B \cap C$, then $x \in B$ and $x \in C$. But this implies that $x \in A \cup B$ and $x \in A \cup C$, so $x \in (A \cup B) \cap (A \cup C)$. Hence $A \cup (B \cap C) \subseteq (A \cup B) \cap (A \cup C)$.

Conversely, if $y \in (A \cup B) \cap (A \cup C)$, then $y \in A \cup B$ and $y \in A \cup C$. There are two cases to consider: when $y \in A$ and when $y \notin A$. If $y \in A$, then $y \in A \cup (B \cap C)$ and this part is done. On the other hand, if $y \notin A$, then since $y \in A \cup B$, we must have $y \in B$. Similarly, since $y \in A \cup C$ and $y \notin A$, we have $y \in C$. Thus $y \in B \cap C$, and this implies that $y \in A \cup (B \cap C)$. Hence $(A \cup B) \cap (A \cup C) \subseteq A \cup (B \cap C)$.

5.15 We wish to prove that $A \backslash (B \cup C) = (A \backslash B) \cap (A \backslash C)$. To this end, let $x \in A \backslash (B \cup C)$. Then $x \in A$ and $x \notin B \cup C$. Since $x \notin B \cup C$, $x \notin B$ and $x \notin C$ (for if it were in either B or C, then it would be in their union). Thus $x \in A$ and $x \notin B$ and $x \notin C$. Hence $x \in A \backslash B$ and $x \in A \backslash C$, which implies that $x \in (A \backslash B) \cap (A \backslash C)$. We conclude that $A \backslash (B \cup C) \subseteq (A \backslash B) \cap (A \backslash C)$.

Conversely, suppose that $x \in (A \backslash B) \cap (A \backslash C)$. Then $x \in A \backslash B$ and $x \in A \backslash C$. But then $x \in A$ and $x \notin B$ and $x \notin C$. This implies that $x \notin (B \cup C)$, so $x \in A \backslash (B \cup C)$. Hence $(A \backslash B) \cap (A \backslash C) \subseteq A \backslash (B \cup C)$, as desired.

5.18 $\{0\}$.

EXERCISES

5.1 Let $A = \{2, 4, 6, 8\}$, $B = \{3, 4, 5, 6\}$, and $C = \{4, 5\}$. Which of the following statements are true?

(a) $\{6, 8\} \subseteq A$ (b) $C \subseteq A \cap B$
(c) $(B \backslash C) \cap A = \{6\}$ (d) $(A \backslash B) \cap C \subseteq B$
(e) $\varnothing \in A$ (f) $C \subseteq B$
(g) $(A \cup B) \backslash C = \{2, 3, 6, 8\}$ (h) $A \cap B \cap C = \{4\}$

5.2 Let $A = \{2, 4, 6, 8\}$, $B = \{1, 2, 3, 4\}$, and $C = \{5, 6, 7\}$. Find the following sets.

(a) $A \cap B$ (b) $A \cup B$
(c) $A \backslash B$ (d) $B \cap C$
(e) $B \backslash C$ (f) $(B \cup C) \backslash A$
(g) $(A \cap C) \backslash B$ (h) $C \backslash (A \cup B)$

5.3 Fill in the blanks in the following proof of Theorem 5.13(a).

> **THEOREM:** Let A be a subset of a universal set U. Then $A \cup (U \backslash A) = U$.
>
> **Proof:** If $x \in A \cup (U \backslash A)$, then $x \in$ __________ or $x \in$ __________. Since both A and $U \backslash A$ are subsets of U, in either case we have __________. Thus __________ $\subseteq$ __________.
>
> On the other hand, suppose that $x \in$ __________. Now either $x \in A$ or $x \notin A$. If $x \notin A$, then $x \in$ __________. In either case $x \in$ __________. Hence __________ $\subseteq$ __________. ▮

5.4 Fill in the blanks in the proof of the following theorem.

> **THEOREM:** $A \subseteq B$ iff $A \cup B = B$.
>
> **Proof:** Suppose that $A \subseteq B$. If $x \in A \cup B$, then $x \in A$ or $x \in$ __________. Since $A \subseteq B$, in either case we have $x \in B$. Thus __________ $\subseteq$ __________. On the other hand, if $x \in$ __________, then $x \in A \cup B$, so __________ $\subseteq$ __________. Hence $A \cup B = B$.
>
> Conversely, suppose that $A \cup B = B$. If $x \in A$, then $x \in$ __________. But $A \cup B = B$, so $x \in$ __________. Thus __________ $\subseteq$ __________. ▮

5.5 Fill in the blanks in the proof of the following theorem.

> **THEOREM:** $A \subseteq B$ iff $A \cap B = A$.
>
> **Proof:** Suppose that $A \subseteq B$. If $x \in A \cap B$, then clearly $x \in A$. Thus $A \cap B \subseteq A$. On the other hand, ____. Thus $A \subseteq A \cap B$, and we conclude that $A \cap B = A$.
>
> Conversely, suppose that $A \cap B = A$. If $x \in A$, then ____. Thus $A \subseteq B$. ▮

5.6 Prove that the empty set is unique. That is, suppose that A and B are empty sets and prove that $A = B$.

5.7 Prove: If $U = A \cup B$ and $A \cap B = \emptyset$, then $A = U \backslash B$.

5.8 Prove: $(A \backslash B) \cup (B \backslash A) = (A \cup B) \backslash (A \cap B)$.

5.9 Prove: $A \cap B$ and $A \backslash B$ are disjoint and $A = (A \cap B) \cup (A \backslash B)$.

5.10 Prove: $A \cap B = A \backslash (A \backslash B)$.

5.11 Use Venn diagrams with three overlapping circles to illustrate the identity in Theorem 5.13(d): $A \cup (B \cap C) = (A \cup B) \cap (A \cup C)$.

5.12 Use Venn diagrams with three overlapping circles to illustrate the identity in Theorem 5.13(f): $A \backslash (B \cup C) = (A \backslash B) \cap (A \backslash C)$.

5.13 Finish the proof of Theorem 5.13.

***5.14** Let $\{A_j : j \in J\}$ be an indexed family of sets and let B be a set. Prove the following generalizations of Theorem 5.13.

(a) $B \cup \left[\bigcap_{j \in J} A_j\right] = \bigcap_{j \in J} (B \cup A_j)$

(b) $B \cap \left[\bigcup_{j \in J} A_j\right] = \bigcup_{j \in J} (B \cap A_j)$

(c) $B \backslash \left[\bigcup_{j \in J} A_j\right] = \bigcap_{j \in J} (B \backslash A_j)$

(d) $B \backslash \left[\bigcap_{j \in J} A_j\right] = \bigcup_{j \in J} (B \backslash A_j)$

5.15 Find $\bigcup_{B \in \mathscr{B}} B$ and $\bigcap_{B \in \mathscr{B}} B$ for each collection $\mathscr{B}$.

(a) $\mathscr{B} = \left\{\left[1, 1 + \frac{1}{n}\right] : n \in \mathbb{N}\right\}$

(b) $\mathscr{B} = \left\{\left(1, 1 + \frac{1}{n}\right) : n \in \mathbb{N}\right\}$

(c) $\mathscr{B} = \{[2, x] : x \in \mathbb{R} \text{ and } x > 2\}$

(d) $\mathscr{B} = \{[0, 3], (1, 5), [2, 4)\}$

Section 6 RELATIONS

Ordered Pairs

In our discussion of sets in Section 5 we noted that the order of the elements is not important. Thus the set $\{a, b\}$ is the same as the set $\{b, a\}$. There are times, however, when order *is* important. For example,

* Exercises that are used in later sections are marked with an asterisk.

in plane analytic geometry the coordinates of a point (x, y) represent an *ordered pair* of numbers. The point $(1, 3)$ is different from the point $(3, 1)$. When we wish to indicate that a set of two elements a and b is *ordered*, we enclose the elements in parentheses: (a, b). Then a is called the first element and b is called the second.† The important property of ordered pairs is that

$$(a, b) = (c, d) \quad \text{iff} \quad a = c \text{ and } b = d.$$

So far we have not really told you what ordered pairs *are*; we have only identified the property that they must satisfy. The question of what they actually are is more subtle. It is possible to give a precise definition of an ordered pair, and we shall do so in a moment. But first one word of caution: Do not expect too much from the formal definition. It will show how the existence of ordered pairs can be derived from set theory, but it will probably not add to your intuitive understanding of the concept at all.

6.1 DEFINITION $(a, b) = \{\{a\}, \{a, b\}\}$.

This definition states that the ordered pair (a, b) is a set consisting of two elements: $\{a\}$ and $\{a, b\}$. The acceptability of the definition depends on the ordered pairs so defined actually having the property expected of them. This we prove in the following theorem.

6.2 THEOREM $(a, b) = (c, d)$ iff $a = c$ and $b = d$.

Proof: If $a = c$ and $b = d$, then

$$(a, b) = \{\{a\}, \{a, b\}\} = \{\{c\}, \{c, d\}\} = (c, d).$$

Conversely, suppose that $(a, b) = (c, d)$. Then by our definition we have $\{\{a\}, \{a, b\}\} = \{\{c\}, \{c, d\}\}$. We wish to conclude that $a = c$ and $b = d$. To this end we consider two cases depending on whether $a = b$ or $a \neq b$.

If $a = b$, then $\{a\} = \{a, b\}$, so $(a, b) = \{\{a\}\}$. Since $(a, b) = (c, d)$, we then have

$$\{\{a\}\} = \{\{c\}, \{c, d\}\}.$$

The set on the left has only one member, $\{a\}$. Thus the set on the right can have only one member, so $\{c\} = \{c, d\}$, and we conclude that $c = d$. But then $\{\{a\}\} = \{\{c\}\}$, so $\{a\} = \{c\}$ and $a = c$. Thus $a = b = c = d$.

† The notation used for an ordered pair (a, b) is the same as that used for an open interval of real numbers $\{x \in \mathbb{R}: a < x < b\}$. The context will make clear which meaning is intended.

On the other hand, if $a \neq b$, then from the preceding argument it follows that $c \neq d$. Since $(a, b) = (c, d)$, we must have

$$\{a\} \in \{\{c\}, \{c, d\}\},$$

which means that $\{a\} = \{c\}$ or $\{a\} = \{c, d\}$. In either case we have $c \in \{a\}$, so $a = c$. Again, since $(a, b) = (c, d)$, we must also have

$$\{a, b\} \in \{\{c\}, \{c, d\}\}.$$

Thus $\{a, b\} = \{c\}$ or $\{a, b\} = \{c, d\}$. But $\{a, b\}$ has two distinct members and $\{c\}$ has only one, so we must have $\{a, b\} = \{c, d\}$. Now $a = c$, $a \neq b$, and $b \in \{c, d\}$, which implies that $b = d$. ∎

6.3 PRACTICE Let $A = \{1, 2\}$. List all the ordered pairs (x, y) such that $x \in A$ and $y \in A$.

Cartesian Products

6.4 DEFINITION If A and B are sets, then the **Cartesian product** (or **cross product**) of A and B, written $A \times B$, is the set of all ordered pairs (a, b) such that $a \in A$ and $b \in B$. In symbols,

$$A \times B = \{(a, b)\colon a \in A \text{ and } b \in B\}.$$

6.5 EXAMPLE If A and B are intervals of real numbers, then using the familiar Cartesian coordinate system with A on the horizontal axis and B on the vertical axis, $A \times B$ is represented by a rectangle. For example, if A is the interval $[1, 4)$ and B is the interval $(2, 4]$, then $A \times B$ is the rectangle shown in Figure 6.1, where the solid line indicates that the edge is included in $A \times B$ and the dashed line indicates that the edge is not included.

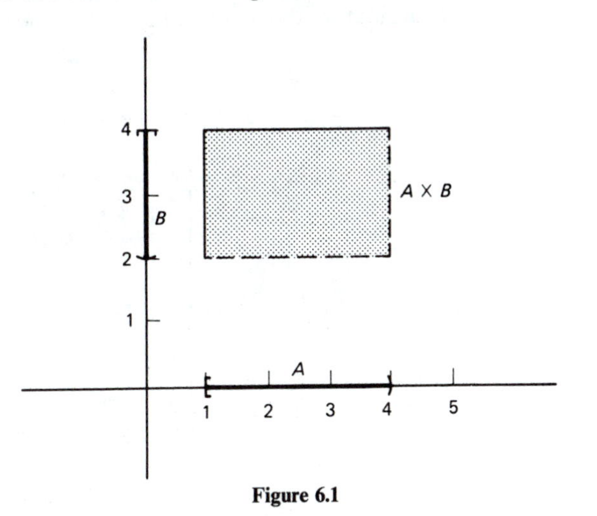

Figure 6.1

6.6 PRACTICE Consider the following intervals:

$$A = (1, 5], \qquad B = [1, 4)$$
$$C = [4, 7), \qquad D = (2, 5).$$

Draw a sketch as in Figure 6.1 to illustate that

$$(A \times B) \cap (C \times D) = (A \cap C) \times (B \cap D).$$

Relations

Intuitively, a relation between two objects a and b is a condition involving a and b that is either true or false. For example, "less than" is a relation between positive integers. We have

$$1 < 3 \quad \text{is true}, \qquad 2 < 7 \quad \text{is true}, \qquad 5 < 4 \quad \text{is false},$$

and in general $a < b$ is either true or false for any $a, b \in \mathbb{N}$.

When considering a relation between two objects, it is necessary to know which object comes first. For instance, $1 < 3$ is true but $3 < 1$ is false. Thus it is natural for the formal definition of a relation to depend on the concept of an ordered pair. Essentially, we collect together all the ordered pairs of objects where the first element is related to the second element and identify this collection of ordered pairs as the relation.

6.7 DEFINITION Let A and B be sets. A **relation between** A **and** B is any subset R of $A \times B$. We say that $a \in A$ and $b \in B$ are **related** by R if $(a, b) \in R$, and we often denote this by writing "aRb." If $B = A$, then we speak of a relation $R \subseteq A \times A$ being a relation on A.

6.8 EXAMPLE Let R be the subset of $\mathbb{N} \times \mathbb{N}$ consisting of the points with positive integral coordinates above the line $y = x$. (See Figure 6.2.) Then $(a, b) \in R$ iff

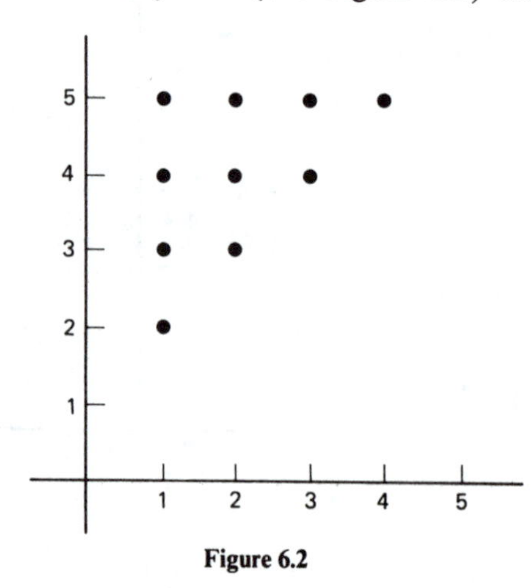

Figure 6.2

$a < b$. That is, R is precisely the relation "less than" defined on the positive integers $\mathbb{N}$.

Equivalence Relations

Certain relations are singled out because they possess the properties naturally associated with the idea of equality. They are called equivalence relations, as we see in the following definition.

6.9 DEFINITION A relation R on a set S is an **equivalence relation** if it has the following properties for all x, y, z in S:

(a) xRx (reflective property)
(b) If xRy, then yRx. (symmetric property)
(c) If xRy and yRz, then xRz. (transitive property)

6.10 EXAMPLES

(a) The relation "$\leqslant$" defined on $\mathbb{N}$ is reflexive and transitive, but not symmetric. Thus it is not an equivalence relation.
(b) When considering lines in the plane, the relation "is parallel to" is reflexive (if we agree that a line is parallel to itself), symmetric, and transitive. Hence it is an equivalence relation.
(c) Let S be the set of all people who live in Chicago, and suppose that two people x and y are related by R if x lives within a mile of y. Then R is reflexive and symmetric, but not transitive.

6.11 PRACTICE Determine which of the three properties (reflexive, symmetric, and transitive) apply to each relation.

(a) Let S be the set of all lines in the plane and let R be the relation "is perpendicular to."
(b) Let S be the set of real numbers and let R be the relation "$>$".
(c) Let S be the set $\mathbb{Z}$ of all integers and let $R = \{(m, n) \in \mathbb{Z} \times \mathbb{Z}: m - n$ is even$\}$.

Given an equivalence relation R on a set S, it is natural to group together all the elements that are related to a particular element. More precisely, we define the **equivalence class** (with respect to R) of $x \in S$ to be the set

$$E_x = \{y \in S: yRx\}.$$

Since R is reflexive, each element of S is in some equivalence class. Furthermore, two different equivalence classes must be disjoint. That is, if two equivalence classes overlap, they must be equal. To see this, suppose that $w \in E_x \cap E_y$. Then for any $x' \in E_x$ we have $x'Rx$. But $w \in E_x$, so wRx and, by symmetry, xRw. Also, $w \in E_y$, so wRy. Using transitivity twice we have $x'Ry$, so that $x' \in E_y$ and $E_x \subseteq E_y$. The reverse inclusion follows in a similar manner.

Thus we see that an equivalance relation R on a set S breaks S up into disjoint pieces in a natural way. These pieces are an example of a partition.

6.12 DEFINITION A **partition** of a set S is a collection $\mathscr{T}$ of nonempty subsets of S such that

(a) Each $x \in S$ belongs to some subset $A \in \mathscr{T}$.
(b) If $A, B \in \mathscr{T}$ and $A \neq B$, then $A \cap B = \varnothing$.

6.13 EXAMPLE Let S be the set of all students in a particular university. For x and y in S, define xRy iff x and y were born in the same calendar year. Then R is an equivalence relation and a typical equivalence class is the set of all students who were born in a particular year. For example, if student x was born in 1967, then E_x consists of all the students that were born in 1967. This relation partitions S into disjoint subsets, where students born in the same year are grouped together.

6.14 PRACTICE In Example 6.13, if $y \in E_x$, does this mean that x and y are the same age?

Not only does an equivalance relation on a set S determine a partition of S, but the partition can be used to determine the relation. We formalize this in the following theorem.

6.15 THEOREM Let R be an equivalance relation on a set S. Then $\{E_x : x \in S\}$ is a partition of S. The relation "belongs to the same piece as" is the same as R. Conversely, if $\mathscr{T}$ is a partition of S, let R be defined by xRy iff x and y are in the same piece of the partition. Then R is an equivalence relation and the corresponding partition into equivalence classes is the same as $\mathscr{T}$.

Proof: Let R be an equivalence relation on S. We have already shown that $\{E_x : x \in S\}$ is a partition. Now suppose that $\tilde{R}$ is the relation "belongs to the same piece (equivalence class) as." Then

$$
\begin{aligned}
x\tilde{R}y \quad &\text{iff} \quad x, y \in E_z \text{ for some } z \in S \\
&\text{iff} \quad xRz \text{ and } yRz \text{ for some } z \in S \\
&\text{iff} \quad xRy.
\end{aligned}
$$

Thus $\tilde{R}$ and R are the same.

Conversely, suppose that $\mathscr{T}$ is a partition of S and let R be defined by xRy iff x and y are in the same piece of the partition. Clearly, R is reflexive and symmetric. To see that R is transitive, suppose that xRy and yRz. Then $y \in E_x \cap E_z$. But this implies that

$E_x = E_z$ by the contrapositive of 6.12(b), so xRz. Finally, the equivalence classes of R correspond to the pieces of $\mathscr{T}$ because of the way R was defined. ▌

6.16 PRACTICE Let S be the set of all atoms in the universe and let $R = \{(x, y) \in S \times S: x$ and y have the same number of protons$\}$. Suppose that x is an atom with one proton. What is a natural name for E_x? Suppose that y is an atom with six protons. What is the natural name for E_y?

ANSWERS TO PRACTICE PROBLEMS

6.3 (1, 1), (1, 2), (2, 1), (2, 2)

6.6

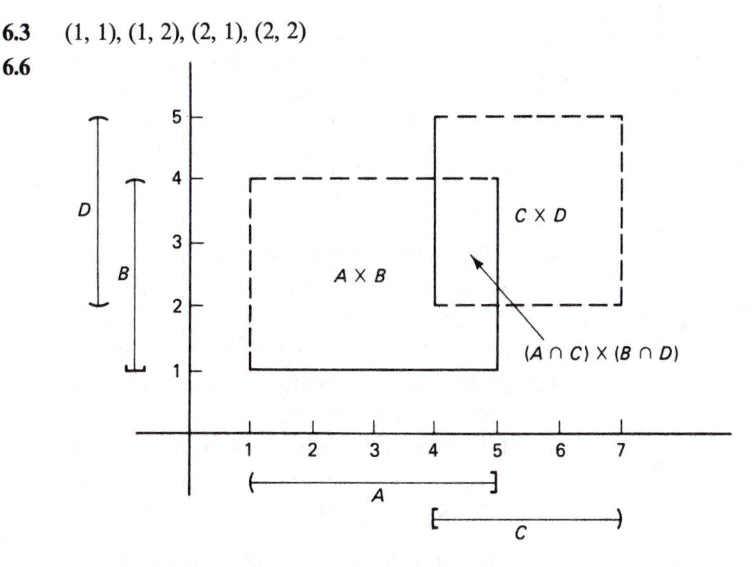

Note: in displaying intervals A and C we have offset them from the axis for the sake of clarity. A similar comment applies to intervals B and D.

6.11 (a) symmetric only; (b) transitive only; (c) all three

6.14 No. Two people born in the same year will be different ages on any date between their birthdays.

6.16 E_x is hydrogen and E_y is carbon.

EXERCISES

6.1 (a) Define an ordered triple (a, b, c) to be equal to $((a, b), c)$. Prove that $(a, b, c) = (d, e, f)$ iff $a = d$, $b = e$, and $c = f$.

(b) On the basis of our definition of an ordered pair (a, b) as $\{\{a\}, \{a, b\}\}$, we might be tempted to define an ordered triple (a, b, c) as $\{\{a\}, \{a, b\}, \{a, b, c\}\}$. Show by means of an example that this will not work. That is, find two different ordered triples that have equivalent representations in this set notation.

6.2 Prove or give a counterexample: $A \times B = B \times A$.

6.3 Fill in the blanks in the proof of the following theorem.

THEOREM: $(A \cap B) \times C = (A \times C) \cap (B \times C)$

Proof: Let $(x, y) \in (A \cap B) \times C$. Then $x \in$ __________ and $y \in$ __________. Since $x \in A \cap B$, $x \in$ __________ and $x \in$ __________. Thus $(x, y) \in$ __________ and $(x, y) \in$ __________. Hence $(x, y) \in (A \times C) \cap (B \times C)$, so __________ $\subseteq$ __________.

On the other hand, suppose that $(x, y) \in$ __________. Then $(x, y) \in$ __________ and $(x, y) \in$ __________. Since $(x, y) \in A \times C$, $x \in$ __________ and $y \in$ __________. Since $(x, y) \in B \times C$, __________ and __________. Thus $x \in A \cap B$, so __________ $\in$ __________ and __________ $\subseteq$ __________. ∎

6.4 Prove or give a counterexample.
(a) $(A \cup B) \times C = (A \times C) \cup (B \times C)$
(b) $(A \times B) \cap (C \times D) = (A \cap C) \times (B \cap D)$
(c) $(A \times B) \cup (C \times D) = (A \cup C) \times (B \cup D)$

6.5 Determine which of the three properties (reflexive, symmetric, and transitive) apply to each relation.
(a) Let R be the relation on $\mathbb{N}$ given by xRy iff x divides y.
(b) Let X be a set and let R be the relation "$\subseteq$" defined on subsets of X.
(c) Let S be the set of people in the school. Define R on S by xRy iff "x likes y."
(d) Let R be the relation on the real numbers given by xRy iff $x - y$ is rational.
(e) Let R be the relation on the real numbers given by xRy iff $x - y$ is irrational.
(f) Let R be the relation on the real numbers given by xRy iff $(x - y)^2 < 0$.
(g) Let R be the relation on the real numbers given by xRy iff $|x - y| \leqslant 2$.

6.6 Find examples of relations with the following properties.
(a) Reflexive, but not symmetric and not transitive.
(b) Symmetric, but not reflexive and not transitive.
(c) Transitive, but not reflexive and not symmetric.
(d) Reflexive and symmetric, but not transitive.
(e) Reflexive and transitive, but not symmetric.
(f) Symmetric and transitive, but not reflexive.

6.7 Let S be the Cartesian coordinate plane $\mathbb{R} \times \mathbb{R}$ and define a relation R on S by $(a, b)R(c, d)$ iff $a = c$. Verify that R is an equivalence relation and describe a typical equivalence class $E_{(a,b)}$.

6.8 Let S be the set $\mathbb{Z}$ of all integers and let $R = \{(m, n) \in \mathbb{Z} \times \mathbb{Z}: m - n \text{ is even}\}$. Verify that R is an equivalence relation and describe the equivalence class E_5. How many distinct equivalence classes are there?

Section 7 FUNCTIONS

We now turn our attention to a consideration of the concept of a function. The reader no doubt thinks of a function f as a formula or a rule that enables us to compute the function value $f(x)$ given any specific value for x. For example, if we have the formula

$$f(x) = x^2 + 3,$$

we can compute $f(5)$ by taking 5^2 and adding 3. Thus the real number 5 is made to correspond to the real number 28, and the function f is seen to establish a correspondence between real numbers and certain other real numbers. Thinking of a function in this way as a formula is helpful in many contexts, but it is inadequate for some of the things we wish to do in analysis.

In searching for a more general understanding of a function, we need to hold on to the idea of a correspondence between sets, but not require that it be described by a formula. What is important is that a given element in the first set cannot correspond to two different elements in the second set. Thus we see that a function is a special kind of a relation. We make this precise in the following definition.

7.1 DEFINITION Let A and B be sets. A **function** between A and B is a nonempty relation $f \subseteq A \times B$ such that if $(a, b) \in f$ and $(a, b') \in f$, then $b = b'$. The **domain** of f is the set of all first elements of members of f and the **range** of f is the set of all second elements of members of f. Symbolically,

$$\text{domain} f = \{a: \exists\, b \in B \ni (a, b) \in f\}$$

$$\text{range} f = \{b: \exists\, a \in A \ni (a, b) \in f\}.$$

The set B is referred to as the **codomain** of f. If it happens that the domain of f is equal to all of A, then we say f is a function **from** A into B and we write $f: A \to B$.

If (x, y) is a member of f, we often say that f maps x onto y or that y is the image of x under f. It is also customary to write $y = f(x)$ instead of $(x, y) \in f$. This agrees with the familiar usage when f is described by a formula, but also applies in more general settings.

When a function consists of just a few ordered pairs, it can be identified simply by listing them. Usually, however, there are too many to list, so the function is identified by specifying the domain and giving a rule for determining the unique second element in the ordered pair that corresponds to any particular first element. When this rule is a formula, we are back to the intuitive notion of a function with which we began the section. Thus to say that a function f is given by the formula $f(x) = x^2 + 3$ means that

$$f = \{(x, y): y = x^2 + 3 \text{ and } x \in \mathbb{R}\}$$

or

$$f = \{(x, f(x)): f(x) = x^2 + 3 \text{ and } x \in \mathbb{R}\}$$

or

$$f = \{(x, x^2 + 3): x \in \mathbb{R}\}.$$

The domain of the function would either be obtained from the context or by stating it explicitly. Unless told otherwise, when a function is given by a formula, the domain is taken to be the largest subset of $\mathbb{R}$ for which the formula will always yield a real number.

7.2 PRACTICE Let $A = \{1, 2, 3\}$ and $B = \{2, 4, 6, 8\}$. Which of the following relations are functions between A and B? Which are functions from A into B?

(a) $\{(1, 2), (2, 6), (3, 4), (2, 8)\}$
(b) $\{(1, 4), (3, 8)\}$
(c) $\{(1, 6), (2, 6), (3, 2)\}$
(d) $\{(1, 8), (2, 2), (3, 4)\}$

7.3 EXAMPLE Let's look more carefully at Practice 7.2(c). Notice that it is permitted for a member of B to appear in more than one ordered pair in the function. It may also be the case that some members of B, such as 4 and 8, do not appear at all. If we call the given function f, we may write $f: A \to B$, where $A = \{1, 2, 3\}$ and $B = \{2, 4, 6, 8\}$. If $C = \mathbb{R}$ and $D = \{2, 6\}$, it is equally correct to write $f: A \to C$ or $f: A \to D$. The latter description is particularly good, since there are no extraneous members in D. That is, D is equal to the range of f.

Properties of Functions

7.4 DEFINITION A function $f: A \to B$ is called **surjective** (or is said to map A **onto** B) if $B = \text{range } f$.

The question of whether or not a function is surjective depends on the choice of codomain. The function can always be made surjective by restricting the codomain to being equal to the range, but sometimes this is not convenient. For the function f given by the formula $f(n) = (n!)^{1/n}$, there is no simple description for the range. It is easier just to write f: $\mathbb{N} \to \mathbb{R}$ and not be more precise.

If it happens, as in Practice 7.2(d), that no member of the codomain appears more than once as a second element in one of the ordered pairs, then we have another important type of function:

7.5 DEFINITION A function $f: A \to B$ is called **injective** (or **one-to-one**) if, for all a and a' in A, $f(a) = f(a')$ implies that $a = a'$.

If a function is both surjective and injective, then it is particularly well behaved.

7.6 DEFINITION A function $f: A \to B$ is called **bijective** if it is both surjective and injective.

7.7 EXAMPLE Consider the function given by the formula $f(x) = x^2$. If we take $\mathbb{R}$ for both the domain and codomain so that $f: \mathbb{R} \to \mathbb{R}$, then f is not surjective because there is no real number that maps onto -1. If we limit the codomain to be the set $[0, \infty)$, then the function $f: \mathbb{R} \to [0, \infty)$ is surjective.

Since $f(-2) = f(2)$, we see that f is not injective when defined on all of $\mathbb{R}$. But by restricting f to be defined only on $[0, \infty)$, it becomes injective. Thus $f: [0, \infty) \to [0, \infty)$ is bijective.

7.8 EXAMPLE Let A be a nonempty set and let S be a subset of A. We may define a function $\chi_S: A \to \{0, 1\}$ by

$$\chi_S(a) = \begin{cases} 1, & \text{if } a \in S \\ 0, & \text{if } a \notin S. \end{cases}$$

This function is called the **characteristic function** (or **indicator function**) of S and is widely used in probability and statistics. If S is a nonempty proper subset of A, then χ_S is surjective. If $S = \varnothing$ or $S = A$, then χ_S is not surjective.

7.9 PRACTICE Let χ_S be given as in Example 7.8. Under what conditions on A and S will χ_S be injective?

Visualizing Functions

There are two helpful ways of picturing a function. The first is by using the familiar Cartesian coordinate system to display the graph of the function. Of course, this technique applies only to functions whose domain and codomain are subsets of $\mathbb{R}$. Thus if $f: \mathbb{R} \to \mathbb{R}$ is the function

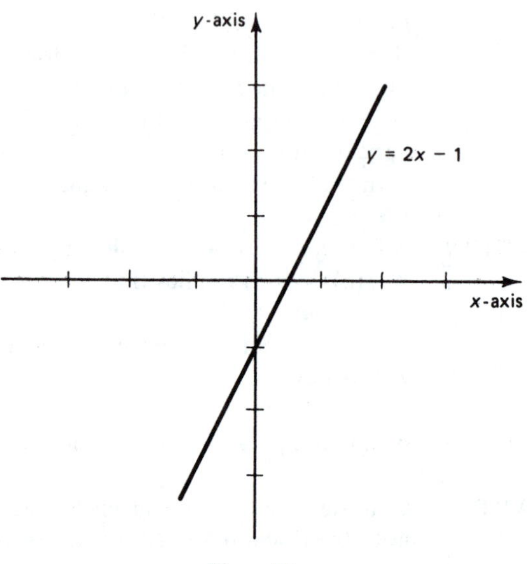

Figure 7.1

defined by $f(x) = 2x - 1$, we obtain the graph pictured in Figure 7.1. In set-theoretic terms, the plane is $\mathbb{R} \times \mathbb{R}$ and a point (x, y) is on the graph of f iff $y = 2x - 1$. That is, the graph of f is the set

$$\{(x, y): y = 2x - 1 \text{ and } x \in \mathbb{R}\}$$

or

$$\{(x, f(x)): f(x) = 2x - 1 \text{ and } x \in \mathbb{R}\}.$$

But this is precisely what f is in terms of Definition 7.1. That is, since we defined functions in terms of ordered pairs, what we ordinarily think of as the graph of a function is really the same thing as the function itself.

If the domain or codomain of a function $f: A \to B$ is not a subset of $\mathbb{R}$, we may visualize f by a diagram as in Figure 7.2. We think of f as

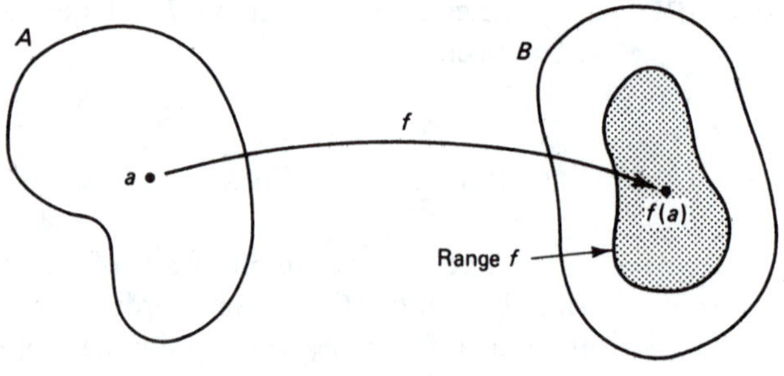

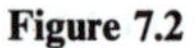
Figure 7.2

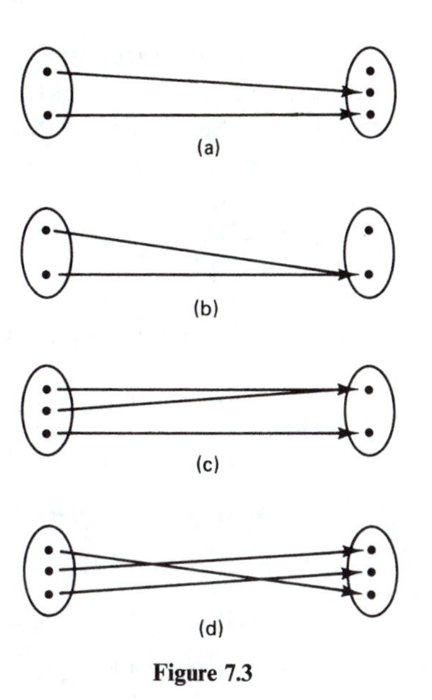

Figure 7.3

transforming its domain A into its range in B. We may even draw arrows from a few points in its domain to their images in B to illustrate its behavior. We often use this kind of geometrical picture even when A and B are not subsets of the plane.

7.10 PRACTICE Consider the four functions pictured in Figure 7.3. For each function, the domain and codomain are sets consisting of two or three points as indicated. Classify each function as being surjective, injective, or bijective.

Functions Acting on Sets

When thinking of a function as transforming its domain into its range, we may wish to consider what happens to certain subsets of the domain. Or we may wish to identify the set of all points in the domain that are mapped into a particular subset of the range. To do this we use the following notation:

7.11 NOTATION Suppose that $f: A \to B$. If $C \subseteq A$, we let $f(C)$ represent the subset $\{f(x): x \in C\}$ of B. The set $f(C)$ is called the **image** of C in B. If $D \subseteq B$, we let $f^{-1}(D)$ represent the subset $\{x \in A: f(x) \in D\}$ of A. The set $f^{-1}(D)$ is called the **pre-image** of D in A (or "f inverse of D").

Before illustrating these ideas with an example, we should comment that the symbol f^{-1} is not to be thought of as an inverse function applied to points in the range of f. In particular, given a point y in B it makes no sense to talk about $f^{-1}(y)$ as a point in A. We shall see a bit later how this idea can be made meaningful *in some cases*; but for now we can only apply f^{-1} to a subset of B and by so doing we obtain a subset of A.

7.12. EXAMPLE Let $f\colon \mathbb{R} \to \mathbb{R}$ be given by $f(x) = x^2$. Then the following hold.

If $C_1 = [0, 2]$, then $f(C_1) = [0, 4]$.

If $C_2 = [-1, 2]$, then $f(C_2) = [0, 4]$.

If $C_3 = [-3, -2] \cup [1, 2]$, then $f(C_3) = [1, 9]$.

If $D_1 = [0, 4]$, then $f^{-1}(D_1) = [-2, 2]$.

If $D_2 = [-5, 4]$, then $f^{-1}(D_2) = [-2, 2]$.

If $D_3 = [1, 4]$, then $f^{-1}(D_3) = [-2, -1] \cup [1, 2]$.

7.13 PRACTICE Let $f\colon \mathbb{R} \to \mathbb{R}$ be given by $f(x) = \sin x$. Find the following: (a) $f([0, \pi])$; (b) $f([0, 8\pi])$; (c) $f^{-1}([-\pi, \pi])$.

Given a function $f\colon A \to B$, there are many relationships that hold between the images and pre-images of subsets of A and B. For example, suppose that we start with a subset C of A. If we map C onto $f(C)$ and then bring it back to $f^{-1}(f(C))$, do we always end up with C again? The answer is no, in general, but $f^{-1}(f(C))$ always contains C. In the following theorem we state a number of relationships like this that apply to images and pre-images. The proofs are all straightforward and provide a good opportunity for you to practice your skills at writing proofs. Thus we sketch only two of the proofs and leave the others for the exercises.

7.14. THEOREM Suppose that $f\colon A \to B$. Let C, C_1, and C_2 be subsets of A and let D, D_1, and D_2 be subsets of B. Then the following hold:

(a) $C \subseteq f^{-1}[f(C)]$
(b) $f[f^{-1}(D)] \subseteq D$
(c) $f(C_1 \cap C_2) \subseteq f(C_1) \cap f(C_2)$
(d) $f(C_1 \cup C_2) = f(C_1) \cup f(C_2)$
(e) $f^{-1}(D_1 \cap D_2) = f^{-1}(D_1) \cap f^{-1}(D_2)$
(f) $f^{-1}(D_1 \cup D_2) = f^{-1}(D_1) \cup f^{-1}(D_2)$
(g) $f^{-1}(B \backslash D) = A \backslash f^{-1}(D)$

7.15 PRACTICE Complete the proofs of the following parts of Theorem 7.14.

(c) Let $y \in f(C_1 \cap C_2)$. Then there exists a point x in $C_1 \cap C_2$ such that __________. Since $x \in C_1 \cap C_2$, $x \in$ __________ and $x \in$ __________. But then $f(x) \in$ __________ and $f(x) \in$ __________, so $y = f(x) \in$ __________.

(f) Let $x \in f^{-1}(D_1 \cup D_2)$. Then $f(x) \in$ __________, so $f(x) \in D_1$ or $f(x) \in D_2$. If $f(x) \in D_1$, then $x \in$ __________. If $f(x) \in D_2$, then __________. In either case, $x \in f^{-1}(D_1) \cup f^{-1}(D_2)$.

Conversely, suppose $x \in$ __________. Then $x \in f^{-1}(D_1)$ or $x \in f^{-1}(D_2)$. If $x \in f^{-1}(D_1)$, then $f(x) \in$ __________. If $x \in f^{-1}(D_2)$, then __________. In either case, $f(x) \in D_1 \cup D_2$, so that $x \in$ __________. ∎

While Theorem 7.14 states the strongest results that hold in general, if we apply certain restrictions on the functions involved, then the containment symbols in parts (a), (b), and (c) may be replaced by equality.

7.16 THEOREM Suppose that $f: A \to B$. Let C, C_1, and C_2 be subsets of A and let D be a subset of B. Then the following hold:

(a) If f is injective, then $f^{-1}[f(C)] = C$.
(b) If f is surjective, then $f[f^{-1}(D)] = D$.
(c) If f is injective, then $f(C_1 \cap C_2) = f(C_1) \cap f(C_2)$.

Proof: Parts (a) and (b) are left to the exercises. To prove (c) we have only to show that $f(C_1) \cap f(C_2) \subseteq f(C_1 \cap C_2)$, since the converse inclusion is Theorem 7.14(c). To this end, let $y \in f(C_1) \cap f(C_2)$. Then $y \in f(C_1)$ and $y \in f(C_2)$. It follows that there exists a point x_1 in C_1 such that $f(x_1) = y$. Similarly, there exists a point x_2 in C_2 such that $f(x_2) = y$. Since f is injective and $f(x_1) = y = f(x_2)$, we must have $x_1 = x_2$. That is, $x_1 \in C_1 \cap C_2$. But then $y = f(x_1) \in f(C_1 \cap C_2)$. ∎

Composition of Functions

If f and g are functions with $f: A \to B$ and $g: B \to C$, then for any $a \in A$, $f(a) \in B$. But B is the domain of g, so g can be applied to $f(a)$. This yields $g(f(a))$, an element of C. (See Figure 7.4.) Thus we have established a correspondence between a in A and $g(f(a))$ in C. This correspondence is called the **composition** function of f and g and is denoted by $g \circ f$. Thus

$$(g \circ f)(a) = g(f(a)).$$

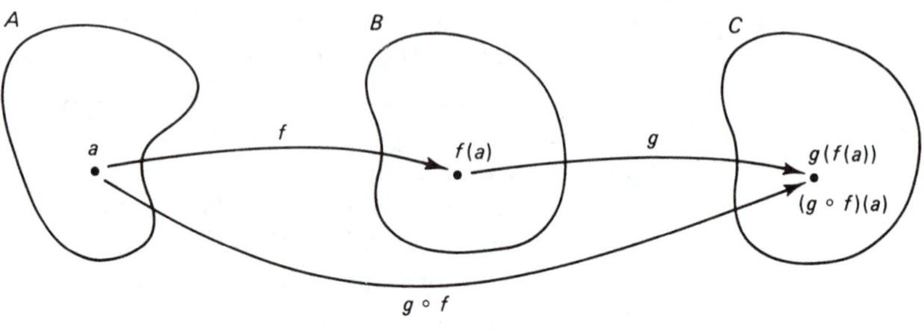

Figure 7.4

In terms of ordered pairs we have

$$g \circ f = \{(a, c) \in A \times C : \exists\, b \in B \ni (a, b) \in f \text{ and } (b, c) \in g\}.$$

7.17 PRACTICE Find an example to show that the composition of two functions need not be commutative. That is, $g \circ f \neq f \circ g$.

Since the composition of two functions is not commutative, we must be careful about the order in which they are written. Unfortunately, the standard notation almost seems backward. That is, the function $g \circ f$ is evaluated by applying f first and then g. This is due to the formula notation

$$(g \circ f)(x) = g(f(x)),$$

in which we evaluate from the "inside out" instead of "left to right."

Fortunately, composition of functions is associative. This is easy to show and is left as an exercise. (See Exercise 7.4.) It is also true that composition preserves the properties of being surjective or injective.

7.18 THEOREM Let $f: A \to B$ and $g: B \to C$. Then

(a) If f and g are surjective, then $g \circ f$ is surjective,
(b) If f and g are injective, then $g \circ f$ is injective,
(c) If f and g are bijective, then $g \circ f$ is bijective.

Proof: (a) Since g is surjective, range $g = C$. That is, for any $c \in C$, there exists $b \in B$ such that $g(b) = c$. Now since f is surjective, there exists $a \in A$ such that $f(a) = b$. But then $(g \circ f)(a) = g(f(a)) = g(b) = c$, so $g \circ f$ is surjective.

(b) See Exercise 7.11.

(c) Follows from parts (a) and (b). ∎

Inverse Functions

Given a function $f: A \to B$, we have seen how f determines a relationship between subsets of B and subsets of A. That is, given $D \subseteq B$, we have the pre-image $f^{-1}(D)$ in A. We would like to be able to extend this idea so that f^{-1} can be applied to a point in B to obtain a point in A. That is, suppose $D = \{y\}$, where $y \in B$. There are two things that can prevent $f^{-1}(D)$ from being a point in A: It may be that $f^{-1}(D)$ is empty, and it may be that $f^{-1}(D)$ has several points instead of just one.

7.19 PRACTICE Given $f: A \to B$ and $y \in B$, under what conditions on f can we assert that there exists an x in A such that $f(x) = y$?

7.20 PRACTICE Given $f: A \to B$ and $y \in B$, under what conditions on f can we assert that there exists a unique x in A such that $f(x) = y$?

Given a bijection $f: A \to B$, we see that each y in B corresponds to exactly one x in A, the unique x such that $f(x) = y$. This correspondence defines a function from B into A called the inverse of f and denoted by f^{-1}. (See Figure 7.5.)

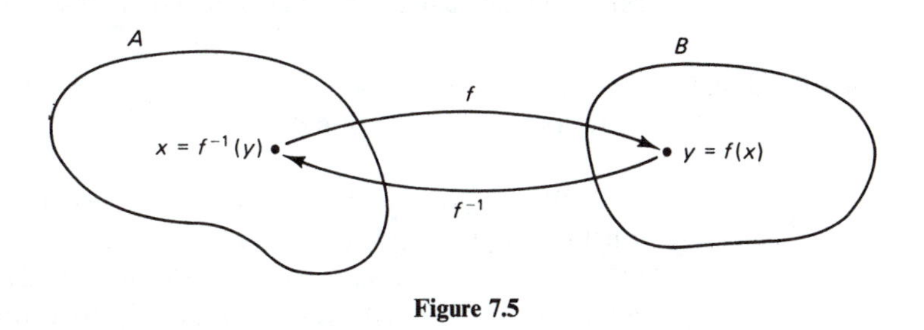

Figure 7.5

7.21 DEFINITION Let $f: A \to B$ be bijective. The **inverse function** of f is the function f^{-1} given by

$$f^{-1} = \{(y, x) \in B \times A: (x, y) \in f\}.$$

If $f: A \to B$ is bijective, then it follows that $f^{-1}: B \to A$ is also bijective. Indeed, since domain $f = A$ and range $f = B$, we have domain $f^{-1} = B$ and range $f^{-1} = A$. Thus f^{-1} is a mapping from B *onto* A. Since f is a function, a given x in A can correspond to only one y in B. This means that f^{-1} is injective, and hence bijective.

When f is followed by f^{-1}, the effect is to map $x \in A$ onto $f(x)$ in B and then back to x in A. That is, $(f^{-1} \circ f)(x) = x$, for every $x \in A$. A function defined on a set A that maps each element in A onto itself is called the **identity function** on A, and is denoted by i_A. Thus we can say that $f^{-1} \circ f = i_A$. Furthermore, if $f(x) = y$, then $x = f^{-1}(y)$, so that

$$(f \circ f^{-1})(y) = f(f^{-1}(y)) = f(x) = y.$$

Thus $f \circ f^{-1} = i_B$.

We summarize these results in the following theorem.

7.22 THEOREM Let $f: A \to B$ be bijective. Then

(a) $f^{-1}: B \to A$ is bijective,
(b) $f^{-1} \circ f = i_A$ and $f \circ f^{-1} = i_B$.

7.23 EXAMPLE Define $f: \mathbb{R} \to [0, \infty)$ by the formula $f(x) = x^2$. Then f is not injective since, for example, $f(3) = 9 = f(-3)$. Thus f does not have an inverse. If, however, we restrict the domain of f to just $[0, \infty)$, then the new function $h: [0, \infty) \to [0, \infty)$ given by $h(x) = x^2$ is injective. Indeed, if $h(a) = h(b)$ so that $a^2 = b^2$, then $a^2 - b^2 = 0$. But then $(a - b)(a + b) = 0$, so that $a = b$ or $a = -b$. Since both a and b are nonnegative, the possibility of $a = -b$ must be ruled out (unless $a = b = 0$). Thus $a = b$, and h is injective.

Furthermore, since each nonnegative number is the square of some nonnegative number, h is bijective. Thus h has an inverse function, usually called the **positive square-root function**, and we write $h^{-1}(y) = \sqrt{y}$.

7.24 PRACTICE Define $f: (-\infty, 0] \to [0, \infty)$ by $f(x) = x^2$. Observe that f is bijective. Describe f^{-1} by a formula.

We are now in a position to prove the main result that relates inverse functions and composition. It will be of particular use to us when we study infinite sets. The proof is based on the original definition of a function as a collection of ordered pairs and the observation that for a bijection f, $(a, b) \in f$ iff $(b, a) \in f^{-1}$.

7.25 THEOREM Let $f: A \to B$ and $g: B \to C$ be bijective. Then the composition $g \circ f: A \to C$ is bijective and $(g \circ f)^{-1} = f^{-1} \circ g^{-1}$.

Proof: We know from Theorem 7.18 that $g \circ f$ is bijective. Thus $g \circ f$ has an inverse. Now

$$g \circ f = \{(a, c): \exists\, b \in B \ni (a, b) \in f \text{ and } (b, c) \in g\},$$

so that

$$\begin{aligned}(g \circ f)^{-1} &= \{(c, a): \exists\, b \in B \ni (a, b) \in f \text{ and } (b, c) \in g\} \\ &= \{(c, a): \exists\, b \in B \ni (b, a) \in f^{-1} \text{ and } (c, b) \in g^{-1}\} \\ &= f^{-1} \circ g^{-1}. \ \blacksquare\end{aligned}$$

ANSWERS TO PRACTICE PROBLEMS

7.2 Part (a) is not a function since 2 is related to both 6 and 8. Part (b) is a function between A and B, but not a function from A into B since 2 in A is not related to anything in B. Parts (c) and (d) are both functions from A into B.

7.9 There are two possibilities for χ_S to be injective. Either A must consist of only two elements and exactly one of these elements is in S, or A must have only one member.

7.10 (a) injective only (b) neither surjective nor injective
(c) surjective only (d) bijective

7.13 (a) $[0, 1]$; (b) $[-1, 1]$; (c) $\mathbb{R}$

7.15 (c) Let $y \in f(C_1 \cap C_2)$. Then there exists a point x in $C_1 \cap C_2$ such that $\underline{f(x) = y}$. Since $x \in C_1 \cap C_2$, $x \in \underline{C_1}$ and $x \in \underline{C_2}$. But then $f(x) \in \underline{f(C_1)}$ and $\underline{f(x) \in f(C_2)}$, so $y = f(x) \in \underline{f(C_1) \cap f(C_2)}$.
(f) Let $x \in f^{-1}(D_1 \cup D_2)$. Then $\underline{f(x) \in D_1 \cup D_2}$, so $f(x) \in D_1$ or $f(x) \in D_2$. If $f(x) \in D_1$, then $x \in \underline{f^{-1}(D_1)}$. If $f(x) \in D_2$, then $\underline{x \in f^{-1}(D_2)}$. In either case, $x \in f^{-1}(D_1) \cup f^{-1}(D_2)$.
Conversely, suppose $x \in \underline{f^{-1}(D_1) \cup f^{-1}(D_2)}$. Then $x \in f^{-1}(D_1)$ or $x \in f^{-1}(D_2)$. If $x \in f^{-1}(D_1)$, $\underline{\text{then } f(x) \in D_1}$. If $x \in f^{-1}(D_2)$, then $\underline{f(x) \in D_2}$. In either case, $f(x) \in D_1 \cup D_2$, so that $\underline{x \in f^{-1}(D_1 \cup D_2)}$. ∎

7.17 There are many possible examples. A simple one is to define f by $f(x) = x^2$ and g by $g(x) = x + 1$. Then $(g \circ f)(x) = x^2 + 1$ but $(f \circ g)(x) = (x + 1)^2$.

7.19 f must be surjective so that range $f = B$.

7.20 f must be surjective to ensure that there is some x, and f must be injective to ensure that there is only one. Thus f must be bijective.

7.24 $f^{-1}(y) = -\sqrt{y}$

EXERCISES

7.1 Find the range of each function $f\colon \mathbb{R} \to \mathbb{R}$.
(a) $f(x) = x^2 + 1$
(b) $f(x) = (x + 3)^2 - 5$
(c) $f(x) = x^2 + 4x + 1$
(d) $f(x) = 2 \cos 3x$

7.2 (a) Let S be the set of all circles in the plane. Define $f\colon S \to [0, \infty)$ by $f(C) =$ the area of C, for all $C \in S$. Is f injective? Is f surjective?
(b) Let T be the set of all circles in the plane that are centered at the origin. Define $g\colon T \to [0, \infty)$ by $g(C) =$ the area of C, for all $C \in T$. Is g injective? Is g surjective?

7.3 Let f and g be functions. Prove that $f = g$ iff domain $f =$ domain g and for every $x \in$ domain f, $f(x) = g(x)$.

7.4 Suppose that $f\colon A \to B$, $g\colon B \to C$, and $h\colon C \to D$. Prove that $h \circ (g \circ f) = (h \circ g) \circ f$.

7.5 Let $f\colon A \to B$ and $g\colon B \to C$. Using the ordered pair definition of the composition $g \circ f$, prove that $g \circ f$ is a function and that $g\colon A \to C$.

7.6 In each part, find a function $f\colon \mathbb{N} \to \mathbb{N}$ that has the desired properties.

(a) surjective, but not injective
(b) injective, but not surjective
(c) neither surjective nor injective
(d) bijective

7.7 (a) Suppose that A has exactly two elements and B has exactly three. How many different functions are there from A to B? How many of these are injective? How many surjective?
(b) Suppose that A has exactly three elements and B has exactly two. How many different functions are there from A to B? How many of these are injective? How many surjective?
(c) Suppose that A has exactly m elements and B has exactly n (where $m, n \in \mathbb{N}$). How many different functions are there from A to B?

7.8 Find examples to show that equality does not hold in parts (a), (b), and (c) of Theorem 7.14. For instance, in part (a) find specific sets A, B, and C with $C \subseteq A$, and a specific function $f\colon A \to B$ such that $C \neq f^{-1}[f(C)]$.

7.9 Prove parts (a), (b), (d), (e), and (g) of Theorem 7.14.

7.10 Prove parts (a) and (b) of Theorem 7.16.

7.11 Prove Theorem 7.18(b). That is, suppose that $f\colon A \to B$ and $g\colon B \to C$ are both injective. Prove that $g \circ f\colon A \to C$ is injective.

7.12 Suppose that $f\colon A \to B$ and suppose that $C \subseteq A$ and $D \subseteq B$.

(a) Prove or give a counterexample: $f(C) \subseteq D$ iff $C \subseteq f^{-1}(D)$.
(b) What condition on f will ensure that $f(C) = D$ iff $C = f^{-1}(D)$? Prove your answer.

7.13 Suppose that $f\colon A \to B$ and let C be a subset of A.

(a) Prove or give a counterexample: $f(A \backslash C) \subseteq f(A) \backslash f(C)$.
(b) Prove or give a counterexample: $f(A) \backslash f(C) \subseteq f(A \backslash C)$.
(c) What condition of f will ensure that $f(A \backslash C) = f(A) \backslash f(C)$? Prove your answer.
(d) What condition on f will ensure that $f(A \backslash C) = B \backslash f(C)$? Prove your answer.

7.14 Find an example of functions $f\colon A \to B$ and $g\colon B \to C$ such that f and $g \circ f$ are both injective, but g is not injective.

7.15 Find an example of functions $f\colon A \to B$ and $g\colon B \to C$ such that g and $g \circ f$ are both surjective, but f is not surjective.

7.16 Suppose that $g\colon A \to C$ and $h\colon B \to C$. Prove that if h is bijective then there exists a function $f\colon A \to B$ such that $g = h \circ f$. *Hint*: Draw a picture.

7.17 Let $f\colon A \to B$ and suppose that there exists a function $g\colon B \to A$ such that $g \circ f = i_A$ and $f \circ g = i_B$.

(a) Prove that f is bijective.
(b) Prove that $g = f^{-1}$.

***7.18** Suppose that $f: A \to B$ is any function. Then a function $g: B \to A$ is called a

left inverse for f if $g(f(x)) = x$ for all $x \in A$,
right inverse for f if $f(g(y)) = y$ for all $y \in B$.

(a) Prove that f has a left inverse iff f is injective.
(b) Prove that f has a right inverse iff f is surjective.

7.19 Let S be a nonempty set and let $\mathscr{F}$ be the set of all functions that map S into S. Suppose that for every f and g in $\mathscr{F}$ we have

$$(f \circ g)(x) = (g \circ f)(x) \qquad \text{for all } x \in S.$$

Prove that S has only one element.

7.20 Suppose that $f: A \to B$. Define a relation R on A by xRy iff $f(x) = f(y)$.
(a) Prove that R is an equivalence relation on A.
(b) For any $x \in A$, let E_x be the equivalence class of x. That is,

$$E_x = \{y \in A: yRx\}.$$

Let E be the collection of all equivalence classes. That is,

$$E = \{E_x: x \in A\}.$$

Prove that the function $g: A \to E$ defined by $g(x) = E_x$ is surjective.
(c) Prove that the function $h: E \to B$ defined by $h(E_x) = f(x)$ is injective.
(d) Prove that $f = h \circ g$. [That is, $f(x) = h(g(x))$ for all $x \in A$.] Thus we conclude that any function can be written as the composition of a surjective function and an injective function.
(c) Let A be the set of all students in the school. Define $f: A \to [0, 200]$ by "$f(x)$ is the age of x." Describe the functions h and g as given above.

Section 8 CARDINALITY

How can we compare the sizes of two sets? If $S = \{x \in \mathbb{R}: x^2 = 9\}$, then $S = \{-3, 3\}$ and we say that S has two elements. If $T = \{1, 7, 11\}$, then T has three elements and we think of T as being "larger" than S. These intuitive ideas are fine for small (finite) sets, but how can we compare the size of (infinite) sets like $\mathbb{N}$ or $\mathbb{R}$?

We shall begin by deciding what it means for two sets to be the same size and then approach the question of comparing size. Certainly, it is reasonable to say that two sets S and T are the same size if there is a bijective function $f: S \to T$, for this function will set up a one-to-one correspondence between the elements of each set.

8.1 DEFINITION Two sets S and T are called **equinumerous**, and we write $S \sim T$, if there exists a bijective function from S onto T.

8.2 PRACTICE If $\mathscr{F}$ is a family of sets, then the concept of being equinumerous is a relation on $\mathscr{F}$. Show that "$\sim$" is an equivalence relation in the sense of Definition 6.9.

Cardinal Numbers

Since "$\sim$" is an equivalence relation, it partitions any family of sets into disjoint equivalence classes. With each equivalence class we associate a cardinal number that we think of as giving the size of the set. Technically, a cardinal number is sometimes defined to *be* an equivalence class determined by the relation $\sim$. But this raises questions about the domain of $\sim$. That is, just what sets do we include in the family $\mathscr{F}$ on which $\sim$ is defined? Since we wish to avoid these complications, it will be simpler merely to associate a cardinal number with each equivalence class. This leaves the question "What *is* a cardinal number?" unanswered, but it will be adequate for our purposes. Given any two sets, it makes sense to ask if they have the same cardinal number. They will iff they are equinumerous.

Using the concept of two sets being equinumerous, we can classify sets according to size.

8.3 DEFINITION A set S is said to be **finite** if $S = \varnothing$ or if there exists, $n \in \mathbb{N}$ and a bijection $f: \{1, 2, \ldots, n\} \to S$. If a set is not finite, it is said to be **infinite**.

It will be convenient to abbreviate the set $\{1, 2, \ldots, n\}$ by I_n. Thus we can say that S is finite iff $S = \varnothing$ or S is equinumerous with I_n for some $n \in \mathbb{N}$. The cardinal number of I_n is just n, and if $S \sim I_n$, we say that S has n elements. The cardinal number of $\varnothing$ is taken to be 0. If a cardinal number is not finite, it is called **transfinite**.

8.4 PRACTICE Suppose that S and T are both sets having n elements. Then from Definition 8.3 there exist bijections $f: I_n \to S$ and $g: I_n \to T$. Show that S and T are equinumerous directly by finding a bijection $h: S \to T$.

8.5 DEFINITION A set S is said to be **denumerable** if there exists a bijection $f: \mathbb{N} \to S$. If a set is finite or denumerable, it is called **countable**. If a set is not countable, it is **uncountable**.

In other words, a set is denumerable iff it is equinumerous with the set of natural numbers $\mathbb{N}$. The cardinal number of a denumerable set is denoted by $\aleph_0$.† Since the set $\mathbb{N}$ is not finite, there exists at least one

† The symbol $\aleph$ (read "aleph") is the first letter of the Hebrew alphabet.

infinite set.† It is not the case, however, that every infinite set has $\aleph_0$ as its cardinal number. That is, not every infinite set is denumerable. It will turn out that there are many different sizes of infinity. Before showing this surprising fact, let us look carefully at some of the properties of countable sets.

Countable Sets

8.6 EXAMPLE It would seem at first glance that the set $\mathbb{N}$ of natural numbers should be "bigger" than the set E of even natural numbers. Indeed, E is a proper subset of $\mathbb{N}$ and, in fact, it contains only "half" of $\mathbb{N}$. But what is "half" of $\aleph_0$? Our experience with finite sets is a poor guide here, for $\mathbb{N}$ and E are actually equinumerous! The function $f(n) = 2n$ is a bijection from $\mathbb{N}$ onto E, so E also has cardinality $\aleph_0$.

8.7 PRACTICE Find a bijection $f: \mathbb{N} \to \mathbb{Z}$, thereby showing that the set $\mathbb{Z}$ of all integers is also denumerable.

If a nonempty set S is finite, then there exists $n \in \mathbb{N}$ and a bijection $f: I_n \to S$. Using the function f, we can count off the members of S as follows: $f(1)$, $f(2)$, $f(3), \ldots, f(n)$. Letting $f(k) = s_k$ for $1 \leqslant k \leqslant n$, we obtain the more familiar notation $S = \{s_1, s_2, \ldots, s_n\}$. The same kind of counting process is possible for a denumerable set, and this is why both kinds of sets are called countable. For example, if T is denumerable, then there exists a bijection $g: \mathbb{N} \to T$, and we may write $T = \{g(1), g(2), g(3), \ldots\}$ or $T = \{t_1, t_2, t_3, \ldots\}$, where $g(n) = t_n$.

This ability to list the members of a set as a first, second, third and so on, characterizes countable sets. If the list terminates, then the set is finite. On the other hand, if $t_1, t_2, t_3, \ldots$ is a nonterminating list of the members of T without repetitions, then T is denumerable since the function $g: \mathbb{N} \to T$ given by $g(n) = t_n$ will be bijective.

8.8 THEOREM Let S be a countable set and let $T \subseteq S$. Then T is countable.

Proof: If T is finite, then we are done. Thus we may assume that T is infinite. This implies (Exercise 8.4) that S is infinite, so S is

† The "obvious" fact that $\mathbb{N}$ is infinite is actually nontrivial to prove. The proof is based on the Dirichlet pigeonhole principle: If $n > m$, then there exists no injection $f: I_n \to I_m$. That is, if n pigeons must fit into m pigeonholes with $n > m$, then at least two pigeons will end up in the same hole. The pigeonhole principle in turn depends on the principle of mathematical induction (Theorem 10.2). See Henkin and others, (1962), page 125, for a complete proof.

denumerable (since it is countable). Therefore, there exists a bijection $f: \mathbb{N} \to S$ and we can write S as a list of distinct members

$$S = \{s_1, s_2, \ldots\}$$

where $f(n) = s_n$. Now let

$$A = \{n \in \mathbb{N}: s_n \in T\}.$$

Since A is a nonempty subset of $\mathbb{N}$, it has a least member, say a_1. Similarly, the set $A \backslash \{a_1\}$ has a least member, say a_2. In general, having chosen $a_1, \ldots, a_k$, let a_{k+1} be the least member in $A \backslash \{a_1, \ldots, a_k\}$. Essentially, if we select from our listing of S those terms that are in T and keep them in the same order, then a_n is the subscript of the nth term in this new list.

Now define a function $g: \mathbb{N} \to \mathbb{N}$ by $g(n) = a_n$. Since T is infinite, g is defined for every $n \in \mathbb{N}$. Since $a_{n+1} \notin \{a_1, \ldots, a_n\}$, g must be injective. Thus the composition $f \circ g$ is also injective. Since each element of T is somewhere in the listing of S, $g(\mathbb{N})$ includes all the subscripts of terms in T. Thus $f \circ g$ is a bijection from $\mathbb{N}$ onto T and T is denumerable. ∎

Using Theorem 8.8 we can derive two very useful criteria for determining when a set is countable.

8.9 THEOREM Let S be a nonempty set. The following three conditions are equivalent.

(a) S is countable.
(b) There exists an injection $f: S \to \mathbb{N}$.
(c) There exists a surjection $g: \mathbb{N} \to S$.

Proof: Suppose that S is countable. Then there exists a bijection $h: J \to S$, where $J = I_n$ for some $n \in \mathbb{N}$ if S is finite and $J = \mathbb{N}$ if S is denumerable. In either case, h^{-1} is a bijection from S onto J and hence an injection (at least) from S to $\mathbb{N}$. Thus (a) implies (b).

Now suppose that there exists an injection $f: S \to \mathbb{N}$. Then f is a bijection from S to $f(S)$, so f^{-1} is a bijection from $f(S)$ back to S. We use f^{-1} to obtain a function g from all of $\mathbb{N}$ onto S as follows: Let p be any fixed member of S. Define $g: \mathbb{N} \to S$ by

$$g(n) = \begin{cases} f^{-1}(n), & \text{if } n \in f(S) \\ p, & \text{if } n \notin f(S). \end{cases}$$

Then $g[f(S)] = f^{-1}[f(S)] = S$ and $g[\mathbb{N} \backslash f(S)] = \{p\}$, so that g is a surjection from $\mathbb{N}$ onto S. Thus (b) implies (c).

Finally, suppose that there exists a surjection $g: \mathbb{N} \to S$. Define $h: S \to \mathbb{N}$ by

$$h(s) \text{ is the smallest } n \in \mathbb{N} \text{ such that } g(n) = s.$$

Then h is an injection from S to $\mathbb{N}$, and hence a bijection from S onto the subset $h(S)$ of $\mathbb{N}$. Since $\mathbb{N}$ is countable, Theorem 8.8 implies that $h(S)$ is countable. Since S and $h(S)$ are equinumerous, S is also countable. ∎

8.10 EXAMPLES (a) Let S and T be nonempty countable sets. We shall show that $S \cup T$ is countable. Since S and T are countable, Theorem 8.9 implies that there exist surjections $f: \mathbb{N} \to S$ and $g: \mathbb{N} \to T$. Define $h: \mathbb{N} \to S \cup T$ by

$$h(n) = \begin{cases} f\left(\dfrac{n+1}{2}\right), & \text{if } n \text{ is odd} \\ g\left(\dfrac{n}{2}\right), & \text{if } n \text{ is even.} \end{cases}$$

Then h is surjective, so $S \cup T$ is countable. Notice how the use of Theorem 8.9 allowed us to consider both the finite and denumerable cases at the same time. It also meant that we did not have to worry about some points being in both S and T. [If $S \cap T \neq \varnothing$, then $f(n) = g(m)$ for some n, $m \in \mathbb{N}$. Then h will not be injective, even if both f and g are.]

(b) Recall that every natural number can be written as a product of primes, and this representation is unique except for the order of the factors. For example, $12 = 2^2 \cdot 3$. Using this fact and Theorem 8.9, we can show that the Cartesian product of two countable sets is countable. Suppose that S and T are nonempty countable sets. Then there exist injections $f: S \to \mathbb{N}$ and $g: T \to \mathbb{N}$. Define $h: S \times T \to \mathbb{N}$ by

$$h(s, t) = 2^{f(s)} \cdot 3^{g(t)}, \qquad \text{where } s \in S \text{ and } t \in T.$$

Then h is injective, for if $h(s, t) = h(u, v)$, then

$$2^{f(s)} \cdot 3^{g(t)} = 2^{f(u)} \cdot 3^{g(v)}.$$

Since the prime factored form of a number is unique, we have $f(s) = f(u)$ and $g(t) = g(v)$. Finally, since f and g are injective, this implies that $s = u$ and $t = v$. Thus, since h is injective, we conclude (by Theorem 8.9 again) that $S \times T$ is countable.

(c) Following the approach used in the preceding example, we can show that the set $\mathbb{Q}$ of rational numbers is countable. To begin with, let $\mathbb{Q}^+$ and $\mathbb{Q}^-$ be the set of positive rationals and negative rationals, respectively. We first show that $\mathbb{Q}^+$ is countable. Any member of $\mathbb{Q}^+$ can be written uniquely as m/n, where $m, n \in \mathbb{N}$, $n \neq 0$, and m and n are relatively prime (have no common prime divisors). Define $f: \mathbb{Q}^+ \to \mathbb{N}$ by

$$f(m/n) = 2^m \cdot 3^n.$$

Then f is injective as in the previous example, so $\mathbb{Q}^+$ is countable by Theorem 8.9. The mapping $g: \mathbb{Q}^+ \to \mathbb{Q}^-$ given by $g(r) = -r$ is clearly bijective, so $\mathbb{Q}^+$ and $\mathbb{Q}^-$ are equinumerous. That is, $\mathbb{Q}^-$ is countable. Since $\mathbb{Q} = \mathbb{Q}^- \cup \{0\} \cup \mathbb{Q}^+$, by applying example (a) twice we see that $\mathbb{Q}$ is countable.

(d) By generalizing on the approach used in Example (a), we can show that the union of a countable family of countable sets is countable. To see this, let $\{S_\alpha: \alpha \in \mathscr{A}\}$ be such a family. Since empty sets contribute nothing to the union, we may assume that all the sets are nonempty. Since the family is countable, we can replace the index set by $\mathbb{N}$ and consider $\{S_n: n \in \mathbb{N}\}$. If the original family had only a finite number of sets, $S_1, \ldots, S_k$, let $S_n = S_1$ for all $n > k$. Now for each set S_n there exists a surjection $f_n: \mathbb{N} \to S_n$, so we can write $S_n = \{S_{n1}, S_{n2}, S_{n3}, \ldots\}$, where $f_n(j) = s_{nj}$. We now arrange the elements of $\bigcup_{n=1}^{\infty} S_n$ in a rectangular array:

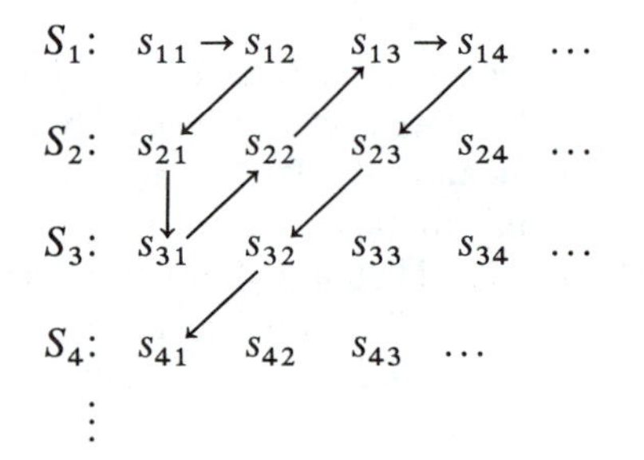

By moving along each diagonal of the array in the manner indicated, we obtain a listing of all the elements in $\bigcup_{n=1}^{\infty} S_n$:

$$s_{11}, \quad s_{12}, \quad s_{21}, \quad s_{31}, \quad s_{22}, \quad s_{13}, \quad s_{14}, \quad \ldots.$$

This listing defines a surjection $f: \mathbb{N} \to \bigcup_{n=1}^{\infty} S_n$, so that the union is countable.†

Having seen several examples of countable sets, the reader may be wondering what sets (if any) are uncountable. We now show that the set of all real numbers is uncountable. Our proof uses the "diagonal process" developed by George Cantor in the late nineteenth century.

8.11 THEOREM The set $\mathbb{R}$ of real numbers is uncountable.

Proof: Since any subset of a countable set is countable (Theorem 8.8), it suffices to show that the interval $J = (0, 1)$ is uncountable. If J were countable, we could list its members and have

$$J = \{x_1, x_2, x_3, \ldots\} = \{x_n: n \in \mathbb{N}\}.$$

† Our argument has used the axiom of choice in a subtle way. See Section 9.

We shall show that this leads to a contradiction by constructing a real number that is in J but is not included in the list of x_n's. Each element of J has an infinite decimal expansion, so we can write

$$x_1 = 0.a_{11}a_{12}a_{13}\cdots,$$
$$x_2 = 0.a_{21}a_{22}a_{23}\cdots,$$
$$x_3 = 0.a_{31}a_{32}a_{33}\cdots,$$
$$\vdots$$

where each $a_{ij} \in \{0, 1, \ldots, 9\}$. (Some numbers, such as $0.5000\cdots = 0.4999\cdots$, have more than one representation, but this will not be a problem.) We now construct a real number $y = 0.b_1b_2b_3\cdots$ by defining

$$b_n = \begin{cases} 2, & \text{if } a_{nn} \neq 2 \\ 3, & \text{if } a_{nn} = 2. \end{cases}$$

Since each digit in the decimal expansion of y is either 2 or 3, $y \in J$. But y is not one of the numbers x_n, since it differs from x_n in the nth decimal place. (Since none of the digits in y are 0 or 9, it is not one of the numbers with two representations.) This contradicts our assumption that J is countable, so J must be uncountable. ∎

8.12 PRACTICE Show that the set of irrational numbers is uncountable.

Ordering of Cardinals†

We conclude this section by returning to our original question about comparing the size of two sets. We would like to make some sense out of the notion that one set is "bigger" or has "more" points than another set. For finite sets we observe that, if S is a proper subset of T, then T certainly has "more" points than S. Unfortunately, this does not hold for infinite sets, as we saw in Example 8.6. In fact, it can be shown (Exercise 8.10) that any inifinite set is equinumerous with a proper subset of itself. Thus the property of being a proper subset is not an adequate basis for comparing the size of sets in general.

A more fruitful approach is built on our definition of equinumerous and our understanding of functions. Intuitively, if $f: S \to T$ is injective, then S can be no larger than T. Not only is this true for finite sets, but we have observed that it also holds for countable sets (Theorem 8.9). Since we think of cardinal numbers as representing the size of a set, we shall use them when comparing sizes.

† The remainder of this section may be omitted on a first reading.

Let us denote the cardinal number of a set S by $|S|$. Then we have $|S| = |T|$ iff S and T are equinumerous. That is, $|S| = |T|$ iff there exists a bijection $f: S \to T$. In light of our discussion above, we define $|S| \leqslant |T|$ to mean that there exists an injection $f: S \to T$. As usual, $|S| < |T|$ means that $|S| \leqslant |T|$ and $|S| \neq |T|$.

The basic properties of our ordering of cardinals are included in Theorem 8.13. The proofs are all straightforward and are left to the exercises. Part (a) corresponds to our intuitive feeling about the relative sizes of subsets. Part (b) and (c) are the reflexive and transitive properties, respectively. Part (d) means that the order of m and n as integers is the same as the order for the finite cardinals m and n. (Recall that $|\{1, 2, \ldots, m\}|$ is denoted by m.)

8.13 THEOREM Let S, T, and U be sets.

(a) If $S \subseteq T$, then $|S| \leqslant |T|$.
(b) $|S| \leqslant |S|$.
(c) If $|S| \leqslant |T|$ and $|T| \leqslant |U|$, then $|S| \leqslant |U|$.
(d) If $m, n \in \mathbb{N}$ and $m \leqslant n$, then $|\{1, 2, \ldots, m\}| \leqslant |\{1, 2, \ldots, n\}|$.
(e) If S is finite, then $|S| < \aleph_0$.

Proof: Exercise 8.6. ∎

It is customary to denote the cardinal number of $\mathbb{R}$ by c, for continuum. Since $\mathbb{Q} \subseteq \mathbb{R}$, we have $\aleph_0 \leqslant c$. In fact, since $\mathbb{Q}$ is countable and $\mathbb{R}$ is uncountable, we have $\aleph_0 < c$. Thus Theorem 8.13(e) implies that $\aleph_0$ and c are unequal transfinite cardinals. Are there any others? The answer is an emphatic yes, as we see in our next theorem.

8.14 NOTATION Given any set S, let $\mathscr{P}(S)$ denote the collection of all the subsets of S. The set $\mathscr{P}(S)$ is called the **power set** of S.

8.15 THEOREM For any set S, we have $|S| < |\mathscr{P}(S)|$.

Proof: The function $g: S \to \mathscr{P}(S)$ given by $g(s) = \{s\}$ is clearly injective, so $|S| \leqslant |\mathscr{P}(S)|$. To prove that $|S| \neq |\mathscr{P}(S)|$, we show that no function from S to $\mathscr{P}(S)$ can be surjective. Suppose that $f: S \to \mathscr{P}(S)$. Then, for each $x \in S$, $f(x)$ is a subset of S. Now for some x in S it may be that x is in the subset $f(x)$ and for others it may not be. Let

$$T = \{x \in S: x \notin f(x)\}.$$

Now $T \subseteq S$, so $T \in \mathscr{P}(S)$. If f were surjective, then $T = f(y)$ for some $y \in S$. Now either $y \in T$ or $y \notin T$, but both possibilities lead to contradictions: If $y \in T$, then $y \notin f(y)$ by the definition of T. But

$f(y) = T$, so $y \notin T$. On the other hand, if $y \notin T$, then $y \notin f(y)$, which implies that $y \in T$.

Thus we conclude that no function from S to $\mathscr{P}(S)$ can be surjective, so $|S| \neq |\mathscr{P}(S)|$. ∎

By applying Theorem 8.15 again and again, we obtain an infinite sequence of transfinite cardinals each larger than the preceding:

$$\aleph_0 = |\mathbb{N}| < |\mathscr{P}(\mathbb{N})| < |\mathscr{P}(\mathscr{P}(\mathbb{N}))| < |\mathscr{P}(\mathscr{P}(\mathscr{P}(\mathbb{N})))| < \cdots.$$

Does the cardinal c fit into this sequence? In exercise 8.16 we sketch the proof that $|\mathscr{P}(\mathbb{N})| = c$. In Exercise 8.9 we show that every infinite set has a denumerable subset. Since (by Theorem 8.8) every infinite subset of a denumerable set is denumerable, we see that $\aleph_0$ is the smallest transfinite cardinal.

What is the first cardinal greater than $\aleph_0$? We know that $c > \aleph_0$, but is there any cardinal number λ such that

$$\aleph_0 < \lambda < c?$$

More specifically, is there any subset of $\mathbb{R}$ with size "in between" $\mathbb{N}$ and $\mathbb{R}$. Experience tells us that there is not, because no such set has ever been found. The conjecture that there is no such set was first made by Cantor and is known as the **continuum hypothesis**. In 1900 it was included as the first of Hilbert's famous 23 unsolved problems. Whether it is true or false is still an unanswered—perhaps unanswerable—question. It is known, however, that the assumption of the continuum hypothesis does not contradict any of the usual axioms of set theory. (This was proved by Kurt Gödel in 1938.) But lest we take too much comfort in this, we should also point out that it has been proved (by Paul Cohen in 1963) that the denial of the continuum hypothesis does not lead to any contradictions either.

Thus the continuum hypothesis is undecidable on the basis of the currently accepted axioms for set theory. (It can neither be proved nor disproved.) It remains to be seen whether new axioms will be found that will enable future mathematicians finally to settle the issue.

ANSWERS TO PRACTICE PROBLEMS

8.2 The identity function is a bijection, so $S \sim S$. Thus "$\sim$" is reflexive. Now, if $S \sim T$, then there exists a bijection $f: S \to T$. By Theorem 7.22, $f^{-1}: T \to S$ is also bijective, so $T \sim S$ and "$\sim$" is symmetric. Finally, suppose that $S \sim T$ and $T \sim U$. Then there exist bijections $f: S \to T$ and $g: T \to U$. By Theorem 7.18, $g \circ f: S \to U$ is bijective, so $S \sim U$. Thus "$\sim$" is transitive.

8.4 Since f is a bijection from I_n onto S, f^{-1} is a bijection from S onto I_n. Thus, if we follow f^{-1} by g, we get a bijection from S onto T. That is, let $h = g \circ f^{-1}$.

8.7 One such function is defined by $f(n) = n/2$ for n even and $f(n) = -(n-1)/2$ for n odd.

8.12 The rationals are countable by Example 8.10(c). If the irrationals were also countable, then $\mathbb{R}$ (their union) would be countable by Example 8.10(a). Since $\mathbb{R}$ is uncountable, so are the irrationals.

EXERCISES

8.1 Show that each of the following pairs of sets S and T are equinumerous by finding a specific bijection between them.

(a) $S = [0, 1]$ and $T = [1, 3]$
(b) $S = [0, 1]$ and $T = [0, 1)$
(c) $S = [0, 1)$ and $T = (0, 1)$
(d) $S = (0, 1)$ and $T = (0, \infty)$
(e) $S = (0, 1)$ and $T = \mathbb{R}$

8.2 (a) Suppose that $m < n$. Prove that the intervals $(0, 1)$ and (m, n) are equinumerous by finding a specific bijection between them.
(b) Use part (a) to prove that any two open intervals are equinumerous.

8.3 Prove: If $(S \backslash T) \sim (T \backslash S)$, then $S \sim T$.

8.4 Prove: Every subset of a finite set is finite.

8.5 Use Example 8.10(d) to prove that $\mathbb{Q}$ is countable.

8.6 Prove Theorem 8.13.

8.7 A real number is said to be **algebraic** if it is a root of a polynomial equation

$$a_n x^n + \cdots + a_1 x + a_0 = 0$$

with integer coefficients. Note that the algebraic numbers include the rationals and all roots of rationals (such as $\sqrt{2}$, $\sqrt[3]{5}$, etc). If a number is not algebraic, it is called **transcendental**.

(a) Show that the set of polynomials with integer coefficients is countable.
(b) Show that the set of algebraic numbers is countable.
(c) Are there more algebraic numbers or transcendental numbers?

8.8 Prove: If S denumerable, then S is equinumerous with a proper subset of itself.

8.9 Prove: Every infinite set has a denumerable subset.

8.10 Prove: Every infinite set is equinumerous with a proper subset of itself.

8.11 Prove that our ordering of cardinal numbers is antisymmetric: if $|S| \leqslant |T|$ and $|T| \leqslant |S|$, then $|S| = |T|$. This result is known as the Schröder–Bernstein theorem and is very useful in proving sets equinumerous. (The proof is hard, but the hint in the back of the book will help.)

8.12 Use the Schröder–Bernstein theorem (Exercise 8.11) to do parts (b) and (c) of Exercise 8.1.

8.13 Suppose that $A \subseteq B \subseteq C$ and $A \sim C$. Prove that $A \sim B$ and $B \sim C$.

8.14 Suppose that we let U denote the "set of all things." Then for any set S we have $S \subseteq U$. In particular, $\mathscr{P}(U) \subseteq U$. Use Theorem 8.13 and Theorem 8.15 to obtain a contradiction. Thus we conclude that a giant universal set that contains "everything" is an impossibility.

8.15 (a) Prove: If $|S| \leqslant |T|$, then $|\mathscr{P}(S)| \leqslant |\mathscr{P}(T)|$.
(b) Prove: If $|S| = |T|$, then $|\mathscr{P}(S)| = |\mathscr{P}(T)|$.

8.16 In this exercise we outline a proof that $|\mathscr{P}(\mathbb{N})| = c$. By Exercise 8.11, it suffices to show that $|\mathscr{P}(\mathbb{N})| \leqslant c$ and $c \leqslant |\mathscr{P}(\mathbb{N})|$.

(a) To show that $|\mathscr{P}(\mathbb{N})| \leqslant c$, we define a function $f\colon \mathscr{P}(\mathbb{N}) \to \mathbb{R}$ by $f(A) = 0.a_1a_2a_3 \cdots a_n \cdots$, where

$$a_n = \begin{cases} 0, & \text{if } n \notin A \\ 1, & \text{if } n \in A. \end{cases}$$

Show that f is injective.

(b) To show that $c \leqslant |\mathscr{P}(\mathbb{N})|$, we use Exercise 8.15 to conclude that $|\mathscr{P}(\mathbb{N})| = |\mathscr{P}(\mathbb{Q})|$ since $|\mathbb{N}| = |\mathbb{Q}|$. Thus it suffices to find an injection $f\colon \mathbb{R} \to \mathscr{P}(\mathbb{Q})$. Define $f(x) = \{y \in \mathbb{Q}\colon y < x\}$. Use the fact (Theorem 12.12) that given $a, b \in \mathbb{R}$ with $a < b$, there exists $r \in \mathbb{Q}$ such that $a < r < b$ to show that f is injective.

8.17 Let α and β be cardinal numbers. The **cardinal sum** of α and β, denoted $\alpha + \beta$, is the cardinal $|A \cup B|$, where A and B are disjoint sets such that $|A| = \alpha$ and $|B| = \beta$.

(a) Prove that the sum is well-defined. That is, if $|A| = |C|$, $|B| = |D|$, $A \cap B = \varnothing$, and $C \cap D = \varnothing$, then $|A \cup B| = |C \cup D|$.
(b) Prove that the sum is commutative and associative. That is, for any cardinals α, β, γ we have $\alpha + \beta = \beta + \alpha$ and $\alpha + (\beta + \gamma) = (\alpha + \beta) + \gamma$.
(c) Show that $n + \aleph_0 = \aleph_0$ for any finite cardinal n.
(d) Show that $\aleph_0 + \aleph_0 = \aleph_0$.
(e) Show that $\aleph_0 + c = c$.
(f) Show that $c + c = c$.

8.18 Let α and β be cardinal numbers. The **cardinal product** $\alpha\beta$ is defined to be the cardinal $|A \times B|$, where $|A| = \alpha$ and $|B| = \beta$.

(a) Prove that the product is well-defined. That is, if $|A| = |C|$ and $|B| = |D|$, then $|A \times B| = |C \times D|$.
(b) Prove that the product is commutative and associative and that the distributive law holds. That is, for any cardinals α, β, γ, we have $\alpha\beta = \beta\alpha$, $\alpha(\beta\gamma) = (\alpha\beta)\gamma$, and $\alpha(\beta + \gamma) = \alpha\beta + \alpha\gamma$.
(c) Show that $0\alpha = 0$ for any cardinal α.
(d) Show that $n\aleph_0 = \aleph_0$ for any finite cardinal n with $n \neq 0$.
(e) Show that $\aleph_0\aleph_0 = \aleph_0$.
(f) Show that $cc = c$.

Section 9 AXIOMS FOR SET THEORY†

Throughout this chapter we have used the concept of a set informally without really saying what sets are or what properties they have. While a definition of "set" is essentially impossible, it is possible to discuss

† This section is optional and may be omitted without loss of continuity.

properties of sets and to indicate things that *cannot* be sets. In this section we present a list of axioms (the Zermelo–Fraenkel axioms) from which set theory can be derived in a formal way. It is not our intent actually to do the formal derivation, but rather to indicate a foundation upon which set theory may be built. We begin by considering two paradoxes.

Paradoxes

In Section 5 we saw the utility of having a "universal" set U that focused our attention on a particular mathematical system. Then any set under consideration was a subset of U. It is natural to ask: Why not let U denote the "set of all things"? Then certainly U will contain everything we might want to consider, and every set will be a subset of this U. In Exercise 8.14 we pointed out that if U contains every set then $\mathscr{P}(U) \subseteq U$, and this leads to the contradiction $|\mathscr{P}(U)| \leqslant |U| < |\mathscr{P}(U)|$. While the notion of a giant universal set that contained "everything" is intuitively plausible, its existence is seen to contradict another established "fact." [If $|\mathscr{P}(U)| < |\mathscr{P}(U)|$, then there can be no bijection $f: \mathscr{P}(U) \to \mathscr{P}(U)$. But the identity map on $\mathscr{P}(U)$ is just such a bijection.] The only way out of this dilemma is to agree that the "set of all things" cannot really be considered a set.

Our second paradox is more subtle. We certainly want to be able to consider sets of sets, that is, sets whose members are themselves sets. [The power set $\mathscr{P}(\mathbb{N})$ is an example of a set of sets.] If sets can be members of sets, then it may be that some sets are members of themselves. That is, there may be some set x such that $x \in x$. On the other hand, there are many sets, such as $x = \mathbb{N}$, for which $x \notin x$. In 1901 Bertrand Russell pointed out the following paradox: Let

$$B = \{x: x \notin x\}.$$

Then it must either be the case that $B \in B$ or $B \notin B$. If $B \in B$, then B must satisfy the defining condition for B; that is, $B \notin B$. On the other hand, if $B \notin B$, then B satisfies the condition for being a member of B and $B \in B$. Thus we are faced with an impasse: $B \in B$ iff $B \notin B$. Once again, the way out of this dilemma is to agree that B is not really a set.

But if some things cannot be sets and other things can, how do we decide what isn't and what is? In an attempt to answer this sort of question, Ernst Zermelo in 1908 proposed a list of axioms from which he built a formal system of set theory. These 10 axioms (as modified slightly by Abraham Fraenkel in 1922) are still one of the most widely accepted foundations for set theory.

The Zermelo–Fraenkel Axioms

Before presenting the axioms and discussing them briefly, we need to point out that the axioms contain two undefined primitive ideas: the concept of a "set" and the concept of "membership" or "belongs to." The latter notion is denoted in the usual way by "$\in$." We also assume the rules of logic as developed in Chapter 1. Since we wish to have as few undefined terms as possible, we shall consider the elements in a set to be sets themselves. Thus there is no distinction in kind between an element and a set. For this reason, we shall use (in this section only) lowercase letters for both. This is not to say that x and $\{x\}$ are equal. Indeed, If $x = \{a, b, c\}$, then x has three members (a, b, and c) and $\{x\}$ has only one member (that is, x). The point is that x and $\{x\}$ are both sets.

AXIOM 1 (The axiom of extension)
Two sets are equal iff they have the same elements.

AXIOM 2 (The axiom of the null set)
There exists a set with no elements, and we denote it by $\varnothing$.

AXIOM 3 (The axiom of pairing)
Given any sets x and y, there exists a set whose elements are x and y.

AXIOM 4 (The axiom of union)
Given any set x, the union of all the elements in x is a set.

AXIOM 5 (The axiom of the power set)
Given any set x, there exists a set consisting of all the subsets of x.

These first five axioms express properties of sets that are very familiar. Using Axiom 3, we obtain sets of the form $\{x, y\}$. By taking $x = y$, we also obtain singleton sets: $\{x\}$. The pair $\{x, y\}$ is not ordered, but the ordered pair (x, y) can be obtained from this in the usual way (Definition 6.1): $(x, y) = \{\{x\}, \{x, y\}\}$. Once we have ordered pairs we can talk about relations and functions.

Our next axiom enables us to define a set (or more precisely, a subset) by identifying a property that its members must satisfy.

AXIOM 6 (The axiom of separation)
Given any set x and any sentence $p(y)$ that is a statement for all $y \in x$, then there exists a set $\{y \in x\colon p(y) \text{ is true}\}$.

It might seem that Axiom 6 will lead to the same problem that was encountered in Russell's paradox. This is not the case, however, since the set constructed is a *subset* of x. To see what happens, let

$$b = \{y \in x \colon y \notin y\}.$$

Then, in order for $w \in b$, it must be the case that $w \in x$ and $w \notin w$. When we suppose $b \in b$, we obtain $b \in x$ and $b \notin b$, which is a contradiction. Now suppose $b \notin b$. If $b \in x$, then b satisfies the defining condition for being a member of b, so $b \in b$, a contradiction. But the possibility $b \notin x$ still remains. Thus, instead of reaching a paradoxical dilemma, we are simply led to conclude that $b \notin x$, and there is no problem.

One of the benefits of Axiom 6 is that it enables us to form the intersection of two sets. If z is a set and we let $p(y)$ be the sentence "$y \in z$," then $\{y \in x \colon y \in z\}$ is just $x \cap z$.

AXIOM 7 (The axiom of replacement)

Given any set x and any function f defined on x, the image $f(x)$ is a set.[†]

It can be shown [see Hamilton (1982)] that Axiom 7 implies Axiom 6. Thus we could have omitted Axiom 6 from our list. It has been included for historical reasons and because it represents our usual method for defining a set. We note that Axiom 6 was one of the original axioms given by Zermelo, and Axiom 7 is essentially a refinement due to Fraenkel.

AXIOM 8 (The axiom of infinity)

There exists a set x such that $\varnothing \in x$, and whenever $y \in x$ it follows that $y \cup \{y\} \in x$.

This axiom looks a bit strange, but it is necessary to guarantee the existence of an infinite set. Indeed, the axiom specifies what elements the set must contain:

$$\varnothing, \quad \varnothing \cup \{\varnothing\}, \quad \varnothing \cup \{\varnothing\} \cup \{\varnothing \cup \{\varnothing\}\}, \ldots$$

As soon as we know that these sets are all distinct, we have an infinite set. The general proof that they are distinct is messy, but the first step is not (Exercise 9.2.).

[†] We are using the term "function" in a more general way than in Definition 7.1. In the present context a function defined on x is a correspondence that associates with each $y \in x$ a unique set $f(y)$. Axiom 7 then asserts that $\{f(y) \colon y \in x\}$ is a set. We do not require that the sets $f(y)$ are all subsets of a given set.

AXIOM 9 (The axiom of regularity)

Given any nonempty set x, there exists $y \in x$ such that $y \cap x = \emptyset$.

The axiom of regularity has the effect of ruling out the possibility that some set is a member of itself. Indeed, suppose that $x \in x$. Now the set $\{x\}$ is nonempty since it contains x. By the axiom of regularity, there exists $y \in \{x\}$ such that $y \cap \{x\} = \emptyset$. Since $y \in \{x\}$, we must have $y = x$. But then $y \in y$ and $y \in \{x\}$, so $y \in y \cap \{x\}$, a contradiction.

The Axiom of Choice

Our final axiom deserves special attention because of the controversy surrounding it. It is the only one of the Zermelo-Fraenkel axioms that has been seriously challenged by mathematicians. There are some (perhaps most) mathematicians who accept it as an axiom. But there are others who feel that it is meaningless (not false, just meaningless). At first glance the axiom certainly seems reasonable:

AXIOM 10 (The axiom of choice)

Given any nonempty set x whose members are pairwise disjoint nonempty sets, there exists a set y consisting of exactly one element taken from each set belonging to x.

The axiom of choice essentially says that we can pick an element out of each set in x and gather all these elements together in a new set y. When x has only a finite number of members, there is no problem (and Axiom 10 is not even needed) since the elements chosen from each set in x can be specified one at a time. But when x contains infinitely many members (sets), there may not be any way of indicating which elements are chosen.

The following example due to Bertrand Russell may make the point more clearly. Suppose that we have a pile S of infinitely many pairs of shoes. Is it possible to construct a set consisting of one shoe from each pair? The answer is yes, and we can do this without using the axiom of choice. Let $L = \{s \in S: s \text{ is a left shoe}\}$. Then L exists by the axiom of separation. But suppose now that sitting beside S we have a pile T consisting of infinitely many pairs of socks. Does there exist a set containing exactly one sock from each pair? Without the axiom of choice we would have to say no, for with socks there is no way to indicate which sock was chosen.

The basic problem with the axiom of choice is one of existence. What does it mean for a set to exist? In Section 5 we adopted the "realistic" point of view that a set must be characterized by some defining property so that it becomes a question of fact whether or not a particular object belongs to the set. If we persist in this point of view, we cannot use the axiom of choice. The "idealistic" point of view, on the other hand, does not attempt to define existence. The notion of existence is taken as an undefined primitive concept. We shall move (somewhat reluctantly) toward this latter position so that we may use the axiom of choice where necessary.

Applications of the Axiom of Choice

We have already had three occasions that called for the use of the axiom of choice (AC). In Exercise 7.18 we were asked to show that, if a function $f: A \to B$ is surjective, then there exists a function $g: B \to A$ such that $f \circ g(y) = y$, for all $y \in B$. If $y \in B$, then $y = f(x)$ for some x in A, but this x may not be unique. If for each y in B we let $g(y)$ be one particular x in $f^{-1}(\{y\})$, then g is the desired function. But to obtain g we have had to select one point from each of possibly infinitely many sets. Thus we have used (AC).

Another way to look at Exercise 7.18 is to realize that $f^{-1} = \{(y, x): (x, y) \in f\}$ defines a relation between B and A having domain $\{y: \exists\, x \in A \ni (x, y) \in f\} = f(A)$. This relation needs to be restricted so that only one x corresponds to each y. Thus we obtain the following alternate (and equivalent) form of the axiom of choice:

> Given any relation R, there exists a function $g \subseteq R$ such that domain $g =$ domain R.

In Example 8.10(d) we used (AC) in a more subtle way. In showing that the union of a countable collection $\{S_n: n \in \mathbb{N}\}$ of countable sets is countable, we needed a surjection $f_n: \mathbb{N} \to S_n$ for each $n \in \mathbb{N}$. Now given any n, the existence of at least one surjection from $\mathbb{N}$ to S_n is guaranteed by Theorem 8.9. But we had to choose exactly one surjection for each natural number, and this requires (AC).

In showing that every infinite set has a denumerable subset (Exercise 8.9), it is also necessary to make infinitely many arbitrary choices without being able to specify which element is being chosen. Again, this can be done only by using (AC). Since Exercise 8.9 is used in the proof of Exercise 8.10, our approach to the latter result (any infinite set is equinumerous with a proper subset) also depends on (AC).

Many other useful results depend on (AC). They occur in various branches of mathematics, including linear algebra, abstract algebra, measure theory, and functional analysis. Most of these results are far beyond the scope of this book, but we mention three that are not:

1. Given any sets S and T, either $|S| \leqslant |T|$ or $|T| \leqslant |S|$.
2. Given any infinite set S, $|S \times S| = |S|$.
3. Given any subset S of $\mathbb{R}$, a point x is an accumulation point of S iff there exists a sequence of points in $S \backslash \{x\}$ that converges to x. (See Exercise 14.11.)

While (AC) is an important axiom for obtaining useful results in many different areas, it also leads to some unexpected (and perhaps unwanted) results. This is another reason why some mathematicians are reluctant to accept (AC) as a valid axiom. Thus we end this section back where we began, with a paradox.

The Banach–Tarski Paradox

In 1942, S. Banach and A. Tarski proved a most disconcerting result using the axiom of choice. It seems to go strongly against our intuition and so is considered a paradox. We say that two subsets of the plane (or three-dimensional space) are **congruent** if one set can be moved rigidly so as to coincide with the other. These rigid motions are just translations and rotations. We say that two sets S and T are equivalent by finite decomposition if there exist sets $S_1, \ldots, S_n$, and $T_1, \ldots, T_n$ such that the following hold:

(a) $S = S_1 \cup \cdots \cup S_n$ and $T = T_1 \cup \cdots \cup T_n$.
(b) $S_i \cap S_j = \varnothing$ and $T_i \cap T_j = \varnothing$ for $i \neq j$.
(c) S_i is congruent to T_i, for each $i = 1, \ldots, n$.

Here now is the paradox.†

The Banach–Tarski Paradox. Let S and T be solid three-dimensional spheres of possibly different radii. Then S and T are equivalent by finite decomposition.

For example, consider a solid sphere S the size of a golf ball and another T the size of a basketball. The paradox implies that S can be decomposed into disjoint pieces $S_1, \ldots, S_n$ in such a way that by rigid

† What we are presenting here is only a special case of their more general theorem, but this is sufficient to make our point. For a good discussion of this and related paradoxes, see the article by Blumethal (1940).

motions these pieces can be fitted together without any holes to fill out all of T. While the number of pieces required in this example is certainly large, in other equally astonishing cases the number is small. It has been shown, for example, that the solid sphere of radius 1 can be decomposed into nine mutually disjoint subsets that can be reassembled by means of rotations and translations to fill out two solid spheres of radius 1! Four of the pieces fit together to make one sphere and the other five pieces make up the second sphere.

One possible response to a paradox like this is to refuse to accept the axiom of choice upon which it is built. But in doing so you would also have to reject the other nonparadoxical results that depend on the axiom of choice for their proofs. In recent years a possible compromise has begun to develop by looking for a new axiom that would be adequate to produce the "good" results but that would not lead to unwanted paradoxes. Although there has been some progress in this direction, the final results are by no means clear. [See Hamilton (1982) for a discussion of some alternative axioms.]

EXERCISES

9.1 Let S and T be nonempty sets. Prove that $|T| \leqslant |S|$ iff there exists a surjection $f: S \to T$.

9.2 Without using the axiom of regularity, show that $\varnothing \neq \{\varnothing\}$, and from this conclude that $\varnothing \neq \varnothing \cup \{\varnothing\}$.

9.3 Let x be a set. Show that $\{y: x \subseteq y\}$ cannot be a set.

9.4 Use the axiom of regularity to show that for any set x, $x \cup \{x\} \neq x$.

9.5 Use the axiom of regularity to show: If $x \in y$, then $y \neq x$.

9.6 Use the axiom of regularity to show that there cannot exist three sets, w, x, and y such that $w \in x$, $x \in y$, and $y \in w$.

9.7 Once upon a time there was a town with a perplexing problem. It seems that some of the men liked to shave themselves and the others preferred going to a barber. The problem was that the town had no barber. To remedy this unfortunate situation, the town began to advertise for a barber who would "shave precisely those men who do not shave themselves." But try as they might, they had great difficultly finding anyone who could fill the job description. There were many good barbers who applied, but none of them could do the job that had been advertised. Then one day a mathematician was passing through town and, hearing of their difficulty, announced "I am qualified to be your barber." Without a moment's hesitation, the townspeople agreed that indeed the mathematician could do the job as advertised. Assuming that this story is true, what was the name of the mathematician? *Hint*: It was either Jim, Dave, or Sam.

9.8 Let S_α $(\alpha \in \mathscr{A})$ be an indexed family of sets. We define the Cartesian product $\times_{\alpha \in \mathscr{A}} S_\alpha$ to be the set of all functions f having domain $\mathscr{A}$ such that $f(\alpha) \in S_\alpha$ for all $\alpha \in \mathscr{A}$.

(a) Show that if $\mathscr{A} = \{1, 2\}$, this definition gives a set that corresponds to the usual Cartesian product of two sets $S_1 \times S_2$ in a natural way.

(b) Show that the axiom of choice is equivalent to the following statement: The Cartesian product of a nonempty family of nonempty sets is nonempty.

9.9 This exercise is designed to give some insight into the Banach–Tarski paradox. Find each of the following without using the axiom of choice.

(a) A subset of $\mathbb{R}$ that is congruent to a proper subset of itself.

(b) A bounded subset of the plane that is congruent to a proper subset of itself.

(c) A bounded subset S of the plane such that S is the disjoint union of two nonempty sets S_1 and S_2, and where S_1 and S_2 are both congruent to S.

3

The Real Numbers

In this chapter we present in some detail many of the important properties of the set $\mathbb{R}$ of real numbers. The real numbers form the background for virtually all the analysis that we shall be doing in this text. Our approach will be axiomatic, but not constructive. That is, we shall not construct the real numbers from some simpler (?) set such as the natural numbers. Instead, we shall assume the existence of $\mathbb{R}$ and postulate the properties that characterize it.†

We begin in Section 10 by looking at the natural numbers and mathematical induction. In Section 11 we consider the field and order axioms that begin to characterize $\mathbb{R}$. The completeness axiom in Section 12 is the final axiom and deserves special attention because of its central role in the rest of analysis. In Sections 13 and 14 we develop some of the topological properties of the reals that will be useful in describing the behavior of sequences and functions. In Section 15 (an optional section) we look at these properties in the more general context of a metric space.

Section 10 NATURAL NUMBERS AND INDUCTION

In Section 5 we agreed to let $\mathbb{N}$ denote the set of positive integers, also called the natural numbers. Thus

$$\mathbb{N} = \{1, 2, 3, 4, \ldots\}.$$

† For a constructive approach to developing the reals from the rationals, the rationals from the natural numbers, and the natural numbers from basic set theory, see Henkin and others (1962), Stewart and Tall (1977), or Hamilton (1982). See also Exercises 10.17, 11.10, and 11.11.

It is possible to develop all the properties (and even the existence) of the natural numbers in a rigorous way from set theory and a few additional axioms. But since our discussion of set theory was not entirely rigorous, there is little to be gained by going though the steps of that development here. Rather, we shall assume that the reader is familiar with the usual arithmetic operations of addition and multiplication and with the notion of what it means for one natural number to be less than another.

There is one additional property of $\mathbb{N}$ that we shall assume as an axiom. (That is, we accept it as true without proof.) It expresses in a precise way the intuitive idea that each nonempty subset of $\mathbb{N}$ must have a least element.

10.1 AXIOM (Well-Ordering Property of $\mathbb{N}$) If S is a nonempty subset of $\mathbb{N}$, then there exists an element $m \in S$ such that $m \leqslant k$ for all $k \in S$.

One important tool to be used when proving theorems about the natural numbers is the principle of mathematical induction. It enables us to conclude that a given statement about natural numbers is true for all the natural numbers without having to verify it for each number one at a time (which would be an impossible task!).

10.2 THEOREM (Principle of Mathematical Induction) Let $P(n)$ be a statement that is either true or false for each $n \in \mathbb{N}$. Then $P(n)$ is true for all $n \in \mathbb{N}$ provided that

(a) $P(1)$ is true, and
(b) for each $k \in \mathbb{N}$, if $P(k)$ is true, then $P(k + 1)$ is true.

Proof: The strategy of our argument will be a proof by contradiction using tautology (g) in Example 3.12. That is, we suppose that (a) and (b) hold but that $P(n)$ is false for some $n \in \mathbb{N}$. Let

$$S = \{n \in \mathbb{N}: P(n) \text{ is false}\}.$$

Then S is not empty and the well-ordering property guarantees the existence of an element $m \in S$ that is a least element of S. Since $P(1)$ is true by hypothesis (a), $1 \notin S$, so that $m > 1$. It follows that $m - 1$ is also a natural number, and since m is the least element in S, we must have $m - 1 \notin S$.

But since $m - 1 \notin S$, it must be that $P(m - 1)$ is true. We now apply hypothesis (b) with $k = m - 1$ to conclude that $P(k + 1) = P(m)$ is true. This implies that $m \notin S$, which contradicts our original choice of m. ▮

It is customary to refer to the verification of part (a) of Theorem 10.2 as the **basis for induction** and part (b) as the **induction step**. The assumption that $P(k)$ is true in verifying part (b) is known as the **induction hypothesis**. It is essential that both parts be verified to have a valid proof using mathematical induction. In practice, it is usually the induction step that is the more difficult part.

10.3 EXAMPLE Prove that $1 + 2 + 3 + \cdots + n = \frac{1}{2}n(n + 1)$ for every natural number n.

Proof: Let $P(n)$ be the statement

$$1 + 2 + 3 + \cdots + n = \frac{1}{2}n(n + 1).$$

Then $P(1)$ asserts that $1 = \frac{1}{2}(1)(1 + 1)$, $P(2)$ asserts that $1 + 2 = \frac{1}{2}(2)(2 + 1)$, and so on. In particular, we see that $P(1)$ is true, and this establishes the basis for induction.

To verify the induction step, we suppose that $P(k)$ is true, where $k \in \mathbb{N}$. That is, we assume

$$1 + 2 + 3 + \cdots + k = \frac{1}{2}k(k + 1).$$

Since we wish to conclude that $P(k + 1)$ is true, we add $k + 1$ to both sides to obtain

$$\begin{aligned} 1 + 2 + 3 + \cdots + k + (k + 1) &= \frac{1}{2}k(k + 1) + (k + 1) \\ &= \frac{1}{2}[k(k + 1) + 2(k + 1)] \\ &= \frac{1}{2}(k + 1)(k + 2) \\ &= \frac{1}{2}(k + 1)[(k + 1) + 1]. \end{aligned}$$

Thus $P(k + 1)$ is true whenever $P(k)$ is true, and by the principle of mathematical induction, we conclude that $P(n)$ is true of all n. ▮

Since the format of a proof using mathematical induction always consists of the same two steps (establishing the basis for induction and verifying the induction step), it is common practice to reduce some of the formalism by omitting explicit reference to the statement $P(n)$. It is also acceptable to omit identifying the steps by name, but we must be certain that they are both actually there.

10.4 EXAMPLE Prove by induction that $7^n - 4^n$ is a multiple of 3, for all $n \in \mathbb{N}$.

Proof: Clearly, this is true when $n = 1$, since $7^1 - 4^1 = 3$. If we suppose that $7^k - 4^k$ is a multiple of 3, then $7^k - 4^k = 3m$ for some $m \in \mathbb{N}$. It follows that

$$\begin{aligned} 7^{k+1} - 4^{k+1} &= 7^{k+1} - 7 \cdot 4^k + 7 \cdot 4^k - 4 \cdot 4^k \\ &= 7(7^k - 4^k) + 3 \cdot 4^k \\ &= 7(3m) + 3 \cdot 4^k \\ &= 3(7m + 4^k). \end{aligned}$$

Since m and k are natural numbers, so is $7m + 4^k$. Thus $7^{k+1} - 4^{k+1}$ is also a multiple of 3, and by induction we conclude that $7^n - 4^n$ is a multiple of 3 for all $n \in \mathbb{N}$. ∎

10.5 PRACTICE Observe that

$$\begin{aligned} 1 &= 1^2 \\ 1 + 3 &= 2^2 \\ 1 + 3 + 5 &= 3^2 \\ 1 + 3 + 5 + 7 &= 4^2. \end{aligned}$$

Figure out a general formula and prove your answer using mathematical induction.

There is a generalization of the principle of mathematical induction that enables us to conclude that a given statement is true for all natural numbers sufficiently large. More precisely, we have:

10.6 THEOREM Let $m \in \mathbb{N}$ and let $P(n)$ be a statement that is either true or false for each $n \geqslant m$. Then $P(n)$ is true for all $n \geqslant m$ provided that

(a) $P(m)$ is true, and
(b) for each $k \geqslant m$, if $P(k)$ is true, then $P(k + 1)$ is true.

Proof: The proof will use the original principle of induction (Theorem 10.2). For each $r \in \mathbb{N}$, let $Q(r)$ be the statement "$P(r + m - 1)$ is true." Then from (a) we know that $Q(1)$ holds. Now let $j \in \mathbb{N}$ and suppose that $Q(j)$ holds. That is, $P(j + m - 1)$ is true. Since $j \in \mathbb{N}$,

$$j + m - 1 = m + (j - 1) \geqslant m,$$

so by (b), $P(j + m)$ must be true. Thus $Q(j + 1)$ holds and the induction step is verified. We conclude that $Q(r)$ holds for all $r \in \mathbb{N}$.

Now if $n \geqslant m$, let $r = n - m + 1$, so that $r \in \mathbb{N}$. Since $Q(r)$ holds, $P(r + m - 1)$ is true. But $P(r + m - 1)$ is the same as $P(n)$, so $P(n)$ is true for all $n \geqslant m$. ▮

ANSWERS TO PRACTICE PROBLEMS

10.5 The general formula is

$$1 + 3 + 5 + \cdots + (2n - 1) = n^2,$$

and we have already seen that this is true for $n = 1$. For the induction step, suppose that $1 + 3 + 5 + \cdots + (2k - 1) = k^2$. Then

$$\begin{aligned} 1 + 3 + 5 + \cdots + (2k - 1) + (2k + 1) &= k^2 + (2k + 1) \\ &= (k + 1)^2. \end{aligned}$$

Since this is the formula for $n = k + 1$, we conclude by induction that the formula holds for all $n \in \mathbb{N}$. ▮

EXERCISES

***10.1** Prove that $1^2 + 2^2 + \cdots + n^2 = \frac{1}{6}n(n + 1)(2n + 1)$ for all $n \in \mathbb{N}$.

***10.2** Prove that $1^3 + 2^3 + \cdots + n^3 = \frac{1}{4}n^2(n + 1)^2$ for all $n \in \mathbb{N}$.

10.3 Prove that $1^3 + 2^3 + \cdots + n^3 = (1 + 2 + \cdots + n)^2$ for all $n \in \mathbb{N}$.

***10.4** Prove that

$$\frac{1}{1 \cdot 2} + \frac{1}{2 \cdot 3} + \frac{1}{3 \cdot 4} + \cdots + \frac{1}{n(n + 1)} = \frac{n}{n + 1}, \qquad \text{for all } n \in \mathbb{N}.$$

***10.5** Prove that $1 + r + r^2 + \cdots + r^n = (1 - r^{n+1})/(1 - r)$ for all $n \in \mathbb{N}$, when $r \neq 1$.

***10.6** Prove that

$$\frac{1}{3} + \frac{1}{15} + \frac{1}{35} + \cdots + \frac{1}{4n^2 - 1} = \frac{n}{2n + 1}, \qquad \text{for all } n \in \mathbb{N}.$$

10.7 Prove that $5^{2n} - 1$ is a multiple of 8 for all $n \in \mathbb{N}$.

10.8 Prove that $9^n - 4^n$ is a multiple of 5 for all $n \in \mathbb{N}$.

10.9 Indicate for which natural numbers n the given inequality is true. Prove your answers by induction.

(a) $n^2 \leqslant n!$

(b) $n^2 \leqslant 2^n$

(c) $2^n \leqslant n!$

***10.10** Use induction to prove Bernoulli's inequality: If $1 + x > 0$, then $(1 + x)^n \geqslant 1 + nx$ for all $n \in \mathbb{N}$.

***10.11** Prove the principle of strong induction: Let $P(n)$ be a statement that is either true or false for each $n \in \mathbb{N}$. Then $P(n)$ is true for all $n \in \mathbb{N}$ provided that

(a) $P(1)$ is true, and
(b) for each $k \in \mathbb{N}$, if $P(j)$ is true for all integers j such that $1 \leqslant j \leqslant k$, then $P(k+1)$ is true.

10.12 Indicate what is wrong with each of the following induction "proofs."

(a) **Theorem:** For each $n \in \mathbb{N}$, let $P(n)$ be the statement "Any collection of n marbles consists of marbles of the same color." Then $P(n)$ is true for all $n \in \mathbb{N}$.

Proof: Clearly, $P(1)$ is a true statement. Now suppose that $P(k)$ is a true statement for some $k \in \mathbb{N}$. Let S be a collection of $k + 1$ marbles. If one marble, call it x, is removed, then the induction hypothesis applied to the remaining k marbles implies that these k marbles all have the same color. Call this color C. Now if x is returned to the set S and a different marble is removed, then again the remaining k marbles must all be of the same color C. But one of these marbles is x, so in fact all $k + 1$ marbles have the same color C. Thus $P(k+1)$ is true, and by induction we conclude that $P(n)$ is true for all $n \in \mathbb{N}$. ▮

(b) **Theorem:** For each $n \in \mathbb{N}$ let $P(n)$ be the statement "$n^2 + 7n + 3$ is an even integer." Then $P(n)$ is true for all $n \in \mathbb{N}$.

Proof: Suppose that $P(k)$ is true for some $k \in \mathbb{N}$. That is, $k^2 + 7k + 3$ is an even integer. But then

$$\begin{aligned}(k+1)^2 + 7(k+1) + 3 &= (k^2 + 2k + 1) + 7k + 7 + 3 \\ &= (k^2 + 7k + 3) + 2(k + 4),\end{aligned}$$

and this number is even since it is the sum of two even numbers. Thus $P(k+1)$ is true. We conclude by induction that $P(n)$ is true for all $n \in \mathbb{N}$. ▮

10.13 In the song "The Twelve Days of Christmas," gifts are sent on successive days according to the following pattern:

First day: A partridge in a pear tree.
Second day: Two turtledoves and another partridge.
Third day: Three French hens, two turtle doves, and a partridge.
And so on.

For each $i = 1, \ldots, 12$, let g_i be the number of gifts sent on the ith day. Then $g_1 = 1$ and for $i = 2, \ldots, 12$ we have

$$g_i = g_{i-1} + i.$$

Now let t_n be the total number of gifts sent during the first n days of Christmas. Find a formula for t_n in the form

$$t_n = \frac{n(n + a)(n + b)}{c},$$

where $a, b, c \in \mathbb{N}$.

10.14 Let $0! = 1$ and for $n \in \mathbb{N}$ define $n!$ (read "n factorial") by

$$n! = n[(n - 1)!].$$

Then let

$$\binom{n}{r} = \frac{n!}{r!(n - r)!} \qquad \text{for } r = 0, 1, 2, \ldots, n.$$

(a) Show that

$$\binom{n}{r} + \binom{n}{r - 1} = \binom{n + 1}{r}, \qquad \text{for } r = 1, 2, \ldots, n.$$

*(b) Use part (a) and mathematical induction to prove the *binomial theorem*:

$$(a + b)^n = \binom{n}{0}a^n + \binom{n}{1}a^{n-1}b + \cdots + \binom{n}{r}a^{n-r}b^r + \cdots + \binom{n}{n}b^n$$

$$= a^n + na^{n-1}b + \frac{1}{2}n(n - 1)a^{n-2}b^2 + \cdots + nab^{n-1} + b^n.$$

10.15 Prove Theorem 10.6 by using the well-ordering property of $\mathbb{N}$ instead of the principle of mathematical induction.

10.16 Use the principle of mathematical induction to prove the well-ordering property of $\mathbb{N}$. Thus we could have taken Theorem 10.2 as an axiom and derived 10.1 as a theorem.

Exercise 10.17 illustrates how the basic properties of addition of natural numbers can be derived from a few simple axioms. These axioms are called the Peano axioms in honor of the Italian mathematician Giuseppe Peano, who developed this approach in the late nineteenth century. We suppose that there exists a set P whose elements are called **natural numbers** and a relation of **successor** with the following properties:

P1. There exists a natural number, denoted by 1, that is not the successor of any other natural number.

P2. Every natural number has a unique successor. If $m \in P$, then we let m' denote the successor of m.

P3. Every natural number except 1 is the successor of exactly one natural number.

P4. If M is a set of natural numbers such that
  (i) $1 \in M$ and
  (ii) for each $k \in P$, if $k \in M$, then $k' \in M$,
then $M = P$.

Axioms P1 to P3 express the intuitive notion that 1 is the first natural number and that we can progress through the natural numbers in succession one at a time. Axiom P4 is the equivalent of the principle of mathematical induction. Using these axioms, we can define what addition means. We begin by defining what it means to add 1.

D1. For every $n \in P$, define $n + 1 = n'$.

That is, $n + 1$ is the unique successor of n whose existence is guaranteed by axiom P2. Following this pattern, it is clear that we want to define $n + 2 = (n + 1) + 1$, $n + 3 = [(n + 1) + 1] + 1$, and so on. To define $n + m$ for all $m \in P$, we use a recursive definition:

D2. Let $n, m \in P$. If $m = k'$ and $n + k$ is defined, then define $n + m$ to be $(n + k)'$.

That is, $n + k' = (n + k)'$ or, equivalently, $n + (k + 1) = (n + k) + 1$. Note that if $m \neq 1$ then the existence of k is assured by axiom P3. Now for the exercise.

10.17 (a) Prove that $n + m$ is defined for all $n, m \in P$.
(b) Prove that $n + 1 = 1 + n$ for all $n \in P$.
(c) Prove that $m' + n = (m + n)'$ for all $m, n \in P$.
(d) Prove that addition is commutative. That is, prove that $n + m = m + n$ for all $m, n \in P$.
(e) Prove that addition is associative. That is, prove that $(m + n) + p = m + (n + p)$ for all $m, n, p \in P$.

For the sake of completeness, we indicate how multiplication can be defined using the Peano axioms. As you would expect by now, the definition is recursive.

D3. For every $n \in P$, define $n \times 1 = n$.
D4. Let $n, m \in P$. If $m = k'$ and $n \times k$ is defined, then define $n \times m$ to be $(n \times k) + n$. That is, $n \times (k + 1) = (n \times k) + n$.

The reader is invited to prove some of the basic properties of multiplication using D3 and D4. For a complete discussion of the development of $\mathbb{N}$ from the Peano axioms, see Henkin and others (1962).

Section 11 ORDERED FIELDS

The set $\mathbb{R}$ of real numbers can be described as a "complete ordered field." In this section we present the axioms of an ordered field and in the next section we give the completeness axiom. The purpose of this development is to identify the basic properties that characterize the real numbers. After stating the axioms of an ordered field, we derive some of the basic algebraic properties that the reader no doubt has used for years without question. It is not our intent to derive *all* these properties, but simply to illustrate how this might be done by giving a few examples. Other properties are left for the reader to prove as exercises. Having done this, we shall subsequently assume familiarity with all the basic algebraic properties (whether we have proved them specifically or not).

We begin by assuming the existence of a set $\mathbb{R}$, called the set of real numbers, and two operations $+$ and $\cdot$, called addition and multiplication, such that the following properties apply:

A1. For any $x, y \in \mathbb{R}$, $x + y \in \mathbb{R}$ and if $x = w$ and $y = z$, then $x + y = w + z$.
A2. For any $x, y \in \mathbb{R}$, $x + y = y + x$.
A3. For any $x, y, z \in \mathbb{R}$, $x + (y + z) = (x + y) + z$.
A4. There is a unique number 0 such that $x + 0 = x$, for all $x \in \mathbb{R}$.
A5. For each $x \in \mathbb{R}$ there is a unique number $-x$ such that $x + (-x) = 0$.
M1. For any $x, y \in \mathbb{R}$, $x \cdot y \in \mathbb{R}$ and if $x = w$ and $y = z$, then $x \cdot y = w \cdot z$.
M2. For any $x, y \in \mathbb{R}$, $x \cdot y = y \cdot x$.
M3. For any $x, y, z \in \mathbb{R}$, $x \cdot (y \cdot z) = (x \cdot y) \cdot z$.
M4. There is a unique number 1 such that $1 \neq 0$ and $x \cdot 1 = x$ for all $x \in \mathbb{R}$.
M5. For each $x \in \mathbb{R}$ with $x \neq 0$, there is a unique number $1/x$ such that $x \cdot (1/x) = 1$. We also write x^{-1} in place of $1/x$.
DL. For any $x, y, z \in \mathbb{R}$, $x \cdot (y + z) = x \cdot y + x \cdot z$.

These first 11 axioms are called the field axioms because they describe a system known as a **field** in the study of abstract algebra. Axioms A2 and M2 are called the **commutative laws** and axioms A3 and M3 are the **associative laws**. Axiom DL is the **distributive law** that shows how addition and multiplication relate to each other. Because of axioms A1 and M1 we can think of addition and multiplication as functions that map $\mathbb{R} \times \mathbb{R}$ into $\mathbb{R}$. When writing multiplication we often omit the raised dot and write xy instead of $x \cdot y$.

In addition to the field axioms, the real numbers also satisfy four order axioms. These axioms identify the properties of the relation $<$. As is common practice, we may write $y > x$ instead of $x < y$, and $x \leqslant y$ as an abbreviation for "$x < y$ or $x = y$." The notation $\geqslant$ is defined analogously. A real number x is called nonnegative if $x \geqslant 0$ and positive if $x > 0$. A pair of simultaneous inequalities such as "$x < y$ and $y < z$" is often written in the shorter form "$x < y < z$."

The relation $<$ satisfies the following properties:

O1. For $x, y \in \mathbb{R}$, exactly one of the relations $x = y$, $x > y$, $x < y$ holds (trichotomy law).

O2. For any $x, y, z \in \mathbb{R}$, if $x < y$ and $y < z$, then $x < z$.
O3. For any $x, y, z \in \mathbb{R}$, if $x < y$, then $x + z < y + z$.
O4. For any $x, y \in \mathbb{R}$, if $x < y$ and $z > 0$, then $xz < yz$.

To illustrate how the axioms may be used to derive familiar algebraic properties, we include the following:

11.1 THEOREM Let x, y, and z be real numbers.

(a) If $x + z = y + z$, then $x = y$.
(b) $x \cdot 0 = 0$.
(c) $(-1) \cdot x = -x$.
(d) $xy = 0$ iff $x = 0$ or $y = 0$.
(e) $x < y$ iff $-y < -x$.
(f) If $x < y$ and $z < 0$, then $xz > yz$.

Proof: (a) If $x + z = y + z$, then

$$\begin{aligned}
(x + z) + (-z) &= (y + z) + (-z) && \text{by A5 and A1,}\\
x + [z + (-z)] &= y + [z + (-z)] && \text{by A3,}\\
x + 0 &= y + 0 && \text{by A5,}\\
x &= y && \text{by A4.}
\end{aligned}$$

(b) For any $x \in \mathbb{R}$ we have

$$\begin{aligned}
x \cdot 0 &= x \cdot (0 + 0) && \text{by A4,}\\
x \cdot 0 &= x \cdot 0 + x \cdot 0 && \text{by DL,}\\
0 + x \cdot 0 &= x \cdot 0 + x \cdot 0 && \text{by A4 and A2,}\\
0 &= x \cdot 0 && \text{by part (a).}
\end{aligned}$$

(c) For any $x \in \mathbb{R}$ we have

$$\begin{aligned}
x + (-1) \cdot x &= x + x \cdot (-1) && \text{by M2,}\\
&= x \cdot 1 + x \cdot (-1) && \text{by M4,}\\
&= x \cdot [1 + (-1)] && \text{by DL,}\\
&= x \cdot 0 && \text{by A5,}\\
&= 0 && \text{by part (b).}
\end{aligned}$$

Thus $(-1) \cdot x = -x$ by the uniqueness of $-x$ in A5.

(d) See Practice 11.2.

(e) Suppose that $x < y$. Then

$$\begin{aligned} x + [(-x) + (-y)] &< y + [(-x) + (-y)] && \text{by O3,} \\ x + [(-x) + (-y)] &< y + [(-y) + (-x)] && \text{by A2,} \\ [x + (-x)] + (-y) &< [y + (-y)] + (-x) && \text{by A3,} \\ 0 + (-y) &< 0 + (-x) && \text{by A5,} \\ -y &< -x && \text{by A2 and A4.} \end{aligned}$$

The converse is similar.

(f) See Practice 11.3. ▮

11.2 PRACTICE Fill in the blanks in the following proof of Theorem 11.1(d).

Proof: If $x=0$ or $y=0$, then $xy=0$ by 11.1(b) and M2. Conversely, suppose that $xy = 0$ and $x \neq 0$. By tautology 3.12(p), it suffices to show that $y = 0$. Since $x \neq 0$, $1/x$ exists by (a) ________. Thus

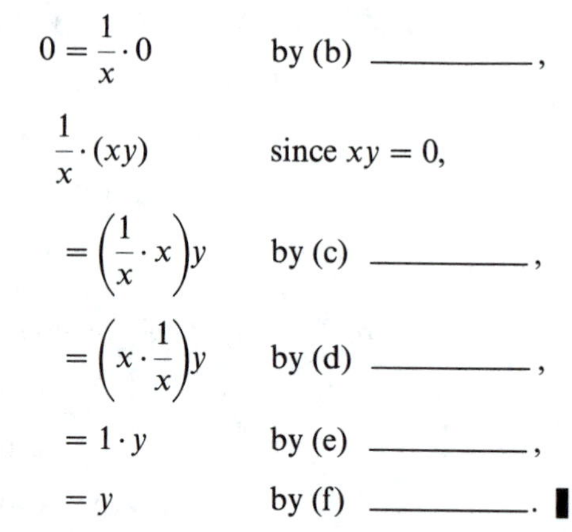

$$\begin{aligned} 0 &= \frac{1}{x} \cdot 0 && \text{by (b) ________,} \\ &\frac{1}{x} \cdot (xy) && \text{since } xy = 0, \\ &= \left(\frac{1}{x} \cdot x\right) y && \text{by (c) ________,} \\ &= \left(x \cdot \frac{1}{x}\right) y && \text{by (d) ________,} \\ &= 1 \cdot y && \text{by (e) ________,} \\ &= y && \text{by (f) ________. ▮} \end{aligned}$$

11.3 PRACTICE Fill in the blanks in the following proof of Theorem 11.1(f).

Proof: If $x < y$ and $z < 0$, then $-z > 0$ by 11.1(e). Thus $x(-z) < y(-z)$ by (a) ________. But

$$\begin{aligned} x(-z) &= x[(-1)(z)] && \text{by (b) ________,} \\ &= [x(-1)]z && \text{by (c) ________,} \\ &= [(-1)(x)]z && \text{by (d) ________,} \\ &= (-1)(xz) && \text{by (e) ________,} \\ &= -xz && \text{by (f) ________.} \end{aligned}$$

Similarly, $y(-z) = -yz$. Thus $-xz < -yz$. But then $yz < xz$ by (g) ________. ▮

We have listed the field axioms and the order axioms as properties of the real numbers. But in fact they are of interest in their own right. Any mathematical system that satisfies these 15 axioms is called an **ordered field**. Thus the real numbers are an example of an ordered field. But there are other examples as well. In particular, the rational numbers $\mathbb{Q}$ are also an ordered field. Recall that

$$\mathbb{Q} = \left\{\frac{m}{n}: m, n \in \mathbb{Z} \text{ and } n \neq 0\right\},$$

where $\mathbb{Z} = \{0, 1, -1, 2, -2, 3, -3, \ldots\}$. Since the rational numbers are a subset of the reals, the commutative and associative laws and the order axioms are automatically satisfied.† Since 0 and 1 are rational numbers, axioms A4 and M4 apply. Since $-(m/n) = -m/n$ and $(m/n)^{-1} = n/m$, axioms A5 and M5 hold. It remains to show that the sum and product of two rationals are also rationals.

11.4 PRACTICE Let a/b and c/d be rational numbers with $a, b, c, d \in \mathbb{Z}$. Show that

$$\frac{a}{b} + \frac{c}{d} \quad \text{and} \quad \frac{a}{b} \cdot \frac{c}{d}$$

are rational.

11.5 EXAMPLE For a more unusual example of an ordered field, let F be the set of all rational functions. That is, F is the set of all quotients of polynomials. A typical element of F looks like

$$\frac{a_n x^n + \cdots + a_1 x + a_0}{b_k x^k + \cdots + b_1 x + b_0},$$

where the coefficients are real numbers and $b_k \neq 0$. Using the usual rules for adding, subtracting, multiplying, and dividing polynomials, it is not difficult to verify that F is a field.

We can define an order on F by saying that a quotient such as above is positive iff a_n and b_k have the same sign; that is, $a_n \cdot b_k > 0$. For example,

$$\frac{3x^2 + 4x - 1}{7x^5 + 5} > 0,$$

since $3 \cdot 7 > 0$. If p/q and f/g are rational functions, then we say that

$$\frac{p}{q} > \frac{f}{g} \quad \text{iff} \quad \frac{p}{q} - \frac{f}{g} > 0.$$

† We have not proved that $\mathbb{Q} \subseteq \mathbb{R}$, but this relationship should come as no surprise to the reader. A rigorous proof may be found in Stewart and Tall (1977).

That is,

$$\frac{p}{q} > \frac{f}{g} \quad \text{iff} \quad \frac{pg - fq}{qg} > 0.$$

The verification that "$>$" satisfies the order axioms is left to the reader (Exercise 11.9). It turns out that the ordered field F has a number of interesting properties, as we shall see later.

There is one more algebraic property of the real numbers to which we give special attention because of its frequent use in proofs in analysis and because it may not be familiar to the reader.

11.6 THEOREM Let $x, y \in \mathbb{R}$ such that $x \leqslant y + \varepsilon$ for every $\varepsilon > 0$. Then $x \leqslant y$.

Proof: We shall establish the contrapositive. By axiom O1, the negation of $x \leqslant y$ is $x > y$. Thus we suppose that $x > y$ and we must show that there exists $\varepsilon > 0$ such that $x > y + \varepsilon$. Let $\varepsilon = (x - y)/2$. Since $x > y$, $\varepsilon > 0$. Furthermore,

$$y + \varepsilon = y + \frac{x - y}{2} = \frac{x + y}{2} < \frac{x + x}{2} = x,$$

as required. ∎

Many of the proofs in analysis involve manipulating inequalities, and one useful tool in working with inequalities is the concept of absolute value. The definition of absolute value was mentioned in Section 4, but we repeat it here for reference.

11.7 DEFINITION If $x \in \mathbb{R}$, then the **absolute value** of x, denoted by $|x|$, is defined by

$$|x| = \begin{cases} x, & \text{if } x \geqslant 0 \\ -x, & \text{if } x < 0. \end{cases}$$

The basic properties of absolute value are summarized in the following theorem.

11.8 THEOREM Let $x, y \in \mathbb{R}$ and let $a \geqslant 0$. Then

(a) $|x| \geqslant 0$,
(b) $|x| \leqslant a$ iff $-a \leqslant x \leqslant a$,
(c) $|xy| = |x| \cdot |y|$,
(d) $|x + y| \leqslant |x| + |y|$.

Proof: (a) There are two cases. If $x \geqslant 0$, then $|x| = x \geqslant 0$. On the other hand, if $x < 0$, then $|x| = -x > 0$. In both cases, $|x| \geqslant 0$.

(b) Since $x = |x|$ or $x = -|x|$, it follows that $-|x| \leqslant x \leqslant |x|$. Now if $|x| \leqslant a$, then we have

$$-a \leqslant -|x| \leqslant x \leqslant |x| \leqslant a.$$

Conversely, suppose that $-a \leqslant x \leqslant a$. If $x \geqslant 0$, then $|x| = x \leqslant a$. And if $x < 0$, then $|x| = -x \leqslant a$. In both cases, $|x| \leqslant a$.

(c) See Exercise 11.3.

(d) As in part (b) we have

$$-|x| \leqslant x \leqslant |x| \quad \text{and} \quad -|y| \leqslant y \leqslant |y|.$$

Adding the inequalities together, we obtain

$$-(|x| + |y|) \leqslant x + y \leqslant |x| + |y|,$$

which implies that $|x + y| \leqslant |x| + |y|$ by part (b). ∎

Part (d) of Theorem 11.8 is referred to as the **triangle inequality**:

$$|x + y| \leqslant |x| + |y|.$$

It is also useful in other forms. For example, letting $x = a - c$ and $y = c - b$, we obtain

$$|a - b| \leqslant |a - c| + |c - b|.$$

If we think of the real numbers as being points on a line, then $|a - b|$ represents the distance from a to b. Thus the distance from a to b is less than or equal to the sum of the distances from a to c and c to b. It is possible to generalize this to higher dimensions where a, b, and c are the vertices of a triangle. It is this more general setting that gives rise to the name "triangle inequality." (See Section 15.)

ANSWERS TO PRACTICE PROBLEMS

11.2 (a) M5; (b) Theorem 11.1(b); (c) M3; (d) M2; (e) M5; (f) M2 and M4.

11.3 (a) O4; (b) Theorem 11.1(c); (c) M3; (d) M2; (e) M3; (f) Theorem 11.1(c); (g) Theorem 11.1(e).

11.4 From the usual rules of arithmetic we have

$$\frac{a}{b} + \frac{c}{d} = \frac{ad + bc}{bd} \quad \text{and} \quad \frac{a}{b} \cdot \frac{c}{d} = \frac{ac}{bd}.$$

Since sums and products of integers are always integers,

$$\frac{ad + bc}{bd} \quad \text{and} \quad \frac{ac}{bd}$$

are both rational.

EXERCISES

11.1 Let x, y and z be real numbers. Prove the following.

(a) $-(-x) = x$.

(b) $(-x)\cdot y = -xy$ and $(-x)\cdot(-y) = xy$.

(c) If $x \neq 0$, then $(1/x) \neq 0$ and $1(1/x) = x$.

(d) If $x \cdot z = y \cdot z$ and $z \neq 0$, then $x = y$.

(e) If $x \neq 0$, then $x^2 > 0$.

(f) $0 < 1$.

(g) If $x > 0$, then $1/x > 0$. If $x < 0$, then $1/x < 0$.

(h) If $0 < x < y$, then $0 < 1/y < 1/x$.

(i) If $xy > 0$, then either (i) $x > 0$ and $y > 0$, or (ii) $x < 0$ and $y < 0$.

(j) For each $n \in \mathbb{N}$, if $0 < x < y$, then $x^n < y^n$.

(k) If $0 < x < y$, then $0 < \sqrt{x} < \sqrt{y}$.

11.2 Prove: If $x \geqslant 0$ and $x \leqslant \varepsilon$ for all $\varepsilon > 0$, then $x = 0$.

11.3 Prove Theorem 11.8(c): $|xy| = |x|\cdot|y|$.

***11.4** Prove: $||x| - |y|| \leqslant |x - y|$.

***11.5** Suppose that $x_1, x_2, \ldots, x_n$ are real numbers. Prove that

$$|x_1 + x_2 + \cdots + x_n| \leqslant |x_1| + |x_2| + \cdots + |x_n|.$$

11.6 Let $P = \{x \in \mathbb{R}: x > 0\}$. Show that P satisfies the following:

(a) If $x, y \in P$, then $x + y \in P$.

(b) If $x, y \in P$, then $x \cdot y \in P$.

(c) For each $x \in \mathbb{R}$, exactly one of the following three statements is true: $x \in P$, $x = 0$, $-x \in P$.

11.7 Let F be a field and suppose that P is a subset of F that satisfies the three properties in Exercise 11.6. Define $x < y$ iff $y - x \in P$. Prove that "$<$" satisfies axioms O1, O2, and O3. Thus in defining an ordered field, we can either begin with the properties of "$<$" as in the text, or we can begin by identifying a certain subset as "positive."

11.8 Prove that in any ordered field F, $a^2 + 1 > 0$ for all $a \in F$. Conclude from this that if the equation $x^2 + 1 = 0$ has a solution in a field, then that field cannot be ordered. (Thus it is not possible to define an order relation on the set of all complex numbers that will make it an ordered field.)

11.9 Let F be the field of rational functions described in Example 11.5.

(a) Show that the ordering given there satisfies the order axioms O1, O2, and O3.

(b) Write the following polynomials in order of increasing size:

$$x^2, \quad -x^3, \quad 5, \quad x + 2, \quad 3 - x.$$

(c) Write the following functions in order of increasing size:

$$\frac{x^2+2}{x-1}, \quad \frac{x^2-2}{x+1}, \quad \frac{x+1}{x^2-2}, \quad \frac{x+2}{x^2-1}.$$

11.10 To actually construct the rationals $\mathbb{Q}$ from the integers $\mathbb{Z}$, let $S = \{(a, b): a, b \in \mathbb{Z} \text{ and } b \neq 0\}$. Define an equivalence relation "$\sim$" on S by $(a, b) \sim (c, d)$ iff $ad = bc$. We then define the set $\mathbb{Q}$ of rational numbers to be the set of equivalence classes corresponding to $\sim$. The equivalence class determined by the ordered pair (a, b) we denote by $[a/b]$. Then $[a/b]$ is what we usually think of as the fraction a/b. For $a, b, c, d \in \mathbb{Z}$ with $b \neq 0$ and $d \neq 0$, we define addition and multiplication in $\mathbb{Q}$ by

$$[a/b] + [c/d] = [(ad + bc)/bd],$$
$$[a/b] \cdot [c/d] = [ac/bd].$$

We say that $[a/b]$ is *positive* if $ab \in \mathbb{N}$. Since $a, b \in \mathbb{Z}$ with $b \neq 0$, this is equivalent to requiring $ab > 0$. The set of positive rationals is denoted by $\mathbb{Q}^+$, and we define an order "$<$" on $\mathbb{Q}$ by

$$x < y \quad \text{iff} \quad y - x \in \mathbb{Q}^+.$$

(a) Verify that $\sim$ is an equivalence relation on S.
(b) Show that addition and multiplication are well-defined. That is, suppose that $[a/b] = [p/q]$ and $[c/d] = [r/s]$. Show that $[(ad + bc)/bd] = [(ps + qr)/qs]$ and $[ac/bd] = [pr/qs]$.
(c) For any $b \in \mathbb{Z}\backslash\{0\}$, show that $[0/b] = [0/1]$ and $[b/b] = [1/1]$.
(d) For any $a, b \in \mathbb{Z}$ with $b \neq 0$, show that $[a/b] + [0/1] = [a/b]$ and $[a/b] \cdot [1/1] = [a/b]$. Thus $[0/1]$ corresponds to zero and $[1/1]$ corresponds to 1.
(e) For any $a, b \in \mathbb{Z}$ with $b \neq 0$, show that $[a/b] + [(-a)/b] = [0/1]$ and $[a/b] \cdot [b/a] = [1/1]$.
(f) Verify that the set $\mathbb{Q}$ with addition, multiplication, and order as given above satisfies the axioms of an ordered field.

11.11 Construct the integers $\mathbb{Z}$ from the natural numbers $\mathbb{N}$ in a method similar to that used in Exercise 11.10 by defining an appropriate equivalence relation on $\mathbb{N} \times \mathbb{N}$.

Section 12 THE COMPLETENESS AXIOM

In the preceding section we presented the field and order axioms of the real numbers. Although these axioms are certainly basic to the real numbers, by themselves they do not characterize $\mathbb{R}$. That is, we have seen that there are other mathematical systems that also satisfy these 15 axioms. In particular, the set $\mathbb{Q}$ of rational numbers is an ordered field. The one additional axiom that distinguishes $\mathbb{R}$ from $\mathbb{Q}$ (and from other ordered fields) is called the completeness axiom. Before presenting this axiom, let us look briefly at why it is needed—at why the rational numbers by themselves are inadequate for analysis.

Consider the graph of the function $f(x) = x^2 - 2$, shown in Figure 12.1. It appears that the graph crosses the horizontal axis at a point between 1 and 2. But does it really? How can we be sure? In other words, how can we be certain that there is a "number" x on the axis such that $x^2 - 2 = 0$? It turns out that if the x-axis consists only of rational numbers, then no such number exists. That is, there is no rational number whose square is 2. In fact, we can easily prove the more general result that $\sqrt{p}$ is irrational (not rational) for any prime number p. (Recall that an integer $p > 1$ is prime iff its only divisors are 1 and p.)

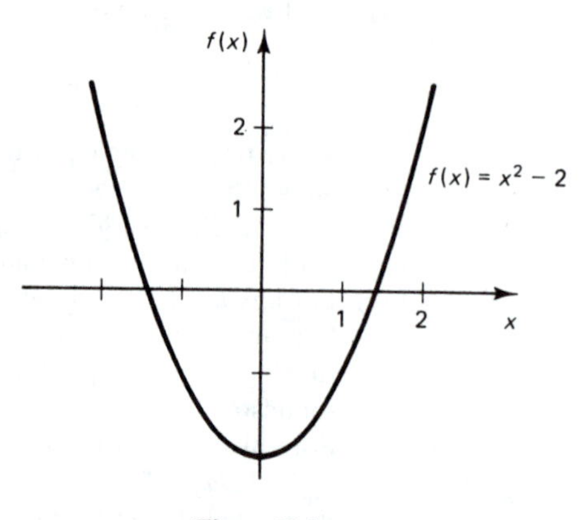

Figure 12.1

12.1 THEOREM Let p be a prime number. Then $\sqrt{p}$ is not a rational number.

Proof: We suppose that $\sqrt{p}$ is rational and obtain a contradiction. If $\sqrt{p}$ is rational, then we can write $\sqrt{p} = m/n$, where m and n are integers with no common factors. Then $pn^2 = m^2$, so m^2 must be a multiple of p. Since p is prime, this implies that m is also a multiple of p. That is, $m = kp$ for some integer k. But then $pn^2 = k^2p^2$, so that $n^2 = k^2p$. Thus n^2 is a multiple of p, and as above we conclude that n is also. Hence m and n are both multiples of p, contradicting the fact that they have no common factors. ∎

There are, of course, many other irrational numbers besides $\sqrt{p}$ for p prime. We saw in Section 8 that there are, in fact, more irrational numbers than there are rational. Thus, if we were to restrict our analysis to rational numbers, our "number line" would have uncountably many "holes" in it. It is these holes in the number line that the completeness axiom fills. To state this final axiom for $\mathbb{R}$, we need some preliminary definitions.

Upper Bounds and Suprema

12.2 DEFINITION Let S be a nonempty subset of $\mathbb{R}$. If there exists a real number m such that $m \geqslant s$ for all $s \in S$, then m is called an **upper bound** for S, and we say that S is bounded above. If $m \leqslant s$ for all $s \in S$, then m is a **lower bound** for S and S is bounded below. The set S is said to be **bounded** if it is bounded above and bounded below.

A set may have upper or lower bounds, or it may have none. If m is an upper bound for S, then any number greater than m is also an upper bound. If an upper bound m for S is a member of S, then m is called the **maximum** (or largest element) of S, and we write

$$m = \max S.$$

Similarly, if a lower bound of S is a member of S, then it is called the **minimum** (or least element) of S, denoted by $\min S$. While a set may have many upper and lower bounds, if it has a maximum or a minimum, then those values are unique. Thus we speak of *an* upper bound and *the* maximum.

12.3 EXAMPLES (a) The set $S = \{2, 4, 6, 8\}$ is bounded above by 8, 9, $8\frac{1}{2}$, π^2, and any other real number greater than or equal to 8. Since $8 \in S$, we have $\max S = 8$. Similarly, S has many lower bounds, including 2, which is the largest of the lower bounds and the minimum of S. It is easy to see that any finite set is bounded and always has a maximum and a minimum.

(b) The interval $[0, \infty)$ is not bounded above. It is bounded below by any nonpositive number, and of these lower bounds, 0 is the largest. Since $0 \in [0, \infty)$, 0 is the minimum of $[0, \infty)$.

(c) The interval $(0, 1]$ has a maximum of 1, and this is the smallest of the upper bounds. It is bounded below by any nonpositive number, and of these lower bounds, 0 is the largest. Since $0 \notin (0, 1]$, the set has no minimum.

12.4 PRACTICE Find upper and lower bounds, the maximum, and the minimum of the set $T = \{q \in \mathbb{Q}: 0 \leqslant q \leqslant \sqrt{2}\}$, if they exist.

Since any number larger than an upper bound is also an upper bound, we have found it useful to identify the smallest or least upper bound in our examples. It is also helpful to know the greatest of the lower bounds.

12.5 DEFINITION Let S be a nonempty set. If S is bounded above, then the least upper bound of S is called its **supremum** and is denoted by sup S. Thus $m =$ sup S iff

(a) $m \geqslant s$, for all $s \in S$, and
(b) if $m' < m$, then there exists $s' \in S$ such that $s' > m'$.

If S is bounded below, then the greatest lower bound of S is called its **infimum** and is denoted by inf S.

12.6 PRACTICE Characterize inf S in a way analogous to that given for sup S.

It is easy to show (Exercise 12.4) that if a set has a supremum, then it is unique. What may not be clear is whether a set that is bounded above must have a least upper bound. Indeed, the set $T = \{q \in \mathbb{Q} : 0 \leqslant q \leqslant \sqrt{2}\}$ in Practice 12.4 does not have a supremum when considered as a subset of $\mathbb{Q}$. The problem is that sup $T = \sqrt{2}$, and $\sqrt{2}$ is one of the "holes" in $\mathbb{Q}$.

When considering subsets of $\mathbb{R}$, it has been true that each set bounded above has had a least upper bound. This supremum may be a member of the set, as in the interval $[0, 1]$, or it may be outside the set, as in the interval $[0, 1)$, but in both cases the supremum *exists* as a real number. This fundamental difference betweeen $\mathbb{Q}$ and $\mathbb{R}$ is the basis for our final axiom of the real numbers, the **completeness axiom**:

> Every nonempty subset S of $\mathbb{R}$ that is bounded above has a least upper bound. That is, sup S exists and is a real number.

While the completeness axiom refers only to sets that are bounded above, the corresponding property for sets bounded below follows readily. Indeed, suppose that S is a nonempty subset of $\mathbb{R}$ that is bounded below. Then the set $-S = \{-s : s \in S\}$ is bounded above and the completeness axiom implies the existence of a supremum, say m. It follows (Exercise 12.5) that $-m$ is the infimum of S. Thus every nonempty subset of $\mathbb{R}$ that is bounded below has a greatest lower bound.

To illustrate the techiques of working with suprema, we include the following two theorems. It goes without saying that analogous results hold for infima.

12.7 THEOREM Given nonempty subsets A and B of $\mathbb{R}$, let C denote the set

$$C = \{x + y : x \in A \text{ and } y \in B\}.$$

If each of A and B has a supremum, then C has a supremum and

$$\sup C = \sup A + \sup B.$$

Proof: Let $\sup A = a$ and $\sup B = b$. If $z \in C$, then $z = x + y$ for some $x \in A$ and $y \in B$. Thus $z = x + y \leqslant a + b$, so $a + b$ is an upper bound for C. By the completeness axiom, C has a least upper bound, say $\sup C = c$. We must show that $c = a + b$. Since c is the *least* upper bound for C, we have $c \leqslant a + b$.

To see that $a + b \leqslant c$, choose any $\varepsilon > 0$. Since $a = \sup A$, $a - \varepsilon$ is not an upper bound for A and there must exist x in A such that $a - \varepsilon < x$. Similarly, since $b = \sup B$, there exists y in B such that $b - \varepsilon < y$. Combining these inequalities, we have

$$a + b - 2\varepsilon < x + y \leqslant c.$$

That is, $a + b < c + 2\varepsilon$ for every $\varepsilon > 0$. Thus by Theorem 11.6, $a + b \leqslant c$.

Finally, since $c \leqslant a + b$ and $c \geqslant a + b$, we conclude that $c = a + b$. ∎

12.8 THEOREM Suppose that D is a nonempty set and that $f: D \to \mathbb{R}$ and $g: D \to \mathbb{R}$. If for every x, $y \in D$, $f(x) \leqslant g(y)$, then $f(D)$ is bounded above and $g(D)$ is bounded below. Furthermore, $\sup f(D) \leqslant \inf g(D)$.

Proof: Given any $y_0 \in D$, we have $f(x) \leqslant g(y_0)$, for all $x \in D$. Thus $f(D)$ is bounded above by $g(y_0)$. It follows that the *least* upper bound of $f(D)$ is no larger than $g(y_0)$. That is, $\sup f(D) \leqslant g(y_0)$. Since this last inequality holds for all $y_0 \in D$, $g(D)$ is bounded below by $\sup f(D)$. Thus the *greatest* lower bound of $g(D)$ is no smaller than $\sup f(D)$. That is, $\sup f(D) \leqslant \inf g(D)$. ∎

The Archimedean Property

One of the important consequences of the completeness axiom is called the Archimedean property. It states that the natural numbers $\mathbb{N}$ are not bounded above in $\mathbb{R}$. Although this property may seem obvious at first, its proof actually depends on the completeness axiom. In fact, there are other ordered fields in which it does not hold. (See Exercise 12.14.)

12.9 THEOREM (Archimedean Property of $\mathbb{R}$) The set $\mathbb{N}$ of natural numbers is unbounded above in $\mathbb{R}$.

Proof: If $\mathbb{N}$ were bounded above, then by the completeness axiom it would have a least upper bound, say $\sup \mathbb{N} = m$. Since m is a *least* upper bound, $m - 1$ is not an upper bound for $\mathbb{N}$. Thus there exists an n_0 in $\mathbb{N}$ such that $n_0 > m - 1$. But then $n_0 + 1 > m$, and since $n_0 + 1 \in \mathbb{N}$, this contradicts m being an upper bound for $\mathbb{N}$. ∎

There are several equivalent forms of the Archimedean property that are useful in different contexts. We establish their equivalence in the following theorem.

12.10 THEOREM Each of the following is equivalent to the Archimedean property.

(a) For each $z \in \mathbb{R}$, there exists $n \in \mathbb{N}$ such that $n > z$.
(b) For each $x > 0$ and for each $y \in \mathbb{R}$, there exists $n \in \mathbb{N}$ such that $nx > y$.
(c) For each $x > 0$, there exists $n \in \mathbb{N}$ such that $0 < 1/n < x$.

Proof: We shall prove that Theorem 12.9 $\Rightarrow$ (a) $\Rightarrow$ (b) $\Rightarrow$ (c) $\Rightarrow$ Theorem 12.9, thereby establishing their equivalence.

If (a) were not true, then for some $z_0 \in \mathbb{R}$ we would have $n \leqslant z_0$ for all $n \in \mathbb{N}$. But then z_0 would be an upper bound for $\mathbb{N}$, contradicting Theorem 12.9. Thus the Archimedean property implies (a).

To see that (a) $\Rightarrow$ (b), let $z = y/x$. Then there exists $n \in \mathbb{N}$ such that $n > y/x$, so that $nx > y$.

(c) follows from (b) by taking $y = 1$ in (b). Then $nx > 1$, so that $1/n < x$. Since $n \in \mathbb{N}$, $n > 0$ and also $1/n > 0$.

Finally, suppose that $\mathbb{N}$ were bounded above by some real number, say m. That is, $n < m$ for all $n \in \mathbb{N}$. But then $1/n > 1/m$, for all $n \in \mathbb{N}$, and this contradicts (c) with $x = 1/m$. Thus (c) implies the Archimedean property. ∎

In Theorem 12.1 we showed that $\sqrt{p}$ is not rational when p is prime. We are now in a position to prove there is a positive *real* number whose square is p, thus illustrating that we actually have filled in the "holes" in $\mathbb{R}$.

12.11 THEOREM Let p be a prime number. Then there exists a positive real number x such that $x^2 = p$.

Proof: Let $S = \{r \in \mathbb{R}: r > 0 \text{ and } r^2 < p\}$. Since $p > 1$, $1 \in S$ and S is nonempty. Furthermore, if $r \in S$, then $r^2 < p < p^2$, so $r < p$. Thus S is bounded above by p, and by the completeness axiom, $\sup S$ exists as a real number. Let $x = \sup S$. It is clear that $x > 0$, and we claim that $x^2 = p$. To prove this, we shall show that neither $x^2 < p$ nor $x^2 > p$ is consistent with our choice of x.

Suppose first that $x^2 < p$. Then $(p - x^2)/(2x + 1) > 0$, so that Theorem 12.10(c) implies the existence of some $n \in \mathbb{N}$ such that†

$$\frac{1}{n} < \frac{p - x^2}{2x + 1}.$$

But then we have

$$\left(x + \frac{1}{n}\right)^2 = x^2 + \frac{2x}{n} + \frac{1}{n^2} = x^2 + \frac{1}{n}\left(2x + \frac{1}{n}\right)$$
$$\leqslant x^2 + \frac{1}{n}(2x + 1) < x^2 + (p - x^2) = p.$$

It follows that $x + 1/n \in S$, which contradicts our choice of x as an upper bound for S.

Now suppose that $x^2 > p$. Then $(x^2 - p)/(2x) > 0$. Again using Theorem 12.10(c), there exists $m \in \mathbb{N}$ such that

$$\frac{1}{m} < \frac{x^2 - p}{2x}.$$

But then we have

$$\left(x - \frac{1}{m}\right)^2 = x^2 - \frac{2x}{m} + \frac{1}{m^2} > x^2 - \frac{2x}{m}$$
$$> x^2 - (x^2 - p) = p.$$

This implies that $x - 1/m > r$, for all $r \in S$, so $x - 1/m$ is an upper bound of S. Since $x - 1/m < x$, this contradicts our choice of x as the *least* upper bound of S.

† In the formal proof above we have mysteriously introduced the inequality $(1/n) < [(p - x^2)/(2x + 1)]$. It is instructive to see how we might come up with such a requirement.

If $x^2 < p$, then we somehow want to contradict the fact that x is an upper bound for S. That is, we want to find some y in S with $y > x$. A simple way to get $y > x$ is to add something positive to x. But if x^2 is close to p, that "something" will have to be small. By taking $y = x + 1/n$, we guarantee that $y > x$, we can make y as close to x as we want, and y is still fairly easy to work with.

Now to make $(x + 1/n) \in S$, we must have $(x + 1/n)^2 = x^2 + 2x/n + 1/n^2 < p$. We can control the size of $1/n$ (by choosing n carefully), so we rewrite the inequality to emphasize the contribution of $1/n$:

$$x^2 + \frac{1}{n}\left(2x + \frac{1}{n}\right) < p.$$

If it were not for the $1/n$ inside the parentheses, we could easily solve for $1/n$. But we can get rid of that $1/n$ by observing that $1/n \leqslant 1$ for all $n \in \mathbb{N}$. Thus we want to choose n so that

$$x^2 + \frac{1}{n}\left(2x + \frac{1}{n}\right) \leqslant x^2 + \frac{1}{n}(2x + 1) < p.$$

Solving the last inequality for $1/n$, we obtain the requirement that appears in the proof.

The inequality that is used in the case $x^2 > p$ is found in a similar manner.

Finally, since neither $x^2 < p$ nor $x^2 > p$ are possibilities, we conclude by the trichotomy law that in fact $x^2 = p$. ▮

The Density of the Rational Numbers

We conclude this section with another property of $\mathbb{R}$ that is probably familiar to the reader: Between any two real numbers there is a rational number. More precisely, we say that the set $\mathbb{Q}$ is **dense** in $\mathbb{R}$. Once again, our proof will ultimately depend on the completeness axiom.

12.12 THEOREM (Density of $\mathbb{Q}$ in $\mathbb{R}$) If x and y are real numbers with $x < y$, then there exists a rational number r such that $x < r < y$.

Proof: We begin by supposing that $x > 0$. Using the Archimedean property 12.10(a), there exists $n \in \mathbb{N}$ such that $n > 1/(y - x)$. That is, $nx + 1 < ny$. Since $nx > 0$, it is not difficult to show (Exercise 12.7) that there exists $m \in \mathbb{N}$ such that $m - 1 \leqslant nx < m$. But then $m \leqslant nx + 1 < ny$, so that $nx < m < ny$. It follows that the rational number $r = m/n$ satisfies $x < r < y$.

Finally, if $x \leqslant 0$, choose an integer k such that $k > |x|$. Then apply the argument above to the positive numbers $x + k$ and $y + k$. If q is a rational satisfying $x + k < q < y + k$, then the rational $r = q - k$ satisfies $x < r < y$. ▮

Using Theorem 12.12, we can easily show that between any two real numbers there is also an irrational number. (Thus the irrationals are also dense in $\mathbb{R}$.) We pause first for you to prove the following preliminary result.

12.13 PRACTICE Let x be a nonzero rational number and let y be irrational. Prove that xy is irrational.

12.14 THEOREM If x and y are real numbers with $x < y$, then there exists an irrational number w such that $x < w < y$.

Proof: Apply Theorem 12.12 to the real numbers $x/\sqrt{2}$ and $y/\sqrt{2}$ to obtain a rational number $r \neq 0$ such that

$$\frac{x}{\sqrt{2}} < r < \frac{y}{\sqrt{2}}.$$

It follows from Practice 12.13 that $w = r\sqrt{2}$ is irrational, and $x < w < y$. ▮

ANSWERS TO PRACTICE PROBLEMS

12.4 Any real number x such that $x^2 \geqslant 2$ is an upper bound for T. The smallest of these upper bounds is $\sqrt{2}$, but since $\sqrt{2} \notin \mathbb{Q}$, set T has no maximum. The minimum of T is 0.

12.6 $m = \inf S$ iff (i) $m \leqslant s$, for all $s \in S$, and (ii) if $m' > m$, then there exists $s' \in S$ such that $s' < m'$.

12.13 Since x is rational and $x \neq 0$, we have $x = m/n$ for some nonzero integers m and n. If xy were rational, then we could write $xy = p/q$ for some p, $q \in \mathbb{Z}$. But then

$$y = \frac{xy}{x} = \frac{p/q}{m/n} = \frac{pn}{mq},$$

so y would have to be rational too, a contradiction.

EXERCISES

12.1 For each subset of $\mathbb{R}$, give its supremum if it has one. Otherwise, write "no sup."

(a) $\{1, 3\}$ (b) $\{\pi, 3\}$

(c) $[0, 4]$ (d) $(0, 4)$

(e) $\left\{\frac{1}{n}: n \in \mathbb{N}\right\}$ (f) $\left\{1 - \frac{1}{n}: n \in \mathbb{N}\right\}$

(g) $\left\{\frac{n}{n+1}: n \in \mathbb{N}\right\}$ (h) $\left\{(-1)^n\left(1 + \frac{1}{n}\right): n \in \mathbb{N}\right\}$

(i) $\left\{n + \frac{(-1)^n}{n}: n \in \mathbb{N}\right\}$ (j) $(-\infty, 4)$

(k) $\bigcap_{n=1}^{\infty}\left(1 - \frac{1}{n}, 1 + \frac{1}{n}\right)$ (l) $\bigcup_{n=1}^{\infty}\left[\frac{1}{n}, 2 - \frac{1}{n}\right]$

(m) $\{r \in \mathbb{Q}: r < 5\}$ (n) $\{r \in \mathbb{Q}: r^2 < 5\}$

12.2 Repeat Exercise 12.1 for the infimum of each set.

12.3 Let S be a nonempty bounded subset of $\mathbb{R}$ and let $m = \sup S$. Prove that $m \in S$ iff $m = \max S$.

12.4 Let S be a nonempty bounded subset of $\mathbb{R}$. Prove that $\sup S$ is unique.

***12.5** Let S be a nonempty bounded subset of $\mathbb{R}$ and let $k \in \mathbb{R}$. Define $kS = \{ks: s \in S\}$. Prove the following:

(a) If $k \geqslant 0$, then $\sup(kS) = k \cdot \sup S$ and $\inf(kS) = k \cdot \inf S$.

(b) If $k < 0$, then $\sup(kS) = k \cdot \inf S$ and $\inf(kS) = k \cdot \sup S$.

12.6 Let S and T be nonempty bounded subsets of $\mathbb{R}$ with $S \subseteq T$. Prove that $\inf T \leqslant \inf S \leqslant \sup S \leqslant \sup T$.

12.7 (a) Prove: If $y > 0$, then there exists $n \in \mathbb{N}$ such that $n - 1 \leqslant y < n$.

(b) Prove that the n in part (a) is unique.

12.8 (a) Prove: If x and y are real numbers with $x < y$, then there are infinitely many rational numbers in the interval $[x, y]$.
(b) Repeat part (a) for irrational numbers.

12.9 Let y be a positive real number. Prove that for every $n \in \mathbb{N}$ there exists a unique positive real number x such that $x^n = y$.

***12.10** Let D be a nonempty set and suppose that $f: D \to \mathbb{R}$ and $g: D \to \mathbb{R}$. Define the function $f + g: D \to \mathbb{R}$ by $(f + g)(x) = f(x) + g(x)$.
(a) If $f(D)$ and $g(D)$ are bounded above, then prove that $(f + g)(D)$ is bounded above and $\sup [(f + g)(D)] \leqslant \sup f(D) + \sup g(D)$.
(b) Find an example to show that a strict inequality in part (a) may occur.
(c) State and prove the analog of part (a) for infima.

12.11 Let $x \in \mathbb{R}$. Prove that $x = \sup \{q \in \mathbb{Q}: q < x\}$.

12.12 Let a/b be a fraction in lowest terms with $0 < a/b < 1$.
(a) Prove that there exists $n \in \mathbb{N}$ such that

$$\frac{1}{n+1} \leqslant \frac{a}{b} < \frac{1}{n}.$$

(b) If n is chosen as in part (a), prove that $a/b - 1/(n + 1)$ is a fraction that in lowest terms has a numerator less than a.
(c) Use part (b) and the principle of strong induction (Exercise 10.11) to prove that a/b can be written as a finite sum of distinct unit fractions:

$$\frac{a}{b} = \frac{1}{n_1} + \cdots + \frac{1}{n_k},$$

where $n_1, \ldots, n_k \in \mathbb{N}$. (As a point of historical interest, we note that in the ancient Egyptian system of arithmetic all fractions were expressed as sums of unit fractions and then manipulated using tables.)

12.13 Prove Euclid's division algorithm: If a and b are natural numbers, then there exist unique numbers q and r, each of which is either 0 or a natural number, such that $r < a$ and $b = qa + r$.

12.14 Let F be the ordered field of rational functions as given in Example 11.5, and note that F contains both $\mathbb{N}$ and $\mathbb{R}$ as subsets.
(a) Show that F does not have the Archimedean property. That is, find a member z in F such that $z > n$ for every $n \in \mathbb{N}$.
(b) Show that the property in Theorem 12.10(c) does not apply. That is, find a positive member z in F such that, for all $n \in \mathbb{N}$, $0 < z \leqslant 1/n$.
(c) Show that F does not satisfy the completeness axiom. That is, find a subset B of F such that B is bounded above but B has no least upper bound. Verify your answer.

12.15 We have said that the real numbers can be characterized as a complete ordered field. This means that any other complete ordered field F is essentially the same as $\mathbb{R}$ in the sense that there exists a bijection $f: \mathbb{R} \to F$ with the following properties for all $a, b \in \mathbb{R}$:

(1) $f(a + b) = f(a) + f(b)$,
(2) $f(a \cdot b) = f(a) \cdot f(b)$,
(3) $a < b$ iff $f(a) < f(b)$.

(Such a function is called an *order isomorphism.*) We can construct the function f by first defining $f(0) = 0_F$ and $f(1) = 1_F$, where 0_F and 1_F are the unique elements of F given in axioms A4 and M4. Then define $f(n + 1) = f(n) + 1_F$ and $f(-n) = -f(n)$ for all $n \in \mathbb{N}$. This extends the domain of f to all of $\mathbb{Z}$.

Next we extend the domain of f to $\mathbb{Q}$ by defining $f(m/n) = f(m)/f(n)$ for $m, n \in \mathbb{Z}$ with $n \neq 0$. Since, for all $x \in \mathbb{R}$,

$$x = \sup \{q \in \mathbb{Q}: q < x\}$$

(Exercise 12.11), we can extend the domain of f to $\mathbb{R}$ by defining

$$f(x) = \sup \{f(q): q \in \mathbb{Q} \text{ and } q < x\}.$$

Verify that the function f so defined is the required order isomorphism. [*Note*: When writing an equation such as $f(a + b) = f(a) + f(b)$, the "$+$" between a and b represents addition in $\mathbb{R}$ and the "$+$" between $f(a)$ and $f(b)$ represents addition in F. Similar comments apply to "$\cdot$" and "$<$."]

Section 13 TOPOLOGY OF THE REALS

Many of the central ideas in analysis are dependent on the notion of two points being "close" to each other. We have seen that the distance between two points x and y in $\mathbb{R}$ is given by the absolute value of their difference: $|x - y|$. Thus, if we are given some positive measure of closeness, say ε, we may be interested in all points y that are less than ε away from x:

$$\{y: |x - y| < \varepsilon\}.$$

We formalize this idea in the following definition.

13.1 DEFINITION Let $x \in \mathbb{R}$ and let $\varepsilon > 0$. A **neighborhood** of x (or an **ε-neighborhood** of x) is a set of the form

$$N(x; \varepsilon) = \{y \in \mathbb{R}: |x - y| < \varepsilon\}.$$

The number ε is referred to as the **radius** of $N(x; \varepsilon)$.[†]

Basically, a neighborhood of x of radius ε is an open interval $(x - \varepsilon, x + \varepsilon)$ of length 2ε centered at x. We prefer to use the term "neighborhood" in subsequent definitions and theorems because this terminology

[†] In some advanced texts the set $N(x; \varepsilon)$ is called an ε-neighborhood of x, and a neighborhood of x is defined to be any set that contains an ε-neighborhood of x for some $\varepsilon > 0$. Since we shall not need this more general notion, we shall use the terms "ε-neighborhood" and "neighborhood' interchangeably.

can be applied in more general settings. In this section we use neighborhoods to define the concepts of open and closed sets. The study of these sets is known as **point set topology**, and this explains the use of the word "topology" in the title of the section.

In some situations, particularly when dealing with limits of functions (Chapter 5), we shall want to consider points y that are close to x but different from x. We can accomplish this by requiring $|x - y| > 0$.

13.2 DEFINITION Let $x \in \mathbb{R}$ and let $\varepsilon > 0$. A **deleted neighborhood** of x is a set of the form

$$N^*(x; \varepsilon) = \{y \in \mathbb{R}: 0 < |x - y| < \varepsilon\}.$$

If $S \subseteq \mathbb{R}$, then a point x in $\mathbb{R}$ can be thought of as being "inside" S, on the "edge" of S, or "outside" S. Saying that x is "outside" S is the same as saying that x is "inside" the complement of S, $\mathbb{R}\backslash S$. Using neighborhoods, we can make the intuitive ideas of "inside" and "edge" more precise.

13.3 DEFINITION Let S be a subset of $\mathbb{R}$. A point x in $\mathbb{R}$ is an **interior** point of S if there exists a neighborhood N of x such that $N \subseteq S$. If for every neighborhood N of x, $N \cap S \neq \varnothing$ and $N \cap (\mathbb{R}\backslash S) \neq \varnothing$, then x is called a **boundary** point of S. The set of all interior points of S is denoted by int S, and the set of all boundary points of S is denoted by bd S.

13.4 EXAMPLES (a) Let S be the open interval $(0, 5)$ and let $x \in S$. If $\varepsilon = \min \{x, 5 - x\}$, then we claim that $N(x; \varepsilon) \subseteq S$. Indeed, for all $y \in N(x; \varepsilon)$ we have $|y - x| < \varepsilon$, so that

$$-x \leqslant -\varepsilon < y - x < \varepsilon \leqslant 5 - x.$$

Thus $0 < y < 5$ and $y \in S$. It follows that every point in S is an interior point of S. Since the inclusion int $S \subseteq S$ always holds, we have $S =$ int S.

The point 0 is not a member of S, but every neighborhood of 0 will contain positive numbers in S. Thus 0 is a boundary point of S. Similarly, $5 \in$ bd S and in fact, bd $S = \{0, 5\}$. Note that none of the boundary of S is contained in S. Of course, there is nothing special about the open interval $(0, 5)$ in this example. Similar comments would apply to any open interval.

(b) Let S be the closed interval $[0, 5]$. The point 0 is still a boundary point of S, since every neighborhood of x will contain negative numbers not in S. We have int $S = (0, 5)$ and bd $S = \{0, 5\}$. This time S contains all of its boundary points, and the same could be said of any other closed interval.

(c) Let S be the interval $[0, 5)$. Then again int $S = (0, 5)$ and bd $S = \{0, 5\}$. We see that S contains some of its boundary, but not all of it.

(d) Let S be the interval $[2, \infty)$. Then int $S = (2, \infty)$ and bd $S = \{2\}$. Note that there is no "point" at ∞ to be included as a boundary point at the right end.

(e) Let $S = \mathbb{R}$. Then int $S = S$ and bd $S = \varnothing$.

13.5 PRACTICE Let $S = (1, 2) \cup (2, 3]$. Find int S and bd S.

Closed Sets and Open Sets

We have seen that a set may contain all of its boundary, part of its boundary, or none of its boundary. Those sets in either the first or last category are of particular interest.

13.6 DEFINITION Let $S \subseteq \mathbb{R}$. If bd $S \subseteq S$, then S is said to be **closed**. If bd $S \subseteq \mathbb{R}\backslash S$, then S is said to be **open**.

If none of the points in S are boundary points of S, then all the points in S must be interior points of S. On the other hand, if S contains its boundary, then since bd $S =$ bd $(\mathbb{R}\backslash S)$, the set $\mathbb{R}\backslash S$ must not contain any of its boundary points. The converse implications also apply, so we obtain the following useful characterizations:

13.7 THEOREM

(a) A set S is open iff $S =$ int S. Equivalently, S is open iff every point in S is an interior point of S.

(b) A set S is closed iff its complement $\mathbb{R}\backslash S$ is open.

13.8 EXAMPLES The interval $(0, 5)$ is open and the interval $[0, 5]$ is closed. Thus our present terminology is consistent with our interval notation in Section 5. That is, an "open interval" (a, b) is an open set and a "closed interval" $[a, b]$ is a closed set. In particular, this means that any neighborhood is an open set, since it is an open interval. The interval $[0, 5)$ is neither open nor closed, and the unbounded interval $[2, \infty)$ is closed.

The entire set $\mathbb{R}$ of real numbers is both open and closed! It is open since int $\mathbb{R} = \mathbb{R}$. It is closed since it contains its boundary: bd $\mathbb{R} = \varnothing$ and $\varnothing \subseteq \mathbb{R}$.

13.9 PRACTICE Is the empty set $\varnothing$ open? Is it closed?

Our next theorem shows how the set operations of intersection and union relate to open sets.

13.10 THEOREM

(a) The union of any collection of open sets is an open set.

(b) The intersection of any finite collection of open sets is an open set.

Proof: (a) Let $\mathscr{A}$ be an arbitrary collection of open sets and let $S = \bigcup \mathscr{A}$. If $x \in S$, then $x \in A$ for some $A \in \mathscr{A}$. Since A is open, x is an interior point of A. That is, there exists a neighborhood N of x such that $N \subseteq A$. But $A \subseteq S$, so $N \subseteq S$ and x is an interior point of S. Hence S is open.

(b) Let $A_1, \ldots, A_n$ be a finite collection of open sets and let $T = \bigcap_{i=1}^{n} A_i$. If $T = \varnothing$, we are done, since $\varnothing$ is open. If $T \neq \varnothing$, let $x \in T$. Then $x \in A_i$ for all $i = 1, \ldots, n$. Since each set A_i is open, there exist neighborhoods $N_i(x; \varepsilon_i)$ of x such that $N_i(x; \varepsilon_i) \subseteq A_i$. Let $\varepsilon = \min \{\varepsilon_1, \ldots, \varepsilon_n\}$. Then $N(x, \varepsilon) \subseteq A_i$ for $i = 1, \ldots, n$, so that $N(x; \varepsilon) \subseteq T$. Thus x is an interior point of T, and T is open. ∎

13.11 COROLLARY (a) The intersection of any collection of closed sets is closed.
(b) The union of any finite collection of closed sets is closed.

Proof: Both parts follow from Theorem 13.10 when combined with Theorem 13.7. Recall (Exercise 5.14) that $\mathbb{R} \backslash (\bigcup_{j \in J} A_j) = \bigcap_{j \in J} (\mathbb{R} \backslash A_j)$ and $\mathbb{R} \backslash (\bigcap_{j \in J} A_j) = \bigcup_{j \in J} (\mathbb{R} \backslash A_j)$. ∎

13.12 EXAMPLE For each $n \in \mathbb{N}$, let $A_n = (-1/n, 1/n)$. Then each A_n is an open set, but $\bigcap_{n=1}^{\infty} A_n = \{0\}$, which is not open. Thus we see that the restriction in Theorem 13.10(b) to intersections of *finitely* many open sets is necessary.

13.13 PRACTICE Find an example of a collection of closed sets whose union is not closed.

Accumulation Points

Our study of open and closed sets so far has been based on the notion of a neighborhood. By using deleted neighborhoods we can consider another property of points and sets.

13.14 DEFINITION Let S be a subset of $\mathbb{R}$. A point x in $\mathbb{R}$ is an **accumulation** point of S if every deleted neighborhood of x contains a point of S. That is, for every $\varepsilon > 0$, $N^*(x; \varepsilon) \cap S \neq \varnothing$.

An equivalent way of handling an accumulation point x of a set S would be to require that each neighborhood of x contain at least one point of S different from x. We shall denote the set of all accumulation points of S by S'. Note that an accumulation point of S may be, but does not have to be, a member of S. If $x \in S$ and $x \notin S'$, then x is called an **isolated** point of S.

13.15 EXAMPLES (a) If S is the interval $(0, 1]$, then $S' = [0, 1]$.
(b) If $S = \{1/n : n \in \mathbb{N}\}$, then $S' = \{0\}$.

(c) If $S = \mathbb{N}$, then $S' = \varnothing$. Thus $\mathbb{N}$ consists entirely of isolated points.
(d) If S is a finite set, then $S' = \varnothing$. Indeed, if $S = \{x_1, \dots, x_n\}$ and $y \in \mathbb{R}$, then let $\varepsilon = \min\{|x_i - y|: x_i \neq y\}$. It follows that $\varepsilon > 0$ and $N^*(y; \varepsilon) \cap S = \varnothing$. Thus y is not an accumulation point of S.

13.16 DEFINITION Let $S \subseteq \mathbb{R}$. Then the **closure** of S, denoted cl S, is defined by

$$\text{cl } S = S \cup S',$$

where S' is the set of all accumulation points of S.

In terms of neighborhoods, a point x is in cl S iff every neighborhood of x intersects S. To see this, let $x \in \text{cl } S$ and let N be a neighborhood of x. If $x \in S$, then $N \cap S$ contains x. If $x \notin S$, then $x \in S'$ and every deleted neighborhood intersects S. Thus in either case, the neighborhood N must intersect S. Conversely, suppose that every neighborhood of x intersects S. If $x \notin S$, then every neighborhood of x intersects S in a point other than x. Thus $x \in S'$, and so $x \in \text{cl } S$.

The basic relationships between accumulation points, closure, and closed sets are presented in the following theorem.

13.17 THEOREM Let S be a subset of $\mathbb{R}$. Then

(a) S is closed iff S contains all of its accumulation points,
(b) cl S is a closed set,
(c) S is closed iff $S = \text{cl } S$.

Proof: (a) suppose that S is closed and let $x \in S'$. We must show that $x \in S$. If $x \notin S$, then x is in the open set $\mathbb{R} \backslash S$. Thus there exists a neighborhood N of x such that $N \subseteq \mathbb{R} \backslash S$. But then $N \cap S = \varnothing$, which contradicts $x \in S'$.

Conversely, suppose that $S' \subseteq S$. We shall show that $\mathbb{R} \backslash S$ is open. To this end, let $x \in \mathbb{R} \backslash S$. Then $x \notin S'$, so there exists a deleted neighborhood $N^*(x; \varepsilon)$ that misses S. Since $x \notin S$, the whole neighborhood $N(x; \varepsilon)$ misses S; that is, $N(x; \varepsilon) \subseteq \mathbb{R} \backslash S$. Thus $\mathbb{R} \backslash S$ is open and S is closed.

(b) By part (a) we must show that if $x \in (\text{cl } S)'$, then $x \in \text{cl } S$. Thus suppose that x is an accumulation point of cl S. Then every deleted neighborhood $N^*(x; \varepsilon)$ intersects cl S. We must show that $N^*(x; \varepsilon)$ intersects S. To this end, let $y \in N^*(x; \varepsilon) \cap \text{cl } S$. (See Figure 13.1.) Since $N^*(x; \varepsilon)$ is an open set (Exercise 13.6), there exists a neighborhood $N(y; \delta)$ contained in $N^*(x; \varepsilon)$. But $y \in \text{cl } S$, so every neighborhood of y intersects S. That is, there exists a point z in

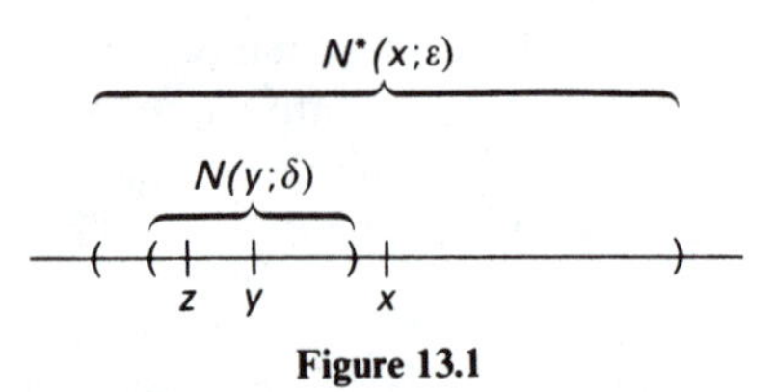

Figure 13.1

$N(y;\ \delta) \cap S$. But then $z \in N(y;\ \delta) \subseteq N^*(x;\ \varepsilon)$, so that $x \in S'$ and $x \in \text{cl } S$.

(c) Exercise 13.12. ▮

ANSWERS TO PRACTICE PROBLEMS

13.5 int $S = (1, 2) \cup (2, 3)$ and bd $S = \{1, 2, 3\}$.

13.9 The empty set $\varnothing$ is both open and closed, since it is the complement of the set $\mathbb{R}$, which is both open and closed. Or, to put it another way, $\varnothing$ is open since int $\varnothing = \varnothing$, and $\varnothing$ is closed since bd $\varnothing = \varnothing \subseteq \varnothing$.

13.13 There are many possibilities. For a simple one, let $A_n = [1/n, 2]$ for all $n \in \mathbb{N}$. Then $\bigcup_{n=1}^{\infty} A_n = (0, 2]$, which is not closed.

EXERCISES

13.1 Find the interior of each set.

(a) $\left\{\frac{1}{n}: n \in \mathbb{N}\right\}$

(b) $[0, 3] \cup (3, 5)$

(c) $\{r \in \mathbb{Q}: 0 < r < \sqrt{2}\}$

(d) $\{r \in \mathbb{Q}: r \geqslant \sqrt{2}\}$

(e) $[0, 2] \cap [2, 4]$

13.2 Find the boundary of each set in Exercise 13.1.

13.3 Classify each of the following sets as open, closed, neither, or both.

(a) $\left\{\frac{1}{n}: n \in \mathbb{N}\right\}$

(b) $\mathbb{N}$

(c) $\mathbb{Q}$

(d) $\bigcap_{n=1}^{\infty} \left(0, \frac{1}{n}\right)$

(e) $\left\{x: |x - 5| \leqslant \frac{1}{2}\right\}$

(f) $\{x: x^2 > 0\}$

13.4 Find the closure of each set in Exercise 13.3.

13.5 If A is open and B closed, prove that $A \backslash B$ is open and $B \backslash A$ is closed.

13.6 Prove: For each $x \in \mathbb{R}$ and $\varepsilon > 0$, $N^*(x, \varepsilon)$ is an open set.

13.7 Prove: $(\text{cl } S) \backslash (\text{int } S) = \text{bd } S$.

13.8 Let S be a bounded infinite set and let $x = \sup S$. Prove: If $x \notin S$, then $x \in S'$.

***13.9** Prove: If x is an accumulation point of the set S, then every neighborhood of x contains infinitely many points of S.

13.10 (a) Prove: $\text{bd } S = (\text{cl } S) \cap [\text{cl } (\mathbb{R} \backslash S)]$.
(b) Prove: $\text{bd } S$ is a closed set.

13.11 Prove: S' is a closed set.

13.12 Prove Theorem 13.17(c).

13.13 Let A be a nonempty open subset of $\mathbb{R}$ and let $\mathbb{Q}$ be the set of rationals. Prove that $A \cap \mathbb{Q} \neq \varnothing$.

13.14 Let S and T be subsets of $\mathbb{R}$. Prove the following.
(a) $\text{cl } (\text{cl } S) = \text{cl } S$
(b) $\text{cl } (S \cup T) = (\text{cl } S) \cup (\text{cl } T)$
(c) $\text{cl } (S \cap T) \subseteq (\text{cl } S) \cap (\text{cl } T)$
(d) Find an example to show that equality need not hold in part (c).

13.15 Let S and T be subsets of $\mathbb{R}$. Prove the following.
(a) $\text{int } S$ is an open set.
(b) $\text{int } (\text{int } S) = \text{int } S$.
(c) $\text{int } (S \cap T) = (\text{int } S) \cap (\text{int } T)$.
(d) $(\text{int } S) \cup (\text{int } T) \subseteq \text{int } (S \cup T)$.
(e) Find an example to show that equality need not hold in part (d).

13.16 For any set $S \subseteq \mathbb{R}$, let $\bar{S}$ denote the intersection of all the closed sets containing S.
(a) Prove that $\bar{S}$ is a closed set.
(b) Prove that $\bar{S}$ is the smallest closed set containing S. That is, show that $S \subseteq \bar{S}$ and if C is any closed set containing S, then $\bar{S} \subseteq C$.
(c) Prove that $\bar{S} = \text{cl } S$.
(d) If S is bounded, prove that $\bar{S}$ is bounded.

13.17 For any set $S \subseteq \mathbb{R}$, let S° denote the union of all the open sets contained in S.
(a) Prove that S° is an open set.
(b) Prove that S° is the largest open set contained in S. That is, show that $S^\circ \subseteq S$ and if U is any open set contained in S, then $U \subseteq S^\circ$.
(c) Prove that $S^\circ = \text{int } S$.

13.18 In this exercise we outline a proof of the following theorem: "A subset of $\mathbb{R}$ is open iff it is the union of countably many disjoint open intervals in $\mathbb{R}$."
(a) Let S be a nonempty open subset of $\mathbb{R}$. For each $x \in S$, let $A_x = \{a \in \mathbb{R}: (a, x] \subseteq S\}$ and let $B_x = \{b \in \mathbb{R}: [x, b) \subseteq S\}$. Use the fact that S is open to show that A_x and B_x are both nonempty.

(b) If A_x is bounded below, let $a_x = \inf A_x$; otherwise, let $a_x = -\infty$. If B_x is bounded above, let $b_x = \sup B_x$; otherwise, let $b_x = \infty$. Show that $a_x \notin S$ and $b_x \notin S$.
(c) Let I_x be the open interval (a_x, b_x). Clearly, $x \in I_x$. Show that $I_x \subseteq S$. (*Hint*: Consider two cases for $y \in I_x$: $y < x$ and $y > x$.)
(d) Show that $S = \bigcup_{x \in S} I_x$.
(e) Show that the intervals $\{I_x \colon x \in S\}$ are pairwise disjoint. That is, suppose $x, y \in S$ with $x \neq y$. If $I_x \cap I_y \neq \varnothing$, show that $I_x = I_y$.
(f) Show that the set of distinct intervals $\{I_x \colon x \in S\}$ is countable.

Section 14 COMPACT SETS

In Section 13 we introduced several important topological concepts in $\mathbb{R}$. Some of these concepts related to points: interior points, boundary points, and accumulation points. Others related to sets: open sets and closed sets. In this section we define another type of set that occurs frequently in applications.

If we require a subset of $\mathbb{R}$ to be both closed and bounded, then it will have a number of special properties not possessed by sets in general. The first such property is called compactness, and although its definition may at first appear strange, it is really a widely used concept of analysis. (For example, see Theorems 22.2, 22.10, 23.5, and 24.9.)

14.1 DEFINITION A set S is said to be **compact** if whenever it is contained in the union of a family $\mathscr{F}$ of open sets, then it is contained in the union of some finite number of the sets in $\mathscr{F}$.

If $\mathscr{F}$ is a family of open sets whose union contains S, then $\mathscr{F}$ is called an **open cover** of S. If $\mathscr{G} \subseteq \mathscr{F}$ and $\mathscr{G}$ is also an open cover of S, then $\mathscr{G}$ is called a **subcover** of S. Thus S is compact iff every open cover of S contains a finite subcover.

14.2 EXAMPLES (a) Let $S = (0, 2)$ and for each $n \in \mathbb{N}$ let $A_n = (1/n, 3)$. If $0 < x < 2$, then by the Archimedean property 12.10(c), there exists $p \in \mathbb{N}$ such that $1/p < x$. Thus $x \in A_p$ and $\mathscr{F} = \{A_n \colon n \in \mathbb{N}\}$ is an open cover for S. However, if $\mathscr{G} = \{A_{n_1}, \ldots, A_{n_k}\}$ is any finite subfamily of $\mathscr{F}$, and if $m = \max\{n_1, \ldots, n_k\}$, then

$$A_{n_1} \cup \cdots \cup A_{n_k} = A_m = \left(\frac{1}{m}, 3\right).$$

It follows that the finite subfamily $\mathscr{G}$ is not an open cover of $(0, 2)$. Since we have exhibited a particular open cover $\mathscr{F}$ that has no finite subcover, we conclude that the interval $(0, 2)$ is not compact.

(b) Let $S = \{x_1, \ldots, x_n\}$ be a finite subset of $\mathbb{R}$, and let $\mathscr{F} = \{A_\alpha : \alpha \in \mathscr{A}\}$ be any open cover of S. For each $i = 1, \ldots, n$, there is a set A_{α_i} from $\mathscr{F}$ that contains x_i since $\mathscr{F}$ is an open cover. It follows that the subfamily $\{A_{\alpha_1}, \ldots, A_{\alpha_n}\}$ also covers S. We conclude that any finite set is compact.

14.3 PRACTICE Show that $[0, \infty)$ is not compact by finding an open cover of $[0, \infty)$ that has no finite subcover.

Notice that in proving a set compact we must show that any open cover (possibly containing uncountably many open sets) has a finite subcover. It is not sufficient to pick a particular open cover and extract a finite subcover. Because of this, the definition is generally difficult to apply in showing a set is compact.

Fortunately, the classical Heine–Borel theorem, which we prove following a preliminary lemma, gives us a much easier characterization to use for subsets of $\mathbb{R}$.

14.4 LEMMA If S is a nonempty closed bounded subset of $\mathbb{R}$, then S has a maximum and a minimum.

Proof: Since S is bounded above, $m = \sup S$ exists by the completeness axiom. If $m \notin S$, then for each $\varepsilon > 0$ there exists x in S such that $m - \varepsilon < x < m$. This would imply that m is an accumulation point of S. But since S is closed, it contains all its accumulation points. Thus we must have $m \in S$, and we conclude that $m = \max S$. Similarly, $\inf S \in S$, so $\inf S = \min S$. ∎

14.5 THEOREM (Heine–Borel) A subset S of $\mathbb{R}$ is compact iff S is closed and bounded.

Proof: First, let us suppose that S is compact. For each $n \in \mathbb{N}$, let $I_n = N(0; n) = (-n, n)$. Then each I_n is open and $S \subseteq \bigcup_{n=1}^{\infty} I_n$. Thus $\{I_n : n \in \mathbb{N}\}$ is an open cover of S. Since S is compact, there exist finitely many integers $n_1, \ldots, n_k$ such that

$$S \subseteq (I_{n_1} \cup \cdots \cup I_{n_k}) = I_m,$$

where $m = \max\{n_1, \ldots, n_k\}$. It follows that $|x| < m$ for all $x \in S$, and S is bounded.

To see that S must be closed, we suppose that it were not closed. Then there would exist a point $p \in (\text{cl } S) \setminus S$. For each $n \in \mathbb{N}$, we let $U_n = \mathbb{R} \setminus \text{cl } N(p; 1/n)$. Now each U_n is an open set and we have

$$\bigcup_{n=1}^{\infty} U_n = \mathbb{R} \setminus \bigcap_{n=1}^{\infty} \text{cl } N(p; 1/n) = \mathbb{R} \setminus \{p\} \supseteq S,$$

by Exercise 5.14(d). Thus $\{U_n: n \in \mathbb{N}\}$ is an open cover of S. Since S is compact there exists $n_1 < n_2 < \cdots < n_k$ in $\mathbb{N}$ such that $S \subseteq \{U_{n_1}, \ldots, U_{n_k}\}$. Furthermore, the U_n's are nested. That is $U_m \subseteq U_n$ if $m \leqslant n$. It follows that $S \subseteq U_{n_k}$. But then $S \cap N(p; 1/n_k) = \varnothing$, contradicting our choice of $p \in (\text{cl } S) \backslash S$.

Conversely, suppose that S is closed and bounded. Let $\mathscr{F}$ be an open cover of S. For each $x \in \mathbb{R}$ let

$$S_x = S \cap (-\infty, x]$$

and let

$$B = \{x: S_x \text{ is covered by a finite subcover of } \mathscr{F}\}.$$

Since S is closed and bounded, Lemma 14.4 implies that S has a minimum, say d. Then $S_d = \{d\}$ and this is certainly covered by a finite subcover of $\mathscr{F}$. Thus $d \in B$ and B is nonempty. If we can show that B is not bounded above, then it will contain a number p greater than sup S. But then $S_p = S$, and since $p \in B$, we can conclude that S is compact.

To this end, we suppose that B is bounded above and let $m = \sup B$. We shall show that $m \in S$ and $m \notin S$ both lead to contradictions.

If $m \in S$, then since $\mathscr{F}$ is an open cover of S, there exists F_0 in $\mathscr{F}$ such that $m \in F_0$. Since F_0 is open, there exists an interval $[x_1, x_2]$ in F_0 such that

$$x_1 < m < x_2.$$

Since $x_1 < m$ and $m = \sup B$, there exist $F_1, \ldots, F_k$ in $\mathscr{F}$ that cover S_{x_1}. But then $F_0, F_1, \ldots, F_k$ cover S_{x_2}, so that $x_2 \in B$. This contradicts $m = \sup B$.

If $m \notin S$, then since S is closed there exists $\varepsilon > 0$ such that $N(m; \varepsilon) \cap S = \varnothing$. But then

$$S_{m-\varepsilon} = S_{m+\varepsilon}.$$

Since $m - \varepsilon \in B$, we have $m + \varepsilon \in B$, which again contradicts $m = \sup B$.

Since the possibility that B is bounded above leads to a contradiction, we must conclude that B is not bounded above, and hence S is compact. ∎

In Example 13.15 we showed that a finite set will have no accumulation points. We also saw that some unbounded sets (such as $\mathbb{N}$) have no accumulation points. As an application of the Heine–Borel theorem, we now derive the classical Bolzano–Weierstrass theorem, which states that these are the only conditions that can allow a set to have no accumulation points.

14.6 THEOREM (Bolzano–Weierstrass) If a bounded subset S of $\mathbb{R}$ contains infinitely many points, then there exists at least one point in $\mathbb{R}$ that is an accumulation point of S.

Proof: Let S be a bounded subset of $\mathbb{R}$ containing infinitely many points. Suppose that S has no accumulation points. Then S is closed by Theorem 13.17(a), so by the Heine–Borel theorem (14.5) S is compact. Since S has no accumulation points, given any $x \in S$, there exists a neighborhood $N(x)$ of x such that $S \cap N(x) = \{x\}$. Now the family $\{N(x): x \in S\}$ is an open cover of S, and since S is compact there exist $x_1, \ldots, x_n$ in S such that $\{N(x_1), \ldots, N(x_n)\}$ covers S. But

$$S \cap [N(x_1) \cup \cdots \cup N(x_n)] = \{x_1, \ldots, x_n\},$$

so $S = \{x_1, \ldots, x_n\}$. This contradicts S having infinitely many points. ▮

We conclude this section with a result that illustrates an important property of compact sets.

14.7 THEOREM Let $\mathscr{F} = \{K_\alpha: \alpha \in \mathscr{A}\}$ be a family of compact subsets of $\mathbb{R}$. Suppose that the intersection of any finite subfamily of $\mathscr{F}$ is nonempty. Then $\bigcap \{K_\alpha: \alpha \in \mathscr{A}\} \neq \varnothing$.

Proof: For each $\alpha \in \mathscr{A}$, let $F_\alpha = \mathbb{R} \backslash K_\alpha$. Choose a member K_1 of $\mathscr{F}$ and suppose that no point of K_1 belongs to every K_α. Then the sets F_α form an open cover of K_1. Since K_1 is compact, there exist finitely many indices $\alpha_1, \ldots, \alpha_n$ such that $K_1 \subseteq (F_{\alpha_1} \cup \cdots \cup F_{\alpha_n})$. But this implies that $K_1 \cap K_{\alpha_1} \cap \cdots \cap K_{\alpha_n} = \varnothing$, a contradiction. Thus some point in K_1 belongs to each K_α, and $\bigcap \{K_\alpha: \alpha \in \mathscr{A}\} \neq \varnothing$. ▮

14.8 COROLLARY (Nested Intervals Theorem) Let $\mathscr{F} = \{A_n: n \in \mathbb{N}\}$ be a family of closed bounded intervals in $\mathbb{R}$ such that $A_{n+1} \subseteq A_n$ for all $n \in \mathbb{N}$. Then $\bigcap_{n=1}^{\infty} A_n \neq \varnothing$.

Proof: Given any $n_1 < n_2 < \cdots < n_k$ in $\mathbb{N}$, we have $\bigcap_{i=1}^{k} A_{n_i} = A_{n_k} \neq \varnothing$. Thus Theorem 14.7 implies that $\bigcap_{n=1}^{\infty} A_n \neq \varnothing$. ▮

ANSWERS TO PRACTICE PROBLEMS

14.3 One possibility is to let $A_n = (-1, n)$ for all $n \in \mathbb{N}$.

EXERCISES

14.1 Show that each subset of $\mathbb{R}$ is not compact by describing an open cover for it that has no finite subcover.

(a) $[1, 3)$
(b) $\mathbb{N}$
(c) $\{1/n: n \in \mathbb{N}\}$
(d) $\{x \in \mathbb{Q}: 0 \leqslant x \leqslant 2\}$

14.2 Prove that the intersection of any collection of compact sets is compact.

14.3 (a) If S_1 and S_2 are compact subsets of $\mathbb{R}$, prove that $S_1 \cup S_2$ is compact.
(b) Find an infinite collection $\{S_n: n \in \mathbb{N}\}$ of compact sets in $\mathbb{R}$ such that $\bigcup_{n=1}^{\infty} S_n$ is not compact.

14.4 Show that compactness is necessary in Corollary 14.8. That is, find a family of intervals $\{A_n: n \in \mathbb{N}\}$ with $A_{n+1} \subseteq A_n$ for all n, $\bigcap_{n=1}^{\infty} A_n = \varnothing$, and such that

(a) The sets A_n are all closed.
(b) The sets A_n are all bounded.

14.5 (a) Let $\mathscr{F}$ be a collection of disjoint open subsets of $\mathbb{R}$. Prove that $\mathscr{F}$ is countable.
(b) Find an example of a collection of disjoint closed subsets of $\mathbb{R}$ that is not countable.

14.6 If S is a compact subset of $\mathbb{R}$ and T is a closed subset of S, then T is compact.

(a) Prove this using the definition of compactness.
(b) Prove this using the Heine–Borel theorem.

14.7 Find an uncountable open cover $\mathscr{F}$ of $\mathbb{R}$ such that $\mathscr{F}$ has no finite subcover. Does $\mathscr{F}$ contain a countable subcover?

14.8 Let $\mathscr{G} = \{N(p; r): p, r \in \mathbb{Q} \text{ and } r > 0\}$.

(a) Prove that $\mathscr{G}$ is countable.
(b) Let A be a nonempty open set and let $\mathscr{G}_A = \{N \in \mathscr{G}: N \subseteq A\}$. Prove that $\bigcup \mathscr{G}_A = A$. What is the cardinality of $\mathscr{G}_A$?
(c) Let $\mathscr{F}$ be any nonempty collection of nonempty open sets. Prove that there is a family $\mathscr{G}_{\mathscr{F}} \subseteq \mathscr{G}$ such that $\bigcup \mathscr{G}_{\mathscr{F}} = \bigcup \mathscr{F}$. Then use $\mathscr{G}_{\mathscr{F}}$ to show that there is a countable subfamily $\mathscr{H} \subseteq \mathscr{F}$ such that $\bigcup \mathscr{H} = \bigcup \mathscr{F}$.
(d) Prove the Lindelöf covering theorem: Let S be a subset of $\mathbb{R}$ and let $\mathscr{F}$ be an open covering of S. Then there is a countable subfamily of $\mathscr{F}$ that also covers S.

14.9 Let I be the interval $[0, 1]$. Remove the open middle third segment $(\frac{1}{3}, \frac{2}{3})$ and let A_1 be the set that remains. That is,

$$A_1 = \left[0, \frac{1}{3}\right] \cup \left[\frac{2}{3}, 1\right].$$

Then remove the open middle third segment from each of the two parts of A_1 and call the remaining set A_2. Thus

$$A_2 = \left[0, \frac{1}{9}\right] \cup \left[\frac{2}{9}, \frac{1}{3}\right] \cup \left[\frac{2}{3}, \frac{7}{9}\right] \cup \left[\frac{8}{9}, 1\right].$$

Continue in this manner. That is, given A_k, remove the open middle third segment from each of the closed segments whose union is A_k, and call the remaining set A_{k+1}. Note that $A_1 \supseteq A_2 \supseteq A_3 \supseteq \cdots$ and that, for each $k \in \mathbb{N}$, A_k is the union of 2^k closed intervals each of length 3^{-k}. The set $C = \bigcap_{k=1}^{\infty} A_k$ is called the Cantor set.

(a) Prove that C is compact.

(b) Let $x = 0.a_1a_2a_3\cdots$ be the base 3 (ternary) expansion of a number $x \in [0, 1]$. Prove that $x \in C$ iff $a_n \in \{0, 2\}$ for all $n \in \mathbb{N}$.

(c) Prove that C is uncountable.

(d) Prove that C contains no intervals.

(e) Prove that $\frac{1}{4} \in C$ but that $\frac{1}{4}$ is not an endpoint of any of the intervals in any of the sets A_k ($k \in \mathbb{N}$).

14.10 Let S be a subset of $\mathbb{R}$. Prove that S is compact iff every infinite subset of S has an accumulation point in S.

14.11 In any ordered field F, we can define absolute value in the usual way: $|x| = x$ if $x \geqslant 0$ and $|x| = -x$ if $x < 0$. Using this we can define neighborhoods, and from neighborhoods obtain the other topological concepts of open sets, closed sets, accumulation points, and so on. Our proof that a closed, bounded set is compact used the completeness of $\mathbb{R}$ in a crucial way. Show that this result does not necessarily hold in an ordered field that is not complete. For an ordered field F as given below, find a subset of F that is bounded and closed in F, but not compact.

(a) Let F be the ordered field of rational numbers $\mathbb{Q}$.

(b) Let F be the ordered field of rational functions described in Example 11.5.

14.12 The Bolzano–Weierstrass theorem (14.6) does not necessarily hold in an arbitrary ordered field F. (See Exercise 14.11.) Find counterexamples when:

(a) F is the ordered field of rational numbers $\mathbb{Q}$.

(b) F is the ordered field of rational functions described in Example 11.5.

Section 15 METRIC SPACES†

The main focus of this text is the study of analysis using the real numbers $\mathbb{R}$. Earlier in this chapter we described two important properties of $\mathbb{R}$: order and completeness. One way to gain a deeper understanding of a property is to study a context in which it does *not* apply. We have already done this briefly for the completeness property by looking at the ordered field of rational functions. (See Exercises 12.14, 14.11, and 14.12.)

In this section we consider a context in which order is not usually available and in which completeness may or may not apply. It will also give us a more general setting in which to consider the topological

† This section is optional. The only later section that depends on this material is Section 24 on continuity in metric spaces.

properties of Sections 13 and 14. While for us the main value of this study will be as a contrast to $\mathbb{R}$, it also gives us an introduction to a topic that is of central importance in higher analysis.

Much of our discussion of topology in the last two sections depended on the concepts of distance and neighborhoods. By identifying the properties of distance that are essential to it, we are able to transfer this concept to a more general setting. In Definition 15.1 we define a general distance function (or **metric**), and in Definition 15.3 we use this to define a general neighborhood.

15.1 DEFINITION Let S be any nonempty set. A function $d: S \times S \to \mathbb{R}$ is called a **metric** on S if it satisfies the following conditions for all $x, y, z \in S$.

(1) $d(x, y) \geqslant 0$
(2) $d(x, y) = 0$ if and only if $x = y$.
(3) $d(x, y) = d(y, x)$.
(4) $d(x, y) \leqslant d(x, z) + d(z, y)$ (**triangle inequality**).

A set S together with a metric d is said to be a **metric space**. Since a set may have more than one metric defined on it, we often identify both and denote the metric space by (S, d). On the other hand, if the particular metric is not important (or if it is otherwise identified), we may simply write S.

15.2 EXAMPLES (a) Let $S = \mathbb{R}$ and define $d: \mathbb{R} \times \mathbb{R} \to \mathbb{R}$ by $d(x, y) = |x - y|$ for all $x, y \in \mathbb{R}$. The fact that d is a metric follows directly from the properties of absolute values. In particular, condition (4) follows from the triangle inequality of Theorem 11.8(d). When we refer to $\mathbb{R}$ as a metric space and do not specify any particular metric, it is understood that we are using this absolute value metric.

(b) Let $S = \mathbb{R} \times \mathbb{R} = \mathbb{R}^2$ and define $d: \mathbb{R}^2 \times \mathbb{R}^2 \to \mathbb{R}$ by

$$d((x_1, x_2), (y_1, y_2)) = \sqrt{(x_1 - y_1)^2 + (x_2 - y_2)^2}$$

for points $\mathbf{x} = (x_1, x_2)$ and $\mathbf{y} = (y_1, y_2)$ in $\mathbb{R}^2$.

This metric is called the **Euclidean metric** on $\mathbb{R}^2$ because it corresponds to our usual measure of distance between two points in the plane. When we refer to $\mathbb{R}^2$ as a metric space and do not specify any other particular metric, it is understood that we are using the Euclidean metric.

The first three conditions for a metric follow easily from the definition of the positive square root function in Example 7.23. The proof of the triangle inequality for this function d will not be included here, but it can be found in most linear algebra books and in more advanced analysis texts.†

† For example, see Rudin (1976), page 16.

We can also see now why condition (4) is called the triangle inequality. If x, y, and z are the vertices of a triangle, then (4) states that the length of one side of a triangle must be less than or equal to the sum of the lengths of the other two sides. (See Figure 15.1.)

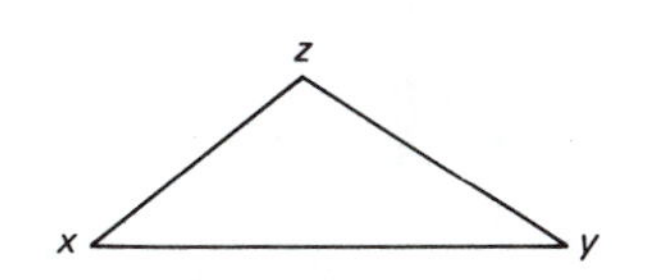

Figure 15.1 $d(x, y) \leqslant d(x, z) + d(z, y)$

(c) Let S be a nonempty set and define the "discrete" metric d on S by

$$d(x, y) = \begin{cases} 0, & \text{if } x = y \\ 1, & \text{if } x \neq y \end{cases}$$

Once again the first three conditions of a metric follow directly from the definition of d. The triangle inequality can easily be established by considering the separate cases when the points x, y, and z are distinct or not. Specifically, if $x \neq z$, then

$$d(x, y) \leqslant 1 \leqslant 1 + d(z, y) = d(x, z) + d(z, y).$$

If $x = z = y$, then

$$d(x, y) = 0 = 0 + 0 = d(x, z) + d(z, y).$$

If $x = z$ but $x \neq y$, then $z \neq y$ so that

$$d(x, y) = 1 = 0 + 1 = d(x, z) + d(z, y).$$

From this example we see that *any* set can be made into a metric space.

15.3 DEFINITION Let (S, d) be a metric space, let $x \in S$, and let $\varepsilon > 0$. The **neighborhood** of x of radius ε is given by

$$N(x; \varepsilon) = \{y \in S : d(x, y) < \varepsilon\}.$$

15.4 EXAMPLES It will be instructive to look again at the examples of metric spaces defined above and see what the neighborhoods look like geometrically.

(a) The metric d defined on $\mathbb{R}$ by $d(x, y) = |x - y|$ is the usual measure of distance in $\mathbb{R}$. The neighborhoods are just open intervals:

$$N(x; \varepsilon) = (x - \varepsilon, x + \varepsilon).$$

(b) The Euclidean metric on $\mathbb{R}^2$ produces neighborhoods that are circular disks. [See Figure 15.2(a).] In particular, the neighborhood of radius 1 centered at the origin θ in $\mathbb{R}^2$ is given by

$$N(\theta; 1) = \{(x_1, x_2): x_1^2 + x_2^2 < 1\}$$

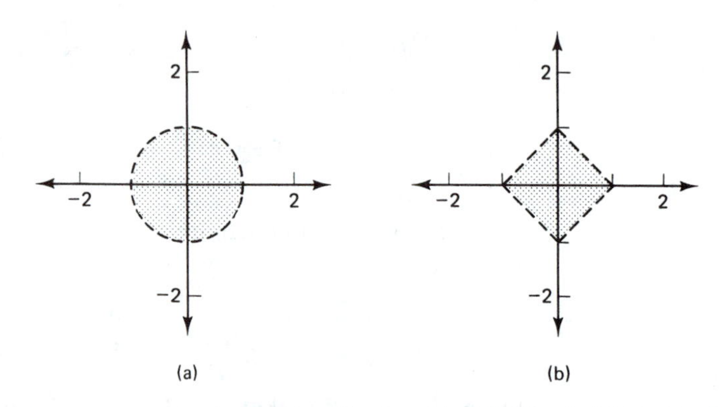

Figure 15.2 (a) $N(\theta; 1)$ for the Euclidean Metric; (b) $N(\theta; 1)$ for the Metric d_1 in Example 15.5

(c) The neighborhoods for the metric defined in Exercise 15.2(c) depend on the size of the radius. If $\varepsilon \leqslant 1$, then the neighborhood contains only the center point itself. If $\varepsilon > 1$, then the neighborhood contains all of S. In particular, if $S = \mathbb{R}^2$, then

$$N(\theta; 1) = \{\theta\} \text{ and } N(\theta; 2) = \mathbb{R}^2.$$

15.5 EXAMPLE For another interesting example, let $S = \mathbb{R}^2$ and define d_1 by

$$d_1((x_1, x_2), (y_1, y_2)) = |x_1 - y_1| + |x_2 - y_2|.$$

It is clear that the first three conditions of a metric are satisfied by d_1. To see that the triangle inequality also holds, let $\mathbf{x} = (x_1, x_2)$, $\mathbf{y} = (y_1, y_2)$, and $\mathbf{z} = (z_1, z_2)$ be arbitrary points in $\mathbb{R}^2$. Then

$$\begin{aligned} d_1(\mathbf{x}, \mathbf{y}) &= |x_1 - y_1| + |x_2 - y_2| \\ &= |x_1 - z_1 + z_1 - y_1| + |x_2 - z_2 + z_2 - y_2| \\ &\leqslant |x_1 - z_1| + |z_1 - y_1| + |x_2 - z_2| + |z_2 - y_2| \\ &= |x_1 - z_1| + |x_2 - z_2| + |z_1 - y_1| + |z_2 - y_2| \\ &= d_1(\mathbf{x}, \mathbf{z}) + d_1(\mathbf{z}, \mathbf{y}), \end{aligned}$$

where the inequality comes from the triangle inequality for absolute value [Theorem 11.8(d)].

Geometrically, the neighborhoods in this metric are diamond shaped. [See Figure 15.2(b).]

If (S, d) is a metric space, then we can use Definition 15.3 for neighborhoods to characterize interior points, boundary points, open sets, and closed sets, just as we did in Section 13 for $\mathbb{R}$. The theorems from Section 13 that relate to these concepts also carry over to this more general setting with little or no change in their proofs. One result that should not be unexpected but that requires a new proof is the following theorem.

15.6 THEOREM Let (S, d) be a metric space. Any neighborhood of a point in S is an open set.

Proof: Let $x \in S$ and let $\varepsilon > 0$. To see that $N(x; \varepsilon)$ is an open set, we shall show that any point y in $N(x; \varepsilon)$ is an interior point of $N(x; \varepsilon)$. If $y \in N(x; \varepsilon)$, then

$$\delta = \varepsilon - d(x, y) > 0.$$

We claim that $N(y; \delta) \subseteq N(x; \varepsilon)$. (See Figure 15.3.)

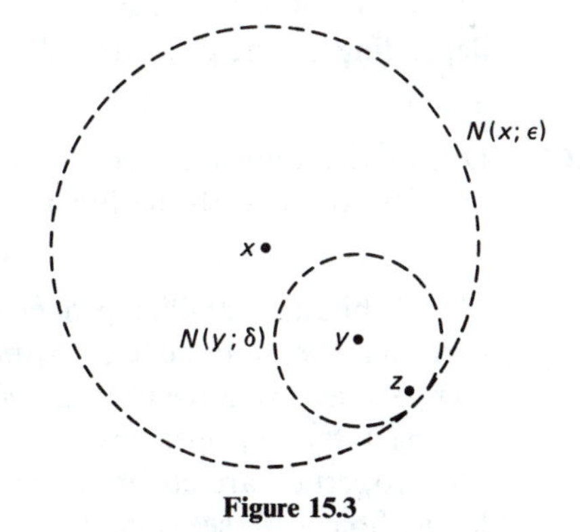

Figure 15.3

If $z \in N(y, \delta)$, then $d(z, y) < \delta$. It follows that

$$\begin{aligned} d(z, x) &\leqslant d(z, y) + d(y, x) \\ &< \delta + d(y, x) \\ &= [\varepsilon - d(x, y)] + d(y, x) = \varepsilon, \end{aligned}$$

so $z \in N(x, \varepsilon)$. Thus $N(y, \delta) \subseteq N(x, \varepsilon)$, and so y is an interior point of $N(x, \varepsilon)$ and $N(x, \varepsilon)$ is open. ∎

15.7 PRACTICE Consider the set $\mathbb{R}^2$ with the Euclidean metric. From our comments prior to Theorem 15.6 we know that the intersection of any finite collection of open sets is an open set. Find an infinite collection of open sets in $\mathbb{R}^2$ whose intersection is not open.

15.8 PRACTICE Consider the set $\mathbb{R}^2$ with the discrete metric of Example 15.2(c). Can you find an infinite collection of open sets whose intersection is not open?

The definition of a deleted neighborhood in a metric space (S, d) is similar to that in $\mathbb{R}$:

$$N^*(x; \varepsilon) = \{y \in S : 0 < d(x, y) < \varepsilon\}.$$

The definition of an accumulation point is also analogous: x is an accumulation point of a set T if for every $\varepsilon > 0$, $N^*(x; \varepsilon) \cap T \neq \varnothing$. Once again, the closure of a set T, denoted cl T, is given by

$$\text{cl } T = T \cup T',$$

where T' is the set of all accumulation points of T.

The properties of closure and closed sets given in Theorem 13.17 continue to apply in a general metric space. Indeed, the proofs given there were all stated in terms of neighborhoods, so they carry over with no change at all. Of course the diagram in Figure 13.1 would be different depending on the geometric shape of the neighborhoods.

15.9 PRACTICE Draw a diagram similar to Figure 13.1 to illustrate the proof of Theorem 13.17(b) for $\mathbb{R}^2$ with the Euclidean metric.

While the familiar properties of open sets and closed sets that we derived for $\mathbb{R}$ continue to apply in a general metric space, when it comes to compact sets we have to be careful. The definition is the same: a set T is compact iff every open cover of T contains a finite subcover. But some of the properties are different. Specifically, the Heine–Borel theorem no longer holds. To see what properties do apply, we first need to define what it means for a set to be bounded in (S, d).

15.10 DEFINITION A set T in a metric space (S, d) is **bounded** if $T \subseteq N(x; \varepsilon)$ for some $x \in S$ and some $\varepsilon > 0$.

15.11 THEOREM Let T be a compact subset of a metric space (S, d). Then

(a) T is closed and bounded.
(b) Every infinite subset of T has an accumulation point in T.

Proof: (a) The first half of the proof of the Heine–Borel Theorem 14.5 was given entirely in terms of neighborhoods and open sets. It applies without any changes in this more general setting.

(b) This proof is the same as the proof of the Bolzano-Weierstrass Theorem 14.6, except that the compactness of T is assumed (instead of using boundedness and the Heine–Borel theorem). We know that the accumulation point of T must be in T, since T is closed. ▮

While a compact subset of (S, d) must be closed and bounded, the converse does not hold in general. The proof of the converse in $\mathbb{R}$ (given in Theorem 14.5) was very dependent on the completeness of $\mathbb{R}$, a property not shared by all metric spaces. We do note, however, that the converse does hold in $\mathbb{R}^d$ for all $d \in \mathbb{N}$, when a generalized Euclidean metric is used.

15.12 EXAMPLE Consider the set $\mathbb{R}^2$ with the discrete metric of Example 15.2(c). In Practice 15.8 we found that *every* subset of $\mathbb{R}^2$ with this metric is open. Since the complement of an open set is closed, this means that every subset is also closed! Let T be the unit square:

$$T = \{(x_1, x_2): 0 \leqslant x_1 \leqslant 1 \text{ and } 0 \leqslant x_2 \leqslant 1\}$$

Clearly, T is bounded since

$$T \subseteq N(\theta; 2) = \mathbb{R}^2,$$

but T is not compact. Indeed, for each point $p \in T$, let $A_p = \{p\}$. Then each A_p is an open set and

$$T \subseteq \bigcup_{p \in T} A_p,$$

so $\mathscr{F} = \{A_p: p \in T\}$ is an open cover for T. But $\mathscr{F}$ contains no finite subcover of T since each set in $\mathscr{F}$ covers only one point in T and there are infinitely many points in T.

We conclude that in this metric space the unit square T is closed and bounded, but not compact.

ANSWERS TO PRACTICE PROBLEMS

15.7 Let $A_n = N(\theta; 1/n)$ for $n \in \mathbb{N}$. Then $\bigcap_{n=1}^{\infty} A_n = \{\theta\}$, which is not open.

15.8 It is not possible to find such a collection. The reason is that *every* subset of $\mathbb{R}^2$ with this metric is an open set. To see this, note that for any point $p \in \mathbb{R}^2$ we have $N(p; 1) = \{p\}$. Since neighborhoods are always open sets (Theorem 15.6), each singleton set consisting of one point is an open set. Since every set is the union of the points in the set, this means that every set is open!

15.9

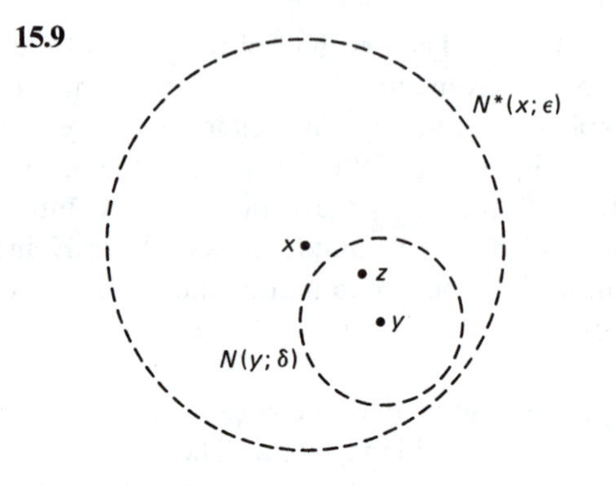

EXERCISES

15.1 Let $S = \mathbb{R}^2$ and define d: $\mathbb{R}^2 \times \mathbb{R}^2 \to \mathbb{R}$ by

$$d_2((x_1, x_2), (y_1, y_2)) = \max \{|x_1 - y_1|, |x_2 - y_2|\}.$$

(a) Verify that d_2 is a metric on $\mathbb{R}^2$.
(b) Draw the neighborhod $N(\theta; 1)$ for d_2.

15.2 Let $S = \mathbb{R} \times \mathbb{R} \times \mathbb{R} = \mathbb{R}^3$ and define the metric d: $\mathbb{R}^3 \times \mathbb{R}^3 \to \mathbb{R}$ by

$$d((x_1, x_2, x_3), (y_1, y_2, y_3)) = \max \{|x_1 - y_1|, |x_2 - y_2|, |x_3 - y_3\}.$$

Describe the neighborhood $N(\theta; 1)$, where θ is the origin in $\mathbb{R}^3$.

15.3 Let $S = \mathbb{R}^2$ and let d be the Euclidean metric. Define d^*: $\mathbb{R}^2 \times \mathbb{R}^2 \to \mathbb{R}$ by $d^*(\mathbf{x}, \mathbf{y}) = \min \{1, d(\mathbf{x}, \mathbf{y})\}$. (This metric is sometimes called the "nearsighted" metric. Up to one unit away, every point is distinguished clearly. But past one, everything gets blurred together.)

(a) Verify that d^* is a metric on $\mathbb{R}^2$.
(b) Draw the neighborhoods $N(\theta; \frac{1}{2})$, $N(\theta; 1)$, and $N(\theta; 2)$ for d^*.

15.4 Let $S = \mathbb{R} \times \mathbb{R} \times \mathbb{R} = \mathbb{R}^3$ and define the metric d: $\mathbb{R}^3 \times \mathbb{R}^3 \to \mathbb{R}$ by

$$d((x_1, x_2, x_3), (y_1, y_2, y_3)) = |x_1 - y_1| + |x_2 - y_2| + |x_3 - y_3|.$$

Describe the neighborhood $N(\theta; 1)$, where θ is the origin in $\mathbb{R}^3$.

15.5 Let $S = \mathbb{R}^2$ and let d be the Euclidean metric. Define w: $\mathbb{R}^2 \times \mathbb{R}^2 \to \mathbb{R}$ by

$$w(\mathbf{x}, \mathbf{y}) = \begin{cases} d(\mathbf{x}, \mathbf{y}) & \text{if } \mathbf{x}, \mathbf{y}, \text{ and } \theta \text{ are colinear} \\ d(\mathbf{x}, \theta) + d(\theta, \mathbf{y}), & \text{otherwise.} \end{cases}$$

(This metric is sometimes called the "Washington metric" because of its similarity to the streets of Washington, D.C.) Let $\mathbf{p} = (4, 5)$ and $\mathbf{q} = (1, 1)$. Draw the following neighborhoods for this metric: (a) $N(\theta; 1)$; (b) $N(\mathbf{p}; 2)$; (c) $N(\mathbf{q}; 2)$.

15.6 If A and B are compact subsets of a metric space (S, d), prove that $A \cup B$ is compact.

15.7 (a) Let x be a point in a metric space (S, d). Prove that the singleton set $\{x\}$ is closed.
(b) Why doesn't part (a) contradict the answer to Practice 15.8?

***15.8** (a) If A is a compact subset of a metric space (S, d) and B is a closed subset of A, prove that B is also compact.
(b) Prove that the intersection of any collection of compact sets in a metric space is compact.

15.9 Let A be a subset of a metric space (S, d). Prove the following:
(a) If A is open, then int (bd A) $= \varnothing$.
(b) If A is closed, then int (bd A) $= \varnothing$.
(c) Find an example of a metric space (S, d) and a subset A such that int (bd A) $= S$.

15.10 In a metric space (S, d), the **closed ball** of radius $\varepsilon > 0$ about the point x in S is the set

$$B(x; \varepsilon) = \{y\colon d(x, y) \leqslant \varepsilon\}.$$

(a) Prove that $B(x; \varepsilon)$ is a closed set.
(b) Prove that cl $N(x; \varepsilon) \subseteq B(x; \varepsilon)$.
(c) Find an example of a metric space (S, d), a point $x \in S$, and a radius $\varepsilon > 0$ such that cl $N(x; \varepsilon) \neq B(x; \varepsilon)$.

15.11 Let $S = \mathbb{R}^2$, let d be the Euclidean metric, let d_1 be the metric of Example 15.5, and let d_2 be the metric of Exercise 15.1. Let A be a subset of $\mathbb{R}^2$. Prove that A is open in $(\mathbb{R}^2, d)$ iff A is open in $(\mathbb{R}^2, d_1)$ iff A is open in $(\mathbb{R}^2, d_2)$. [Two metrics for a set are said to be (topologically) **equivalent** if a subset is open with respect to one metric iff it is open with respect to the other. Thus in this exercise you are to show that d, d_1, and d_2 are equivalent metrics.]

Exercises 15.12 to 15.16 relate to the following definition: Let (S, d) be a metric space. A subset D of S is said to be **dense** in S if cl $D = S$.

15.12 Let D be a subset of a metric space (S, d).
(a) Prove that D is dense in S iff every nonempty open subset of S intersects D.
(b) Prove that D is dense in S iff for every $x \in S$ and every $\varepsilon > 0$ there exists a point z in D such that $d(x, z) < \varepsilon$.

15.13 Prove that $\mathbb{Q}$ is dense in $\mathbb{R}$ with the usual absolute value metric. [If a metric space (S, d) has a countable subset that is dense, then S is said to be **separable**. Thus in this exercise you are to show that $\mathbb{R}$ is separable.]

15.14 Let S be a nonempty set and let d be the metric of Example 15.2(c). Prove that no proper subset of S is dense with this metric.

15.15 Prove that $\mathbb{Q} \times \mathbb{Q} = \{x_1, x_2)\colon x_1 \in \mathbb{Q}$ and $x_2 \in \mathbb{Q}\}$ is dense in $\mathbb{R}^2$ with the Euclidean metric. (Since $\mathbb{Q} \times \mathbb{Q}$ is countable, this means that $\mathbb{R}^2$ is separable. See Exercise 15.13.)

15.16 Prove that $\{(x_1, x_2)\colon x_1 \neq 0$ and $x_2 \neq 0\}$ is dense in $\mathbb{R}^2$ with the usual Euclidean metric.

4

Sequences

Having laid a solid foundation by looking carefully at the properties of real numbers, we now move to a more dynamic topic: the study of sequences. We shall find that sequences play a crucial role throughout analysis, so it is important to gain a thorough understanding of what they are and how they may be used. After discussing the covergence of sequences in Section 16, we devote Section 17 to several theorems that enable us to find the limit of a sequence more easily. In Section 18 we develop some of the properties of monotone sequences and Cauchy sequences, and in the final section we look at subsequences.

Section 16 CONVERGENCE

A **sequence** is a function whose domain is the set $\mathbb{N}$ of natural numbers. If s is a sequence, we usually denote its value at n by s_n instead of $s(n)$. We may refer to the sequence s as (s_n) or by listing the elements $(s_1, s_2, s_3, \ldots)$. We call s_n the nth term of the sequence and we often describe a sequence by giving a formula for the nth term. Thus $(1/n)$ is an abbreviation for the sequence

$$\left(1, \frac{1}{2}, \frac{1}{3}, \frac{1}{4}, \ldots\right).$$

Sometimes we may wish to change the domain of a sequence from $\mathbb{N}$ to $\mathbb{N} \cup \{0\}$ or $\{n \in \mathbb{N} : n \geqslant m\}$. That is, we may want to start with s_0 or s_m, for some $m \in \mathbb{N}$. In this case we write $(s_n)_{n=0}^{\infty}$ or $(s_n)_{n=m}^{\infty}$, respectively. If no mention is made to the contrary, we assume that the domain is just $\mathbb{N}$.

16.1 EXAMPLES (a) Consider the sequence (s_n) given by $s_n = 1 + (-1)^n$. Writing out the first few terms of the sequence, we obtain $(0, 2, 0, 2, 0, \ldots)$, and the pattern to be followed for the rest of the terms is clear. Formally, this sequence is a function

$$s(n) = 1 + (-1)^n = \begin{cases} 0, & \text{if } n \text{ is odd} \\ 2, & \text{if } n \text{ is even,} \end{cases}$$

but it is often more helpful to visualize the sequence as a listing $(0, 2, 0, 2, \ldots)$. Notice that the terms in a sequence do not have to be distinct. We consider s_2 and s_4 to be different terms even though their values are both equal to 2. The range of the sequence is just the set of values obtained, $\{0, 2\}$. Thus, while a sequence will always have infinitely many terms, the set of values in the sequence may be finite.

(b) For any denumerable set S, there exists a bijection from $\mathbb{N}$ onto S. This bijection may be thought of as a sequence that lists the members of S in a particular order. For example, the sequence given by $s_n = 2n$; that is,

$$(2, 4, 6, 8, 10, \ldots),$$

is precisely the function we used in Example 8.6 to show that the set of positive even integers is denumerable. Since $s_1 = 2$, we think of 2 as the "first" even number. Since $s_2 = 4$, 4 is the "second" even number, and so on. Since this function is injective, the terms are all distinct. Thus the range of the sequence is the set $\{2, 4, 6, 8, \ldots\}$. In general, we may think of any denumerable set as the range of a sequence of distinct terms. This is what we mean when we say that the elements of a denumerable set can be listed in a sequence.

(c) The sequence given by $s_n = 1 + 1/2^n$ can be written as

$$\left(\frac{3}{2}, \frac{5}{4}, \frac{9}{8}, \frac{17}{16}, \ldots\right).$$

The graph of this sequence (thinking of it as a function) is shown in Figure 16.1. Sometimes we reduce the graph by displaying only the range, as in Figure 16.2. This can be helpful when the terms of the sequence are distinct, but it can be misleading when they repeat. Notice that the "farther" we go in the sequence, the "closer" the terms appear to get to 1. This prompts us to say that the limit of the sequence is equal to 1. We make this more precise in the following definition.

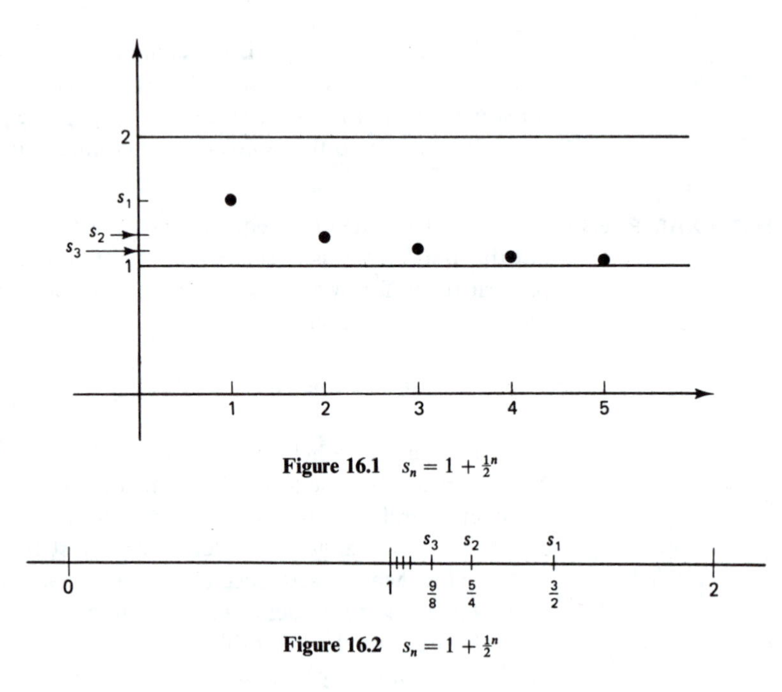

Figure 16.1 $s_n = 1 + \frac{1}{2}^n$

Figure 16.2 $s_n = 1 + \frac{1}{2}^n$

16.2 DEFINITION A sequence (s_n) is said to **converge** to the real number s provided that

> for each $\varepsilon > 0$ there exists a real number N such that $n > N$ implies that $|s_n - s| < \varepsilon$.

If (s_n) converges to s, then s is called the **limit** of the sequence (s_n), and we write $\lim_{n\to\infty} s_n = s$, $\lim s_n = s$, or $s_n \to s$. If a sequence does not converge to a real number, it is said to **diverge**.

It is important to note the order of the quantifiers in Definition 16.2. In trying to show that $s_n \to s$, the N that must be found may depend on the positive number ε. For each ε there must exist an N, but it is not necessary to find one N that works for all ε. We illustrate this in the following examples.

16.3 EXAMPLE Let us show that $\lim 1/n = 0$. Given any particular $\varepsilon > 0$, we want to make $|1/n - 0| < \varepsilon$. Now $|1/n - 0| = 1/n$, and $1/n < \varepsilon$ whenever $n > 1/\varepsilon$. Thus it suffices to let $N = 1/\varepsilon$. We can organize this in a formal proof as follows:

> Given $\varepsilon > 0$, let $N = 1/\varepsilon$. Then for any $n > N$ we have $|1/n - 0| = 1/n < 1/N = \varepsilon$. Thus $\lim 1/n = 0$. ▌

16.4 PRACTICE To show that $\lim 1/\sqrt{n} = 0$, given any $\varepsilon > 0$ we have to find N such that $n > N$ implies that $1/\sqrt{n} < \varepsilon$. What can we take for N?

16.5 EXAMPLE In Example 16.1(c) we observed that $1 + 1/2^n$ seemed to approach 1 as n got large. We can now prove that conjecture. Given any $\varepsilon > 0$, we want $|(1 + 1/2^n) - 1| = 1/2^n < \varepsilon$. Instead of solving directly for n this time, we observe that $1/2^n < 1/n$ for all $n \in \mathbb{N}$. (This can easily be proved using induction, but we omit the details.) Thus once again it suffices to let $N = 1/\varepsilon$. Here is the formal argument:

> Given $\varepsilon > 0$, let $N = 1/\varepsilon$. Then for any $n > N$ we have $|(1 + 1/2^n) - 1| = 1/2^n < 1/n < 1/N = \varepsilon$. Thus $\lim (1 + 1/2^n) = 1$. ∎

16.6 EXAMPLE For a more complicated example let us show that $\lim (n^2 + 2n)/(n^3 - 5) = 0$. Given any $\varepsilon > 0$, we want to make $|(n^2 + 2n)/(n^3 - 5)| < \varepsilon$. By considering only $n \geqslant 2$, we can remove the absolute value signs since $n^3 - 5$ will be positive. Thus we want to know how big n has to be in order to make $(n^2 + 2n)/(n^3 - 5) < \varepsilon$. Since this inequality would be very messy to solve for n, we shall try to find some estimate of how large the left side can be. To do this we seek an upper bound for the numerator and a lower bound for the denominator. Since $n^2 + 2n$ behaves like n^2 for large values of n, we shall try to find an upper bound on $n^2 + 2n$ of the sort bn^2. Similarly, we seek a lower bound for $n^3 - 5$ that is a multiple of n^3, say cn^3. Then we have

$$\frac{n^2 + 2n}{n^3 - 5} \leqslant \frac{bn^2}{cn^3} = \frac{b}{c}\left(\frac{1}{n}\right),$$

and it is relatively easy to make the latter expression small.

Now $n^2 + 2n \leqslant n^2 + n^2 = 2n^2$ when $n \geqslant 2$. And $n^3 - 5 \geqslant n^3/2$ when $n^3/2 \geqslant 5$ or $n^3 \geqslant 10$ or $n \geqslant 3$. Thus for $n \geqslant 3$ we have

$$\frac{n^2 + 2n}{n^3 - 5} \leqslant \frac{2n^2}{n^3 - 5} \leqslant \frac{2n^2}{\frac{1}{2}n^3} = \frac{4}{n}.$$

To make this less than ε, we want $n > 4/\varepsilon$. Thus there are two conditions to be satisfied: we want $n \geqslant 3$ and $n > 4/\varepsilon$. We can accomplish this by letting $N = \max\{3, 4/\varepsilon\}$. We are now ready to organize this into a formal proof.

> Given $\varepsilon > 0$, let $N = \max\{3, 4/\varepsilon\}$. Then $n > N$ implies that $n > 3$ and $n > 4/\varepsilon$. Since $n > 3$ we have $n^2 + 2n \leqslant 2n^2$ and $n^3 - 5 \geqslant n^3/2$. Thus for $n > N$ we have
>
> $$\left|\frac{n^2 + 2n}{n^3 - 5} - 0\right| = \frac{n^2 + 2n}{n^3 - 5} \leqslant \frac{2n^2}{\frac{1}{2}n^3} = \frac{4}{n} < \varepsilon.$$
>
> Hence $\lim (n^2 + 2n)/(n^3 - 5) = 0$. ∎

16.7 PRACTICE Find $k > 0$ and $m \in \mathbb{N}$ so that $5n^3 + 7n \leqslant kn^3$ for all $n \geqslant m$.

The technique involved in our last example can be used in many settings. The amount of work involved can be reduced somewhat by means of the following general theorem.

16.8 THEOREM Let (s_n) and (a_n) be sequences of real numbers and let $s \in \mathbb{R}$. If for some $k > 0$ and some $m \in \mathbb{N}$, we have

$$|s_n - s| \leqslant k|a_n|, \qquad \text{for all } n > m,$$

and if $\lim a_n = 0$, then it follows that $\lim s_n = s$.

Proof: Given any $\varepsilon > 0$, since $\lim a_n = 0$ there exists $N_1 \in \mathbb{R}$ such that $n > N_1$ implies that $|a_n| < \varepsilon/k$. Now let $N = \max\{m, N_1\}$. Then for $n > N$ we have $n > m$ and $n > N_1$, so that

$$|s_n - s| \leqslant k|a_n| < k\left(\frac{\varepsilon}{k}\right) = \varepsilon.$$

Thus $\lim s_n = s$. ∎

16.9 EXAMPLE To illustrate the use of Theorem 16.8, we shall prove that $\lim (4n^2 - 3)/(5n^2 - 2n) = 4/5$. To apply the theorem, we need to find an upper bound for

$$\left|\frac{4n^2 - 3}{5n^2 - 2n} - \frac{4}{5}\right| = \left|\frac{8n - 15}{5(5n^2 - 2n)}\right|$$

when n is sufficiently large. The numerator is easy since $|8n - 15| < 8n$ for all n. For the denominator we want to make $5n^2 - 2n \geqslant kn^2$ for some $k > 0$. If we try $k = 4$, then $5n^2 - 2n \geqslant 4n^2$ or $n^2 \geqslant 2n$ or $n \geqslant 2$. Writing this as a formal proof, we have the following:

If $n \geqslant 2$, then $n^2 \geqslant 2n$ and $5n^2 - 2n \geqslant 4n^2$, so that

$$\left|\frac{4n^2 - 3}{5n^2 - 2n} - \frac{4}{5}\right| = \left|\frac{8n - 15}{5(5n^2 - 2n)}\right| < \frac{8n}{5(4n^2)} = \frac{2}{5}\left(\frac{1}{n}\right).$$

Since $\lim (1/n) = 0$, Theorem 16.8 implies that

$$\lim \frac{4n^2 - 3}{5n^2 - 2n} = \frac{4}{5}. \; ∎$$

16.10 PRACTICE Find $k > 0$ and $m \in \mathbb{N}$ so that $n^3 - 7n \geqslant kn^3$ for all $n \geqslant m$.

16.11 EXAMPLE Let us prove that $\lim n^{1/n} = 1$. Since $n^{1/n} \geqslant 1$ for all n, the number $b_n = n^{1/n} - 1$ is nonnegative. Since $1 + b_n = n^{1/n}$, we have $n = (1 + b_n)^n$.

From the binomial theorem (Exercise 10.14), when $n \geqslant 2$ we obtain

$$n = (1 + b_n)^n = 1 + nb_n + \frac{1}{2}n(n-1)b_n^2 + \cdots + b_n^n \geqslant 1 + \frac{1}{2}n(n-1)b_n^2.$$

It follows that $n - 1 \geqslant \frac{1}{2}n(n-1)b_n^2$, so that $b_n^2 \leqslant 2/n$ and $b_n \leqslant \sqrt{2/n}$. Hence for $n \geqslant 2$ we have

$$|n^{1/n} - 1| = b_n \leqslant \sqrt{2}\left(\frac{1}{\sqrt{n}}\right)$$

Since $\lim (1/\sqrt{n}) = 0$ by Practice 16.4, Theorem 16.8 implies that $\lim n^{1/n} = 1$.

In our next example we show that the sequence given by $s_n = 1 + (-1)^n$ as in Example 16.1(a) is not convergent. Since $(s_n) = (0, 2, 0, 2, \ldots)$, if the limit existed it would have to be close to both 0 and 2. Since no number is less than 1 away from both 0 and 2, we can use $\varepsilon = 1$ in the definition of convergence to obtain a contradiction. This is the reasoning behind our argument.

16.12 EXAMPLE To prove that the sequence $s_n = 1 + (-1)^n$ is divergent, let us suppose that s_n converges to some real number s. Letting $\varepsilon = 1$ in the definition of convergence, we find that there exists N such that $n > N$ implies that $|1 + (-1)^n - s| < 1$. If $n > N$ and n is odd, then we obtain $|s| < 1$ so that $-1 < s < 1$. On the other hand, if $n > N$ and n is even, then $|2 - s| < 1$ and we must have $1 < s < 3$. Since s cannot satisfy both inequalities, we have reached a contradiction. Thus the sequence (s_n) is divergent.

We conclude this section by deriving two important properties of convergent sequences. A sequence (s_n) is said to be **bounded** if the range $\{s_n : n \in \mathbb{N}\}$ is a bounded set, that is, if there exists $M \geqslant 0$ such that $|s_n| \leqslant M$ for all $n \in \mathbb{N}$.

16.13 THEOREM Every convergent sequence is bounded.

Proof: Let (s_n) be a convergent sequence and let $\lim s_n = s$. From the definition of convergence with $\varepsilon = 1$, we obtain $N \in \mathbb{R}$ such that $|s_n - s| < 1$ whenever $n > N$. Thus for $n > N$ the triangle inequality [11.8(d)] implies that $|s_n| < |s| + 1$. If we let

$$M = \max\{|s_1|, |s_2|, \ldots, |s_N|, |s| + 1\}.$$

then we have $|s_n| \leqslant M$ for all $n \in \mathbb{N}$, so (s_n) is bounded. ∎

16.14 THEOREM If a sequence converges, its limit is unique.

Proof: Let (s_n) be a sequence and suppose that (s_n) converges to both s and t. Then, given any $\varepsilon > 0$, there exists $N_1 \in \mathbb{R}$ such that

$$|s_n - s| < \frac{\varepsilon}{2}, \qquad \text{for every } n > N_1.$$

Similarly, there exists $N_2 \in \mathbb{R}$ such that

$$|s_n - t| < \frac{\varepsilon}{2}, \qquad \text{for every } n > N_2.$$

Therefore, if $n > \max\{N_1, N_2\}$, then from the triangle inequality [Theorem 11.8(d)] we have

$$\begin{aligned} |s - t| &= |s - s_n + s_n - t| \\ &\leqslant |s - s_n| + |s_n - t| \\ &< \frac{\varepsilon}{2} + \frac{\varepsilon}{2} = \varepsilon. \end{aligned}$$

Since this holds for all $\varepsilon > 0$, we must have $s = t$. (See Theorem 11.6.) ▮

ANSWERS TO PRACTICE PROBLEMS

16.4 We can take $N = 1/\varepsilon^2$. Any large N will also work.

16.7 There are many possible answers. For example, take $k = 6$ and $m = 3$. Then for $n \geqslant m$ we have $n^2 \geqslant 7$, so that $5n^3 + 7n \leqslant 5n^3 + n^2 n = 6n^3$. As another example, take $k = 12$ and $m = 1$. Then for $n \geqslant m$ we have $n^3 \geqslant n$, so that $5n^3 + 7n \leqslant 5n^3 + 7n^3 = 12n^3$.

16.10 If $7n \leqslant \frac{1}{2}n^3$, then $n^3 - 7n \geqslant \frac{1}{2}n^3$. Now $7n \leqslant \frac{1}{2}n^3$ when $n^2 \geqslant 14$ or $n \geqslant 4$. Thus we can take $k = \frac{1}{2}$ and $m = 4$. Then for $n \geqslant m$ we have $n^2 \geqslant 14$, so that $\frac{1}{2}n^3 - 7n \geqslant 0$. It follows that $n^3 - 7n = \frac{1}{2}n^3 + (\frac{1}{2}n^3 - 7n) \geqslant \frac{1}{2}n^3$. Once again, other estimates are also possible.

EXERCISES

16.1 Write out the first seven terms of each sequence.

(a) $a_n = n^2$ (b) $b_n = \dfrac{(-1)^n}{n}$

(c) $c_n = \cos\dfrac{n\pi}{3}$ (d) $d_n = \dfrac{2n + 1}{3n - 1}$

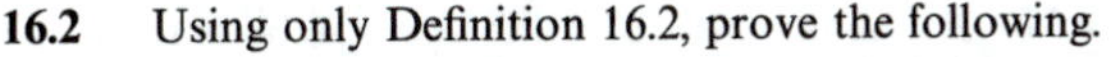

16.2 Using only Definition 16.2, prove the following.

(a) For any real number k, $\lim_{n\to\infty} (k/n) = 0$.

*(b) For any real number $k > 0$, $\lim_{n\to\infty} (1/n^k) = 0$.

(c) $\lim \dfrac{3n+1}{n+2} = 3$

(d) $\lim \dfrac{\sin n}{n} = 0$

(e) $\lim \dfrac{n+2}{n^2-3} = 0$

16.3 Using any of the results in this section, prove the following.

(a) $\lim \dfrac{1}{1+3n} = 0$ (b) $\lim \dfrac{4n^2-7}{2n^3-5} = 0$

(c) $\lim \dfrac{6n^2+5}{2n^2-3n} = 3$ (d) $\lim \dfrac{\sqrt{n}}{n+1} = 0$

(e) $\lim \dfrac{n^2}{n!} = 0$ *(f) If $|x| < 1$, then $\lim_{n\to\infty} x^n = 0$.

16.4 Show that each of the following sequences is divergent.

(a) $a_n = 2n$ (b) $b_n = (-1)^n$

(c) $c_n = \cos \dfrac{n\pi}{3}$ (d) $d_n = (-n)^2$

16.5 For each of the following, prove or give a counterexample.

*(a) If (s_n) converges to s, then $(|s_n|)$ converges to $|s|$.

(b) If $(|s_n|)$ is convergent, then (s_n) is convergent.

(c) $\lim s_n = 0$ iff $\lim |s_n| = 0$.

16.6 Find an example of each of the following.

(a) A convergent sequence of rational numbers having an irrational limit.

(b) A convergent sequence of irrational numbers having a rational limit.

***16.7** Given a sequence (s_n) and given $k \in \mathbb{N}$, let (t_n) be the sequence defined by $t_n = s_{n+k}$. That is, the terms in (t_n) are the same as the terms in (s_n) after the first k terms have been skipped. Prove that (t_n) converges iff (s_n) converges, and if they converge, show that $\lim t_n = \lim s_n$. Thus the convergence of a sequence is not affected by omitting (or changing) a finite number of terms.

***16.8** Suppose that $\lim s_n = 0$. If (t_n) is a bounded sequence, prove that $\lim (s_n t_n) = 0$.

16.9 Suppose that (a_n), (b_n), and (c_n) are sequences such that $a_n \leqslant b_n \leqslant c_n$ for all $n \in \mathbb{N}$ and such that $\lim a_n = \lim c_n = b$. Prove that $\lim b_n = b$.

16.10 Suppose that $\lim s_n = s$, with $s > 0$. Prove that there exists $N \in \mathbb{R}$ such that $s_n > 0$ for all $n > N$.

***16.11** (a) Prove that x is an accumulation point of a set S iff there exists a sequence (s_n) of points in $S \backslash \{x\}$ such that (s_n) converges to x.

(b) Prove that a set S is closed iff whenever (s_n) is a convergent sequence of points in S, it follows that $\lim s_n$ is in S.

***16.12** Recall that $N(s; \varepsilon) = \{x: |x - s| < \varepsilon\}$ is the neighborhood of x of radius ε. Prove the following.

(a) $s_n \to s$ iff for each $\varepsilon > 0$ there exists $M \in \mathbb{N}$ such that $n \geqslant M$ implies that $s_n \in N(s; \varepsilon)$.

(b) $s_n \to s$ iff for each $\varepsilon > 0$, all but finitely many s_n are in $N(s; \varepsilon)$.

(c) $s_n \to s$ iff given any open set U with $s \in U$, all but finitely many s_n are in U.

Section 17 LIMIT THEOREMS

In Section 16 we saw that the definition of convergence can sometimes be messy to use even for sequences given by relatively simple formulas (see Example 16.6). In this section we derive some basic results that will greatly simplify our work. We also introduce the notion of an infinite limit. Our first theorem is a very important result showing that algebraic operations are compatible with taking limits.

17.1 THEOREM Suppose that (s_n) and (t_n) are convergent sequences with $\lim s_n = s$ and $\lim t_n = t$. Then

(a) $\lim (s_n + t_n) = s + t$.

(b) $\lim (ks_n) = ks$ and $\lim (k + s_n) = k + s$, for any $k \in \mathbb{R}$.

(c) $\lim (s_n t_n) = st$.

(d) $\lim (s_n/t_n) = s/t$, provided that $t_n \neq 0$ for all n and $t \neq 0$.

Proof: (a) To show that $\lim (s_n + t_n) = s + t$, we need to make the difference $|(s_n + t_n) - (s + t)|$ small. Using the triangle inequality [Theorem 11.8(d)], we have

$$\begin{aligned} |(s_n + t_n) - (s + t)| &= |(s_n - s) + (t_n - t)| \\ &\leqslant |s_n - s| + |t_n - t|. \end{aligned}$$

Now given any $\varepsilon > 0$, since $s_n \to s$, there exists N_1 such that $n > N_1$ implies that $|s_n - s| < \varepsilon/2$. Similarly, since $t_n \to t$, there exists N_2 such that $n > N_2$ implies that $|t_n - t| < \varepsilon/2$. Thus, if we let $N = \max\{N_1, N_2\}$, then $n > N$ implies that

$$|(s_n + t_n) - (s + t)| \leqslant |s_n - s| + |t_n - t| < \frac{\varepsilon}{2} + \frac{\varepsilon}{2} = \varepsilon.$$

Therefore, we conclude that $\lim (s_n + t_n) = s + t$.

(b) Exercise 17.2(a).

(c) This time we use the inequality

$$\begin{aligned} |s_n t_n - st| &= |(s_n t_n - s_n t) + (s_n t - st)| \\ &\leqslant |s_n t_n - s_n t| + |s_n t - st| \\ &= |s_n| \cdot |t_n - t| + |t| \cdot |s_n - s|. \end{aligned}$$

We know from Theorem 16.13 that the convergent sequence (s_n) is bounded. Thus there exists $M_1 > 0$ such that $|s_n| \leqslant M_1$ for all n. Letting $M = \max\{M_1, |t|\}$, we obtain the inequality

$$|s_n t_n - st| \leqslant M|t_n - t| + M|s_n - s|.$$

Now, given any $\varepsilon > 0$, there exists N_1 and N_2 such that

$$|t_n - t| < \frac{\varepsilon}{2M} \text{ when } n > N_1 \quad \text{and} \quad |s_n - s| < \frac{\varepsilon}{2M} \text{ when } n > N_2.$$

Let $N = \max\{N_1, N_2\}$. Then $n > N$ implies that

$$\begin{aligned} |s_n t_n - st| &\leqslant M|t_n - t| + M|s_n - s| \\ &< M\left(\frac{\varepsilon}{2M}\right) + M\left(\frac{\varepsilon}{2M}\right) = \frac{\varepsilon}{2} + \frac{\varepsilon}{2} = \varepsilon. \end{aligned}$$

Thus $\lim(s_n t_n) = st$.

(d) Since $s_n/t_n = s_n(1/t_n)$, it suffices from part (c) to show that $\lim(1/t_n) = 1/t$. That is, given $\varepsilon > 0$ we must make

$$\left|\frac{1}{t_n} - \frac{1}{t}\right| = \left|\frac{t - t_n}{t_n t}\right| < \varepsilon$$

for all n sufficiently large. To get a lower bound on how small the denominator can be, we note that since $t \neq 0$ there exists N_1 such that $n > N_1$ implies that $|t_n - t| < |t|/2$. Thus for $n > N_1$ we have

$$|t_n| = |t - (t - t_n)| \geqslant |t| - |t - t_n| > |t| - \frac{|t|}{2} = \frac{|t|}{2}$$

by Exercise 11.4. There also exists N_2 such that $n > N_2$ implies that $|t_n - t| < \frac{1}{2}\varepsilon|t|^2$. Let $N = \max\{N_1, N_2\}$. Then $n > N$ implies that

$$\left|\frac{1}{t_n} - \frac{1}{t}\right| = \left|\frac{t - t_n}{t_n t}\right| < \frac{2}{|t|^2}|t - t_n| < \varepsilon.$$

Hence $\lim(1/t_n) = 1/t$. ▮

To illustrate the usefulness of Theorem 17.1, let us return to the sequence used in Example 16.9.

17.2 EXAMPLE To prove that $\lim(4n^2 - 3)/(5n^2 - 2n) = 4/5$, we note that

$$s_n = \frac{4n^2 - 3}{5n^2 - 2n} = \frac{4 - 3/n^2}{5 - 2/n}.$$

Now $\lim (1/n^2) = 0$ by Exercise 16.2(b), so $\lim [(-3)/n^2] = 0$ by Theorem 17.1(b). Thus $\lim [4 - (3/n^2)] = 4$ by 17.1(b). Similarly,

$$\lim \left(5 - \frac{2}{n}\right) = 5 - 2\left(\lim \frac{1}{n}\right) = 5 - 2(0) = 5.$$

Finally, from 17.1(d) we conclude that $\lim s_n = \frac{4}{5}$.

17.3 PRACTICE Show that $\left(\dfrac{n+3}{n^2 - 5n}\right)$ converges and find its limit.

Another useful fact is that the order relation "$\leqslant$" is preserved when taking limits.

17.4 THEOREM Suppose that (s_n) and (t_n) are convergent sequences with $\lim s_n = s$ and $\lim t_n = t$. If $s_n \leqslant t_n$ for all $n \in \mathbb{N}$, then $s \leqslant t$.

Proof: Suppose that $s > t$. Then $\varepsilon = (s - t)/2 > 0$. Thus there exists N_1 such that $n > N_1$ implies that

$$s - \varepsilon < s_n < s + \varepsilon.$$

Similarly, there exists N_2 such that $n > N_2$ implies that

$$t - \varepsilon < t_n < t + \varepsilon.$$

Let $N = \max \{N_1, N_2\}$. Then for $n > N$ we have

$$t_n < t + \varepsilon = s - \varepsilon < s_n,$$

which contradicts the assumption that $s_n \leqslant t_n$ for all n. Thus we conclude that $s \leqslant t$. ▌

17.5 COROLLARY If (t_n) converges to t and $t_n \geqslant 0$ for all $n \in \mathbb{N}$, then $t \geqslant 0$.

Proof: Exercise 17.2(b). ▌

17.6 EXAMPLE Suppose that (t_n) converges to t and that $t_n \geqslant 0$ for all $n \in \mathbb{N}$. To illustrate how algebraic manipulations can be useful in evaluating limits, let us show that $\lim (\sqrt{t_n}) = \sqrt{t}$. First, we note that Corollary 17.5 implies that $t \geqslant 0$, so that $\sqrt{t}$ is defined. Our argument consists of two cases, depending on whether t is positive or zero.

Suppose that $t > 0$. To get a bound on the difference $|\sqrt{t_n} - \sqrt{t}|$ in terms of $|t_n - t|$, we multiply and divide by the conjugate $|\sqrt{t_n} + \sqrt{t}|$. Thus

$$|\sqrt{t_n} - \sqrt{t}| = \frac{|\sqrt{t_n} - \sqrt{t}| \cdot |\sqrt{t_n} + \sqrt{t}|}{|\sqrt{t_n} + \sqrt{t}|} = \frac{|t_n - t|}{|\sqrt{t_n} + \sqrt{t}|}.$$

Since $\sqrt{t_n} + \sqrt{t} \geqslant \sqrt{t} > 0$, we obtain

$$|\sqrt{t_n} - \sqrt{t}| \leqslant \frac{1}{\sqrt{t}} |t_n - t|.$$

Now $\lim (t_n - t) = 0$ since $\lim t_n = t$. Thus from Theorem 16.8 we may conclude that $\lim \sqrt{t_n} = \sqrt{t}$.

The proof of the case when $t = 0$ is similar to Practice 16.4 and is left to the reader.

Our next theorem gives a "ratio test" that can be used to show that certain sequences converge to zero.

17.7 THEOREM Suppose that (s_n) is a sequence of positive terms and that the limit $L = \lim (s_{n+1}/s_n)$ exists. If $L < 1$, then $\lim s_n = 0$.

Proof: Since $L < 1$, there exists a real number c such that $L < c < 1$. Let $\varepsilon = c - L$ so that $\varepsilon > 0$. Then there exists an integer N such that $n > N$ implies that

$$\left| \frac{s_{n+1}}{s_n} - L \right| < \varepsilon.$$

Let $k = N + 1$. Then for all $n > k$ we have $n - 1 > N$, so that

$$\frac{s_n}{s_{n-1}} < L + \varepsilon = L + (c - L) = c.$$

It follows that, for all $n > k$,

$$0 < s_n < s_{n-1}c < s_{n-2}c^2 < \cdots < s_k c^{n-k}.$$

Letting $M = s_k/c^k$, we obtain $0 < s_n < Mc^n$ for all $n > k$. Since $0 < c < 1$, Exercise 16.3(f) implies that $\lim c^n = 0$. Thus $\lim s_n = 0$ by Theorem 16.8. ▮

17.8 PRACTICE Suppose that $0 < x < 1$. Apply Theorem 17.7 to the sequence given by $s_n = nx^n$.

Infinite Limits

The sequence given by $s_n = n$ is certainly not convergent since it is not bounded (Theorem 16.13). But its behavior is not the least erratic: the terms get larger and larger. Although there is no real number that the terms "approach," we would like to be able to say that s_n "goes to ∞." We make this precise in the following definition.

17.9 DEFINITION A sequence (s_n) is said to **diverge to $+\infty$**, and we write $\lim s_n = +\infty$ provided that

> for every $M \in \mathbb{R}$ there exists a number N such that $n > N$ implies that $s_n > M$.

Similarly, (s_n) is said to **diverge to $-\infty$**, and we write $\lim s_n = -\infty$, provided that

> for every $M \in \mathbb{R}$ there exists a number N such that $n > N$ implies that $s_n < M$.

It is important to note that the symbols $+\infty$ and $-\infty$ do not represent real numbers. They are simply part of the notation that is used to describe the behavior of certain sequences. When $\lim s_n = +\infty$ (or $-\infty$), we shall say that the limit exists, but this does not mean that the sequence converges; in fact, it diverges. Thus a sequence converges iff its limit exists *as a real number*. Since Theorems 17.1 and 17.4 refer to convergent sequences, they cannot be used with infinite limits.

17.10 PRACTICE Show that $\lim n^2 = +\infty$.

17.11 EXAMPLE The technique of developing proofs for infinite limits is similar to that for finite limits. To illustrate, let us show that $\lim (4n^2 - 3)/(n + 2) = +\infty$. This time we want to get a *lower* bound on the numerator. We find that

$$4n^2 - 3 \geqslant 4n^2 - n^2 = 3n^2, \qquad \text{when } n > 1.$$

For an *upper* bound on the denominator, we have

$$n + 2 \leqslant n + n = 2n, \qquad \text{when } n > 1.$$

Thus for $n > 1$ we obtain

$$\frac{4n^2 - 3}{n + 2} \geqslant \frac{3n^2}{2n} = \frac{3n}{2}.$$

To make this greater than any particular M, we want $n > 2M/3$. Thus there are two conditions to be satisfied: $n > 1$ and $n > 2M/3$. Here is the proof written out formally:

> Given any $M \in \mathbb{R}$, let $N = \max\{1, 2M/3\}$. Then $n > N$ implies that $n > 1$ and $n > 2M/3$. Since $n > 1$ we have $4n^2 - 3 \geqslant 4n^2 - n^2 = 3n^2$ and $n + 2 \leqslant n + n = 2n$. Thus for $n > N$ we have
>
> $$\frac{4n^2 - 3}{n + 2} \geqslant \frac{3n^2}{2n} = \frac{3n}{2} > M.$$
>
> Hence $\lim (4n^2 - 3)/(n + 2) = +\infty$. ∎

As an analog of Theorem 17.4, we have the following result for infinite limits.

17.12 THEOREM Suppose that (s_n) and (t_n) are sequences such that $s_n \leqslant t_n$ for all $n \in \mathbb{N}$.

(a) If $\lim s_n = +\infty$, then $\lim t_n = +\infty$.
(b) If $\lim t_n = -\infty$, then $\lim s_n = -\infty$.

Proof: Exercise 17.7. ▌

For our final theorem in this section we show the relationship between infinite limits and zero limits.

17.13 THEOREM Let (s_n) be a sequence of positive numbers. Then $\lim s_n = +\infty$ iff $\lim (1/s_n) = 0$.

Proof: Suppose that $\lim s_n = +\infty$. Given any $\varepsilon > 0$, let $M = 1/\varepsilon$. Then there exists N such that $n > N$ implies that $s_n > M = 1/\varepsilon$. Since each s_n is positive we have

$$\left| \frac{1}{s_n} - 0 \right| < \varepsilon, \qquad \text{whenever } n > N.$$

Thus $\lim (1/s_n) = 0$.

The converse is analogous and is left to the reader (Exercise 17.8). ▌

ANSWERS TO PRACTICE PROBLEMS

17.3 We have

$$\lim \left(\frac{n+3}{n^2-5n}\right) = \lim \left(\frac{1/n + 3/n^2}{1 - 5/n}\right)$$
$$= \frac{(\lim 1/n) + 3(\lim 1/n^2)}{1 - 5(\lim 1/n)} = \frac{0 + 3(0)}{1 - 5(0)} = 0.$$

17.8 $\dfrac{s_{n+1}}{s_n} = \dfrac{(n+1)x^{n+1}}{nx^n} = x\left(1 + \dfrac{1}{n}\right) \to x < 1$. Hence $\lim nx^n = 0$.

17.10 Given $M \in \mathbb{R}$, let $N = |M|$. Then for $n > N$ we have $n^2 \geqslant n > N \geqslant M$. Thus $\lim n^2 = +\infty$.

EXERCISES

17.1 Use Theorem 17.1 to find the following limits. Justify your answers.

(a) $\lim \dfrac{3n^2 + 4n}{7n^2 - 5n}$ (b) $\lim \dfrac{n^4 + 13}{2n^5 + 3}$

17.2 (a) Prove Theorem 17.1(b).
(b) Prove Corollary 17.5.

17.3 For s_n given by the following formulas, determine the convergence or divergence of the sequence (s_n). Find any limits that exist.

(a) $s_n = \dfrac{3 - 2n}{1 + n}$ (b) $s_n = \dfrac{(-1)^n}{n + 3}$

(c) $s_n = \dfrac{(-1)^n n}{2n - 1}$ (d) $s_n = \dfrac{2^{3n}}{3^{2n}}$

(e) $s_n = \dfrac{n^2 - 2}{n + 1}$ (f) $s_n = \dfrac{3 + n - n^2}{1 + 2n}$

(g) $s_n = \dfrac{1 - n}{2^n}$ (h) $s_n = \dfrac{3^n}{n^3 + 5}$

(i) $s_n = \dfrac{n!}{2^n}$ (j) $s_n = \dfrac{n!}{n^n}$

(k) $s_n = \dfrac{n^2}{2^n}$ (l) $s_n = \dfrac{n^2}{n!}$

17.4 For each of the following, prove or give a counterexample.
(a) If (s_n) and (t_n) are divergent sequences, then $(s_n + t_n)$ diverges.
(b) If (s_n) and (t_n) are divergent sequences, then $(s_n t_n)$ diverges.
(c) If (s_n) and $(s_n + t_n)$ are convergent sequences, then (t_n) converges.
(d) If (s_n) and $(s_n t_n)$ are convergent sequences, then (t_n) converges.

17.5 Give an example of an unbounded sequence that does not diverge to $+\infty$ or to $-\infty$.

17.6 (a) Give an example of a convergent sequence (s_n) of positive numbers such that $\lim (s_{n+1}/s_n) = 1$.
(b) Give an example of a divergent sequence (t_n) of positive numbers such that $\lim (t_{n+1}/t_n) = 1$.

17.7 Prove Theorem 17.12.

17.8 Prove the converse part of Theorem 17.13.

17.9 Prove: If $\lim s_n = 0$, then for any $k > 0$, $\lim_{n\to\infty} s_n^k = 0$. This finishes the proof in Example 17.6.

17.10 Suppose that (s_n) converges to s. Prove that (s_n^2) converges to s^2 directly without using the product formula of Theorem 17.1(c).

17.11 Write an alternative proof of Theorem 17.1(c) that does not use Theorem 16.13 by using the identity $s_n t_n - st = (s_n - s)(t_n - t) + s(t_n - t) + t(s_n - s)$.

17.12 Prove that $\lim \left(\dfrac{1}{n} - \dfrac{1}{n+1}\right) = 0$.

17.13 Prove the following.

(a) $\lim (\sqrt{n+1} - \sqrt{n}) = 0$
(b) $\lim (\sqrt{n^2+1} - n) = 0$
(c) $\lim (\sqrt{n^2+n} - n) = \frac{1}{2}$

17.14 Let (s_n) be a sequence of positive terms such that $L = \lim (s_{n+1}/s_n)$ exists. Prove that if $L > 1$ then $\lim s_n = +\infty$.

***17.15** (a) Show that $\lim_{n\to\infty} k^n/n! = 0$ for all $k \in \mathbb{R}$.
(b) What can be said about $\lim_{n\to\infty} n!/k^n$?

***17.16** Suppose that (s_n) is a convergent sequence with $a \leqslant s_n \leqslant b$ for all $n \in \mathbb{N}$. Prove that $a \leqslant \lim s_n \leqslant b$.

17.17 Prove the following.

(a) If $\lim s_n = +\infty$ and $k > 0$, then $\lim ks_n = +\infty$.
(b) If $\lim s_n = +\infty$ and $k < 0$, then $\lim ks_n = -\infty$.
(c) $\lim s_n = +\infty$ iff $\lim (-s_n) = -\infty$.
(d) If $\lim s_n = +\infty$ and if (t_n) is a bounded sequence, then $\lim (s_n + t_n) = +\infty$.

17.18 Let (s_n), (t_n), and (u_n) be sequences such that $s_n \leqslant t_n \leqslant u_n$ for all $n \in \mathbb{N}$. Suppose that (s_n) and (u_n) both converge to the real number s. Prove that (t_n) also converges to s.

Section 18 MONOTONE SEQUENCES AND CAUCHY SEQUENCES

In the preceding two sections we have seen a number of results that enable us to show that a sequence converges. Unfortunately, most of these techniques depend on our knowing (or guessing) what the limit of the sequence is before we begin. Often in applications it is desirable to be able to show that a given sequence is convergent without knowing precisely the value of the limit. In this section we obtain two important theorems (18.3 and 18.12) that enable us to do just that.

Monotone Sequences

18.1 DEFINITION A sequence (s_n) of real numbers is **increasing** if $s_n \leqslant s_{n+1}$ for all $n \in \mathbb{N}$ and is **decreasing** if $s_n \geqslant s_{n+1}$ for all $n \in \mathbb{N}$. A sequence is **monotone** if it is either increasing or decreasing.†

18.2 EXAMPLE The sequences given by $a_n = n$, $b_n = 2^n$, and $c_n = 2 - 1/n$ are all increasing. The sequence $(d_n) = (1, 1, 2, 2, 3, 3, \ldots)$ is also called increasing even

† Some authors refer to an increasing sequence as "nondecreasing" and reserve the term "increasing" to apply to a "strictly increasing" sequence: $s_n < s_{n+1}$ for all $n \in \mathbb{N}$.

though some adjacent terms are equal. The sequences given by $s_n = 2/n$ and $t_n = -3n$ are decreasing. A constant sequence $(u_n) = (1, 1, 1, 1, \ldots)$ is both increasing and decreasing. The sequences given by $x_n = (-1)^n/n$ and $y_n = \cos(n\pi/3)$ are not monotone.

Of the monotone examples given above, the sequences (c_n), (s_n), and (u_n) are bounded, while (a_n), (b_n), (d_n), and (t_n) are not bounded. We also note that (c_n), (s_n), and (u_n) are convergent, while the unbounded monotone sequences diverge. It turns out that this is not just a coincidence.

18.3 THEOREM (Monotone Convergence Theorem) A monotone sequence is convergent iff it is bounded.

Proof: Suppose that (s_n) is a bounded increasing sequence. Let S denote the nonempty bounded set $\{s_n : n \in \mathbb{N}\}$. By the completeness axiom (see Section 12) S has a least upper bound, and we let $s = \sup S$. We claim that $\lim s_n = s$. Given any $\varepsilon > 0$, $s - \varepsilon$ is not an upper bound for S. Thus there exists N such that $s_N > s - \varepsilon$. Furthermore, since (s_n) is increasing and s is an upper bound for S, we have

$$s - \varepsilon < s_N \leqslant s_n \leqslant s$$

for all $n > N$. Hence (s_n) converges to s.

In the case when the sequence is decreasing, let $s = \inf S$ and proceed in a similar manner. (See Exercise 18.5.)

The converse implication has already been proved as Theorem 16.13. ∎

18.4 EXAMPLE Let (S_n) be the sequence defined by $s_1 = 1$ and $s_{n+1} = \sqrt{1 + s_n}$ for $n \geqslant 1$. We shall show that (s_n) is a bounded increasing sequence. Computing the next three terms of the sequence, we find

$$s_2 = \sqrt{2} \approx 1.414$$

$$s_3 = \sqrt{1 + \sqrt{2}} \approx 1.554$$

$$s_4 = \sqrt{1 + \sqrt{1 + \sqrt{2}}} \approx 1.598,$$

where the decimals have been rounded off. It appears that the sequence is bounded above by 2. To see if this conjecture is true, let us try to prove it using induction. Certainly, $s_1 = 1 < 2$. Now suppose that $s_k < 2$ for some $k \in \mathbb{N}$. Then $s_{k+1} = \sqrt{1 + s_k} < \sqrt{1 + 2} = \sqrt{3} < 2$. Thus we may conclude by induction that $s_n < 2$ for all $n \in \mathbb{N}$.

To verify that (s_n) is an increasing sequence, we also argue by induction. Since $s_1 = 1$ and $s_2 = 2$, we have $s_1 < s_2$, which establishes the basis for induction. Now suppose that $s_k < s_{k+1}$ for some $k \in \mathbb{N}$. Then we have

$$s_{k+1} = \sqrt{1 + s_k} < \sqrt{1 + s_{k+1}} = s_{k+2}.$$

Thus the induction step holds and we conclude that $s_n < s_{n+1}$ for all $n \in \mathbb{N}$.

Thus (s_n) is an increasing sequence and it is bounded by the interval $[1, 2]$. We concluded from the monotone convergence theorem (18.3) that (s_n) is convergent. The only question that remains is to find the value s to which it converges. Since $\lim s_{n+1} = \lim s_n$ (Exercise 16.7), we see that s must satisfy the equation

$$s = \sqrt{1 + s}.$$

(Here we have used Theorem 17.1 and Example 17.6.) Solving algebraically for s, we obtain $s = (1 \pm \sqrt{5})/2$. Since $s_n \geqslant 1$ for all n, $(1 - \sqrt{5})/2$ cannot be the limit. We conclude that $\lim s_n = s = (1 + \sqrt{5})/2$.

18.5 EXAMPLE Consider the sequence (t_n) defined by $t_1 = 1$ and $t_{n+1} = (t_n + 1)/4$. The first four terms are $t_1 = 1, t_2 = \frac{1}{2}, t_3 = \frac{3}{8}$, and $t_4 = \frac{11}{32}$. In Practice 18.6 you are asked to show that the sequence is decreasing. Assuming this to be true, we have $t_n \leqslant t_1 = 1$ for all n. Since each t_n is clearly positive, we see that (t_n) is a bounded monotone sequence, and hence is convergent. In Practice 18.7 you are asked to find the value t of the limit.

18.6 PRACTICE Use induction to show $t_n > t_{n+1}$ for all n. We have already established the basis for induction: $t_1 > t_2$. The induction step remains.

18.7 PRACTICE Use the fact that $t = \lim t_n = \lim t_{n+1}$ to find t.

While unbounded monotone sequences do not converge, they do have limits.

18.8 THEOREM (a) If (s_n) is an unbounded increasing sequence, then $\lim s_n = +\infty$.
(b) If (s_n) is an unbounded decreasing sequence, then $\lim s_n = -\infty$.

Proof: (a) Let (s_n) be an increasing sequence and suppose that the set $S = \{s_n \colon n \in \mathbb{N}\}$ is unbounded. Since (s_n) is increasing, S is bounded below by s_1. Hence S must be unbounded above. Thus, given any $M \in \mathbb{R}$, there exists $N \in \mathbb{N}$ such that $s_n > M$. But then for any $n > N$ we have $s_n \geqslant s_N > M$, so $\lim s_n = +\infty$. The proof of (b) is similar (Exercise 18.6). ▮

Cauchy Sequences

When a sequence (s_n) is convergent, the terms all get close to the value of the limit for large n. By so doing, they also get close to each other. It turns out that the latter property (called the Cauchy property) is actually sufficient to imply convergence. We prove this after a preliminary definition and two lemmas.

18.9 DEFINITION A sequence (s_n) of real numbers is said to be a **Cauchy sequence** if

for each $\varepsilon > 0$ there exists a number N such that $m, n > N$ implies that $|s_n - s_m| < \varepsilon$.

18.10 LEMMA Every convergent sequence is a Cauchy sequence.

Proof: Suppose that (s_n) converges to s. To show that s_n is close to s_m, we use the fact that they are both close to s. A clever use of the triangle inequality gives us the following estimate:

$$|s_n - s_m| = |s_n - s + s - s_m| \leqslant |s_n - s| + |s - s_m|.$$

Thus, given any $\varepsilon > 0$, we choose N so that $k > N$ implies that $|s_k - s| < \varepsilon/2$. (We can do this since $\lim s_n = s$.) Then for $m, n > N$ we have

$$|s_n - s_m| \leqslant |s_n - s| + |s - s_m| < \frac{\varepsilon}{2} + \frac{\varepsilon}{2} = \varepsilon.$$

Thus (s_n) is a Cauchy sequence. ▮

18.11 LEMMA Every Cauchy sequence is bounded.

Proof: The proof is similar to that of Theorem 16.13 and is included as Exercise 18.7. ▮

18.12 THEOREM (Cauchy Convergence Criterion) A sequence of real numbers is convergent iff it is a Cauchy sequence.

Proof: We have already shown (Lemma 18.10) that a convergent sequence is a Cauchy sequence. For the converse we suppose that (s_n) is a Cauchy sequence and let $S = \{s_n : n \in \mathbb{N}\}$ be the range of the sequence. We consider two cases, depending on whether S is finite or infinite.

If S is finite, then the minimum distance ε between distinct points of S is positive. Since (s_n) is Cauchy, there exists N such that $m, n > N$ implies that $|s_n - s_m| < \varepsilon$. Let n_0 be the smallest integer greater than N. Given any $m > N$, s_m and s_{n_0} are both in S, so if the distance between them is less than ε, it must be zero (since ε is the minimum distance between *distinct* points in S). Thus $s_m = s_{n_0}$ for all $n > N$. It follows that $\lim s_n = s_{n_0}$.

Now suppose that S is infinite. From Lemma 18.11 we know that S is bounded. Thus from the Bolzano-Weierstrass theorem (14.6) there exists a point s in $\mathbb{R}$ that is an accumulation point of S. We claim that (s_n) converges to s. Given any $\varepsilon > 0$, there exists N such that $|x_n - x_m| < \varepsilon/2$ whenever $m, n > N$. Since s is an accumulation point of S, the neighborhood $N(s; \varepsilon/2) = (s - \varepsilon/2, s + \varepsilon/2)$ contains infinitely many points of S. (See Exercise 13.9.) Thus in particular there exists $m > N$ such that $s_m \in N(s; \varepsilon/2)$. (See Figure 18.1.) Hence for any $n > N$ we have

$$\begin{aligned} |s_n - s| &= |s_n - s_m + s_m - s| \\ &\leqslant |s_n - s_m| + |s_m - s| < \frac{\varepsilon}{2} + \frac{\varepsilon}{2} = \varepsilon. \end{aligned}$$

Therefore, $\lim s_n = s$. ∎

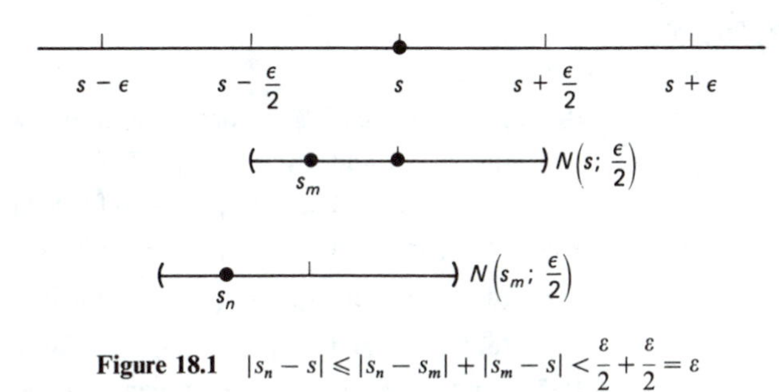

Figure 18.1 $|s_n - s| \leqslant |s_n - s_m| + |s_m - s| < \frac{\varepsilon}{2} + \frac{\varepsilon}{2} = \varepsilon$

It is important to note that the Cauchy convergence criterion depends on the completeness of $\mathbb{R}$ since the proof uses the Bolzano-Weierstrass theorem. In fact, it can be shown that an Archimedean ordered field is complete iff the Cauchy convergence criterion holds. [See Olmsted (1962), page 203.] The property of being a Cauchy sequence can be defined in any setting in which there is a notion of distance. (See Sections 15 and 24.) In this more general setting, a Cauchy sequence may not necessarily converge, although it will be bounded.

18.13 EXAMPLE We illustrate the use of the Cauchy criterion by showing that the sequence given by

$$s_n = 1 + \frac{1}{2} + \frac{1}{3} + \cdots + \frac{1}{n}$$

is divergent. If $m > n$, then

$$s_m - s_n = \frac{1}{n+1} + \frac{1}{n+2} + \cdots + \frac{1}{m}$$
$$> \underbrace{\frac{1}{m} + \frac{1}{m} + \cdots + \frac{1}{m}}_{m-n \text{ terms}} = \frac{m-n}{m} = 1 - \frac{n}{m}.$$

In particular, when $m = 2n$ we have $s_{2n} - s_n > \frac{1}{2}$. Thus the sequence (s_n) cannot be Cauchy and hence it is not convergent.

ANSWERS TO PRACTICE PROBLEMS

18.6 Suppose that $t_k < t_{k+1}$ for some $k \in \mathbb{N}$. Then $t_{k+1} = (t_k + 1)/4 > (t_{k+1} + 1)/4 = t_{k+2}$.

18.7 Since $t = (t + 1)/4$, we obtain $t = \frac{1}{3}$.

EXERCISES

18.1 Prove that each sequence is monotone and bounded. Then find the limit.

(a) $s_1 = 1$ and $s_{n+1} = \frac{1}{4}(s_n + 5)$ for $n \in \mathbb{N}$.

(b) $s_1 = 2$ and $s_{n+1} = \frac{1}{4}(s_n + 5)$ for $n \in \mathbb{N}$.

(c) $s_1 = 1$ and $s_{n+1} = \frac{1}{4}(2s_n + 5)$ for $n \in \mathbb{N}$.

(d) $s_1 = 2$ and $s_{n+1} = \sqrt{2s_n + 1}$ for $n \in \mathbb{N}$.

(e) $s_1 = 3$ and $s_{n+1} = \sqrt{10s_n - 17}$ for $n \in \mathbb{N}$.

18.2 Find an example of a sequence of real numbers satisfying each set of properties.

(a) Cauchy, but not monotone

(b) Monotone, but not Cauchy

(c) Bounded, but not Cauchy

18.3 Suppose that $x > 0$. Define a sequence (s_n) by $s_1 = k$ and $s_{n+1} = (s_n^2 + x)/(2s_n)$ for $n \in \mathbb{N}$. Prove that, for any $k > 0$, $\lim s_n = \sqrt{x}$.

18.4 (a) Suppose that $|r| < 1$. Recall from Exercise 10.5 that $1 + r + r^2 + \cdots + r^n = (1 - r^{n+1})/(1 - r)$. Find $\lim_{n\to\infty}(1 + r + r^2 + \cdots + r^n)$.

(b) If we let the infinite repeating decimal $0.9999\cdots$ stand for the limit

$$\lim_{n\to\infty}\left(\frac{9}{10} + \frac{9}{10^2} + \cdots + \frac{9}{10^n}\right),$$

show that $0.9999\cdots = 1$.

18.5 Finish the proof of Theorem 18.3 for a bounded decreasing sequence.

18.6 Prove Theorem 18.8(b).

18.7 Prove Lemma 18.11.

***18.8** Let (s_n) be the sequence defined by $s_n = (1 + 1/n)^n$. Use the binomial theorem (Exercise 10.14) to show that (s_n) is an increasing sequence with $s_n < 3$ for all n. Conclude that (s_n) is convergent. The limit of (s_n) is referred to as e and is used as the base for natural logarithms. The approximate value of e is 2.71828.

18.9 A sequence (s_n) is said to be **contractive** if there exists a constant k, $0 < k < 1$, such that $|s_{n+2} - s_{n+1}| \leqslant k|s_{n+1} - s_n|$ for all $n \in \mathbb{N}$. Prove that every contractive sequence is a Cauchy sequence, and hence is convergent.

Section 19 SUBSEQUENCES

19.1 DEFINITION Let $(s_n)_{n=1}^{\infty}$ be a sequence and let $(n_k)_{k=1}^{\infty}$ be any sequence of natural numbers such that $n_1 < n_2 < n_3 < \cdots$. The sequence $(s_{n_k})_{k=1}^{\infty}$ is called a **subsequence** of $(s_n)_{n=1}^{\infty}$.

If we delete a finite number of the terms of a sequence and renumber the remaining ones in the same order, we obtain a subsequence. In fact, we may delete infinitely many of the terms in the original sequence as long as there are still infinitely many terms left. Thus the sequence

$$(s_n) = \left(1, \frac{1}{2}, \frac{1}{3}, \frac{1}{4}, \ldots\right)$$

has, for example,

$$(t_k) = \left(\frac{1}{5}, \frac{1}{6}, \frac{1}{7}, \frac{1}{8}, \ldots\right) \quad \text{and} \quad (u_k) = \left(\frac{1}{2}, \frac{1}{4}, \frac{1}{8}, \frac{1}{16}, \ldots\right)$$

as subsequences. Of course, it has many other subsequences, including (s_n) itself. We note, however, that

$$(v_n) = \left(\frac{1}{3}, \frac{1}{2}, \frac{1}{5}, \frac{1}{4}, \frac{1}{7}, \frac{1}{6}, \ldots\right)$$

is not a subsequence of (s_n) since the order of the terms is not preserved. If we use the notation of Definition 19.1 and write $t_k = s_{n_k}$, then we have $n_k = k + 4$. Sometimes we shorten the notation and simply refer to the subsequence (t_k) as (s_{n+4}).

19.2 PRACTICE If $u_k = s_{n_k}$ as given above, what is n_k?

If a sequence is convergent, we would expect that any subsequence is also convergent. This is easy to prove once we have the following simple result.

19.3 PRACTICE Let $(n_k)_{k=1}^{\infty}$ be a sequence of natural numbers such that $n_k < n_{k+1}$ for all $k \in \mathbb{N}$. Use induction to show that $n_k \geqslant k$ for all $k \in \mathbb{N}$.

19.4 THEOREM If a sequence (s_n) converges to a real number s, then every subsequence of (s_n) also converges to s.

Proof: Let (s_{n_k}) be any subsequence of (s_n). Given any $\varepsilon > 0$, there exists N such that $n > N$ implies that $|s_n - s| < \varepsilon$. Thus when $k > N$, we apply Practice 19.3 to obtain $n_k \geqslant k > N$, so that $|s_{n_k} - s| < \varepsilon$. Hence $\lim_{k \to \infty} s_{n_k} = s$. ∎

19.5 EXAMPLE One application of Theorem 19.4 is in finding the value of the limit of a convergent sequence. Suppose that $0 < x < 1$ and consider the sequence (s_n) defined by $s_n = x^{1/n}$. Since $0 < x^{1/n} < 1$ for all n, (s_n) is bounded. Since

$$x^{1/(n+1)} - x^{1/n} = x^{1/(n+1)}(1 - x^{1/[n(n+1)]}) > 0, \quad \text{for all } n,$$

(s_n) is an increasing sequence. Thus, by the monotone convergence theorem (18.3), (s_n) converges to some number, say s. Now for each n, $s_{2n} = x^{1/(2n)} = (x^{1/n})^{1/2} = \sqrt{s_n}$. But by Theorem 19.4, $\lim s_{2n} = \lim s_n$ and by Example 17.6, $\lim \sqrt{s_n} = \sqrt{\lim s_n}$. Thus we have

$$s = \lim s_n = \lim s_{2n} = \lim \sqrt{s_n} = \sqrt{\lim s_n} = \sqrt{s}.$$

It follows that $s^2 = s$, so that $s = 0$ or $s = 1$. But $s_1 = x > 0$ and the sequence is increasing, so $s \neq 0$. Hence $s = 1$ and $\lim x^{1/n} = 1$.

19.6 EXAMPLE Theorem 19.4 can also be useful in showing that a sequence is divergent. For example, if the sequence $s_n = (-1)^n$ were convergent to some number s, then every subsequence would also converge to s. But (s_{2n}) converges to $+1$ and (s_{2n-1}) converges to -1. We conclude that (s_n) is not convergent.

If a sequence is divergent, the behavior of its subsequences can be quite varied. For example, we just saw that $s_n = (-1)^n$ has subsequences converging to two different numbers. On the other hand, none of the subsequences of $(1, 2, 3, 4, \ldots)$ are convergent. If, however, a given sequence is bounded, it will have at least one convergent subsequence. This result is sometimes known as the Bolzano–Weierstrass theorem for sequences.

19.7 THEOREM Every bounded sequence has a convergent subsequence.

Proof: Let (s_n) be a sequence whose range $S = \{s_n \colon n \in \mathbb{N}\}$ is bounded. Suppose first that S is finite. Then there is some number x in S that is equal to s_n for infinitely many values of n. That is, there exists $n_1 < n_2 < \cdots < n_k < \cdots$ such that $s_{n_k} = x$ for all $k \in \mathbb{N}$. It follows that the subsequence (s_{n_k}) converges to x.

On the other hand, suppose that S is infinite. Then the Bolzano–Weierstrass theorem (14.6) implies that S has an accumulation point, say y, in $\mathbb{R}$. We now construct a subsequence of (s_n) that converges to y. For each $k \in \mathbb{N}$, let $A_k = (y - 1/k, y + 1/k)$ be the neighborhood about y of radius $1/k$. Since y is an accumulation point of S, given any $k \in \mathbb{N}$, there are infinitely many values of n such that $s_n \in A_k$. Thus we can pick $s_{n_1} \in A_1$. Then we can choose $n_2 > n_1$ with $s_{n_2} \in A_2$. In general we choose $s_{n_k} \in A_k$ with $n_k > n_{k-1}$. By so doing we obtain a subsequence (s_{n_k}) of (s_n) for which $|s_{n_k} - y| < 1/k$ for all $k \in \mathbb{N}$. It follows from Theorem 16.8 that $\lim_{k \to \infty} s_{n_k} = y$. ∎

While an unbounded sequence may not have any convergent subsequence, it will contain a subsequence that has an infinite limit. In fact, we can prove the following slightly stronger result.

19.8 THEOREM Every unbounded sequence contains a monotone subsequence that has either $+\infty$ or $-\infty$ as a limit.

Proof: Suppose that (s_n) is unbounded above. We shall construct an unbounded increasing subsequence of (s_n). Given any $M \in \mathbb{R}$, there must be infinitely many terms of (s_n) larger than M. (Otherwise, the maximum of the finite number of terms would be an upper bound.) In particular, there exists $n_1 \in \mathbb{N}$ such that $s_{n_1} > 1$. Then there exists $n_2 > n_1$ such that $s_{n_2} > \max\{2, s_{n_1}\}$. In general, given $n_1, \ldots, n_k$ there exists $n_{k+1} > n_k$ such that $s_{n_{k+1}} > \max\{k, s_{n_k}\}$. It follows that the subsequence (s_{n_k}) is unbounded and increasing. By Theorem 18.8, $\lim_{k \to \infty} s_{n_k} = +\infty$.

Finally, if (s_n) is not unbounded above, then it must be unbounded below and a similar argument produces an unbounded decreasing subsequence having limit $-\infty$. ∎

lim sup and lim inf

19.9 DEFINITION Let (s_n) be a bounded sequence. A **subsequential limit** of (s_n) is any real number that is the limit of some subsequence of (s_n). If S is the set of all subsequential limits of (s_n), then we define the **limit superior** (or **upper limit**) of (s_n) to be

$$\limsup s_n = \sup S.$$

Similarly, we define the **limit inferior** (or **lower limit**) of (s_n) to be

$$\liminf s_n = \inf S.$$

We should note that in Definition 19.9 we require (s_n) to be bounded. Thus Theorem 19.7 implies that (s_n) contains a convergent subsequence, so the set S of subsequential limits will be nonempty. It will also be bounded since (s_n) is bounded. The completeness axiom then implies that $\sup S$ and $\inf S$ both exist as real numbers.

It should be clear that we always have $\liminf s_n \leqslant \limsup s_n$. Now, if (s_n) is convergent to some number s, then all its subsequences converge to s, so we have $\liminf s_n = \limsup s_n = s$. The converse of this is also true (Exercise 19.5). If it happens that $\liminf s_n < \limsup s_n$, then we say that (s_n) **oscillates**.

19.10 EXAMPLE Let $s_n = (-1)^n + 1/n$. We see that $|s_n| \leqslant |(-1)^n| + |1/n| \leqslant 2$ for all n, so the sequence (s_n) is bounded. The first few terms are

$$0, \frac{3}{2}, -\frac{2}{3}, \frac{5}{4}, -\frac{4}{5}, \frac{7}{6}, -\frac{6}{7}, \ldots.$$

The subsequence (s_{2n}) is seen to converge to 1 and the subsequence (s_{2n-1}) converges to -1. Since these are the only possible subsequential limits, we have $\limsup s_n = 1$ and $\liminf s_n = -1$.

There are some occasions when we wish to generalize the notion of the limit superior and the limit inferior to apply to unbounded sequences. There are two cases to consider for the lim sup, with analogous definitions applying to the lim inf.

1. Suppose that (s_n) is unbounded above. Then the proof of Theorem 19.1 implies that there exists a subsequence having $+\infty$ as its limit. This prompts us to define $\limsup s_n = +\infty$.
2. Suppose that (s_n) is bounded above but not bounded below. If some subsequence converges to a finite number, we define $\limsup s_n$ to be the supremum of the set of subsequential limits. Essentially, this coincides with Definition 19.9. If no subsequence converges to a finite number, we must have $\lim s_n = -\infty$, so we define $\limsup s_n = -\infty$.

Thus for any sequence (s_n), $\limsup s_n$ always exists as either a real number or $+\infty$ or $-\infty$. When $k \in \mathbb{R}$ and $\alpha = \limsup s_n$, then writing $\alpha > k$ means that α is a real number greater than k or that $\alpha = +\infty$. Similarly, $\alpha < k$ means α is a real number less than k or $\alpha = -\infty$. Sometimes we write $k < \alpha < +\infty$ to indicate that α is a real number greater than k and thereby explicitly rule out the possibility that $\alpha = +\infty$. The only times we shall use this extended meaning for the inequality sign is when we are referring to the value of a limit, a lim sup, or a lim inf. In all other cases, the use of an inequality implies a comparison of real numbers.

If $m = \lim \sup s_n$ is a real number, then some special properties apply. In particular, no number larger than m can be a subsequential limit of (s_n). Thus, given any $\varepsilon > 0$, there can only be finitely many terms s_n as large as $m + \varepsilon$. (If there were infinitely many terms as large as $m + \varepsilon$, then a subsequence of these terms would have a limit greater than m.) On the other hand, if we consider $m - \varepsilon$, then there must be infinitely many terms greater than $m - \varepsilon$. (For otherwise no subsequence could have a limit greater than $m - \varepsilon$, and $m - \varepsilon$ would be an upper bound for the set of subsequential limits.) We summarize these results in our next theorem.

19.11 THEOREM Let (s_n) be a sequence and suppose that $m = \lim \sup s_n$ is a real number. Then the following properties hold:

(a) For every $\varepsilon > 0$ there exists N such that $n > N$ implies that $s_n < m + \varepsilon$.
(b) For every $\varepsilon > 0$ and for every $i \in \mathbb{N}$, there exists an integer $k > i$ such that $s_k > m - \varepsilon$.

Furthermore, if m is a real number satisfying properties (a) and (b), then $m = \lim \sup s_n$.

Proof: The only thing left to prove is the final statement. If m satisfies (a), then (s_n) is bounded above and no number larger than m can be a subsequential limit. If m satisfies (b), then no number smaller than m is an upper bound for the set S of subsequential limits. Hence $m = \sup S = \lim \sup s_n$. ∎

19.12 COROLLARY Let (s_n) be a sequence and suppose that $m = \lim \sup s_n$ is a real number. Then $m \in S$, where S is the set of subsequential limits of (s_n). That is, there exists a subsequence of (s_n) that converges to m.

Proof: Taken together, parts (a) and (b) of Theorem 19.11 imply the existence of a subsequence (s_{n_k}) of (s_n) such that

$$m - \frac{1}{k} < s_{n_k} < m + \frac{1}{k}.$$

Clearly, (s_{n_k}) converges to m. ∎

19.13 PRACTICE Let $s_n = n \sin^2 (n\pi/2)$. Find the set S of subsequential limits, the lim sup, and the lim inf of (s_n).

We conclude this section with a particular result that will be useful later in working with power series (see Theorems 34.3 and 37.2).

19.14 THEOREM Suppose that (r_n) converges to a positive number r and (s_n) is a bounded sequence. Then

$$\limsup r_n s_n = r \cdot \limsup s_n.$$

Proof: Let $s = \limsup s_n$ and $t = \limsup r_n s_n$. By Corollary 19.12 there exists a subsequence (s_{n_k}) of (s_n) such that $\lim_{k\to\infty} s_{n_k} = s$. Now $\lim_{k\to\infty} r_{n_k} = r$ by Theorem 19.4, so $\lim_{k\to\infty} r_{n_k} s_{n_k} = rs$. Thus $rs \leqslant \limsup r_n s_n = t$.

Similarly, let $(r_{n_k} s_{n_k})$ be a subsequence of $(r_n s_n)$ that converges to t. Then since $t > 0$,

$$\lim_{k\to\infty} s_{n_k} = \lim_{k\to\infty} \frac{r_{n_k} s_{n_k}}{r_{n_k}} = \frac{t}{r},$$

so that $t/r \leqslant s$. That is, $t \leqslant rs$. Since $rs \leqslant t$ and $t \leqslant rs$, we conclude that $t = rs$. ∎

ANSWERS TO PRACTICE PROBLEMS

19.2 $n_k = 2^k$

19.3 Since $n_1 \in \mathbb{N}$, $n_1 \geqslant 1$. Now suppose that $n_k \geqslant k$ for some $k \in \mathbb{N}$. Then $n_{k+1} > n_k \geqslant k$, so that $n_{k+1} \geqslant k + 1$. Thus $n_k \geqslant k$ for all $k \in \mathbb{N}$.

19.13 $S = \{0, +\infty\}$, $\limsup s_n = +\infty$, $\liminf s_n = 0$

EXERCISES

19.1 For each sequence, find the set S of subsequential limits, the lim sup, and the lim inf.

(a) $s_n = (-1)^n$

(b) $(t_n) = \left(\frac{1}{2}, 1, \frac{1}{4}, \frac{1}{3}, \frac{1}{6}, \frac{1}{5}, \ldots\right)$

(c) $u_n = n^2[-1 + (-1)^n]$

(d) $v_n = n \sin \dfrac{n\pi}{2}$

19.2 For each sequence, find the set S of subsequential limits, the lim sup, and the lim inf.

(a) $w_n = \dfrac{(-1)^n}{n}$

(b) $(x_n) = (0, 1, 2, 0, 1, 3, 0, 1, 4, \ldots)$
(c) $y_n = n[2 + (-1)^n]$
(d) $z_n = (-n)^n$

19.3 Use Exercise 18.8 to find the limit of each sequence.

(a) $s_n = \left(1 + \frac{1}{2n}\right)^{2n}$ (b) $s_n = \left(1 + \frac{1}{n}\right)^{2n}$

(c) $s_n = \left(1 + \frac{1}{n}\right)^{n-1}$ (d) $s_n = \left(1 + \frac{1}{n}\right)^{-n}$

(e) $s_n = \left(1 + \frac{1}{2n}\right)^{n}$ (f) $s_n = \left(\frac{n+2}{n+1}\right)^{n+3}$

19.4 If (s_n) is a subsequence of (t_n) and (t_n) is a subsequence of (s_n), can we conclude that $(s_n) = (t_n)$? Prove or give a counterexample.

19.5 Let (s_n) be a bounded sequence and suppose that lim inf $s_n =$ lim sup $s_n = s$. Prove that (s_n) is convergent and that lim $s_n = s$.

***19.6** Suppose that $x > 1$. Prove that lim $x^{1/n} = 1$.

19.7 Let (s_n) be a bounded sequence and let S denote the set of subsequential limits of (s_n). Prove that S is closed.

19.8 Let $A = \{x \in \mathbb{Q}: 0 \leqslant x < 2\}$. Since A is denumerable, there exists a bijection $s: \mathbb{N} \to A$. Letting $s(n) = s_n$, find the set of subsequential limits of the sequence (s_n).

19.9 Let (s_n) and (t_n) be bounded sequences.
(a) Prove that lim sup $(s_n + t_n) \leqslant$ lim sup s_n + lim sup t_n.
(b) Find an example to show that equality may not hold in part (a).

19.10 State and prove the analog of Theorem 19.11 for lim inf.

19.11 Let (s_n) and (t_n) be bounded sequences.
(a) Prove that lim inf s_n + lim inf $t_n \leqslant$ lim inf $(s_n + t_n)$.
(b) Find an example to show that equality may not hold in part (a).

19.12 Let (s_n) be a bounded sequence.
(a) Prove that lim sup $s_n = \lim_{N\to\infty}$ sup $\{s_n: n > N\}$.
(b) Prove that lim inf $s_n = \lim_{N\to\infty}$ inf $\{s_n: n > N\}$.

***19.13** Prove that if lim sup $s_n = +\infty$ and $k > 0$ then lim sup $(ks_n) = +\infty$.

19.14 Let C be a nonempty subset of $\mathbb{R}$. Prove that C is compact iff every sequence in C has a subsequence that converges to a point in C.

5

Limits and Continuity

The study of limits is central to the development of analysis. We have already encountered the notion of a limit of a convergent sequence in Chapter 4. In Section 20 we discuss the limit of a function. We shall find that limits of functions are closely related to limits of sequences, and we shall exploit this relationship in our proofs. In Section 21 we introduce the concept of a continuous function and show its relationship both to limits of functions and to sequences. Some of the important properties of continuous functions are developed in Section 22, and in Section 23 we discuss uniform continuity.

Section 20 LIMITS OF FUNCTIONS

In examining the limit of a function f at a point c, we wish to know what the values of $f(x)$ are getting close to as x gets close to c. To do this, f must be defined at points arbitrarily close to c, although not necessarily at c itself. Of course, the notion of "close" must be made more precise, and in so doing we shall follow the pattern developed for limits of sequences. Throughout this chapter we will be considering functions whose domain D is a nonempty subset of $\mathbb{R}$.

20.1 DEFINITION Let $f: D \to \mathbb{R}$ and let c be an accumulation point of D. We say that a real number L is a **limit of f at** c, and we write $\lim_{x \to c} f(x) = L$, if

> for each $\varepsilon > 0$ there exists a $\delta > 0$ such that $|f(x) - L| < \varepsilon$ whenever $x \in D$ and $0 < |x - c| < \delta$.

Requiring $0 < |x - c| < \delta$ in the definition of limit has the effect of making x "close" to c but deleting the point $x = c$ from consideration. Thus the function f need not be defined at c, and even if $f(c)$ is defined, it is not necessarily equal to the limit of f at c.

If we translate the definition of a limit into the terminology of neighborhoods, we obtain the following characterization.

20.2 THEOREM Let $f: D \to \mathbb{R}$ and let c be an accumulation point of D. Then $\lim_{x \to c} f(x) = L$ iff

for each neighborhood V of L there exists a deleted neighborhood U of c such that $f(U \cap D) \subseteq V$. (See Figure 20.1.)

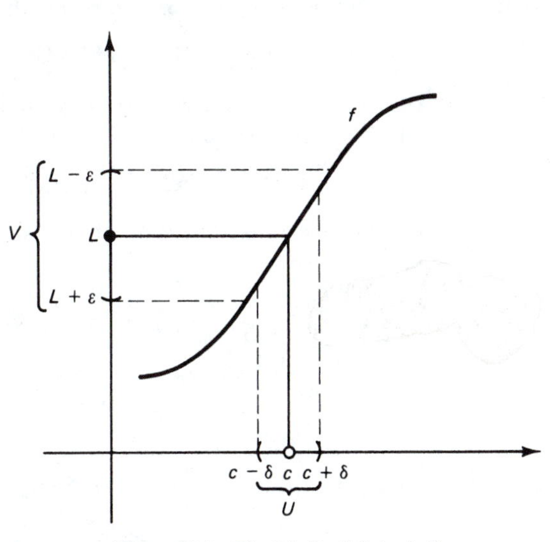

Figure 20.1 The Limit of f at c is L

20.3 EXAMPLE Let $k \in \mathbb{R}$. Define the constant function $f: \mathbb{R} \to \mathbb{R}$ by $f(x) = k$ for all $x \in \mathbb{R}$. Then for any $c \in \mathbb{R}$, $\lim_{x \to c} f(x) = k$. Indeed, given any $\varepsilon > 0$, let $\delta = 1$ (or any other positive number). Then $|f(x) - k| = |k - k| = 0 < \varepsilon$ whenever $0 < |x - c| < 1$. Thus $\lim_{x \to c} k = k$.

20.4 PRACTICE Consider the function $f(x) = x$ for all $x \in \mathbb{R}$. We claim that $\lim_{x \to c} f(x) = c$ for all $c \in \mathbb{R}$. Given any $\varepsilon > 0$, if we want to make $|f(x) - c| < \varepsilon$, how big can δ be?

20.5 EXAMPLE Let $f(x) = (2x^2 - 3x + 1)/(x - 1)$ for $x \neq 1$ and $f(1) = 5$. To find $\lim_{x \to 1} f(x)$, we note that

$$\frac{2x^2 - 3x + 1}{x - 1} = \frac{(2x - 1)(x - 1)}{x - 1} = 2x - 1$$

when $x \neq 1$. Thus as x approaches 1 we guess that $f(x)$ is approaching $2(1) - 1 = 1$. To prove this, given any $\varepsilon > 0$, let $\delta = \varepsilon/2$. Then, whenever $0 < |x - 1| < \varepsilon/2$, we have

$$|f(x) - 1| = \left| \frac{2x^2 - 3x + 1}{x - 1} - 1 \right| = |(2x - 1) - 1|$$

$$= |2(x - 1)| = 2|x - 1| < 2\left(\frac{\varepsilon}{2}\right) = \varepsilon.$$

Thus we have $\lim_{x \to 1} f(x) = 1$. Notice that having $f(1) = 5$ did not affect the value of the limit at 1 at all.

20.6 EXAMPLE Let $f(x) = x^2 + 2x + 6$. To prove that $\lim_{x \to 3} f(x) = 21$, we write

$$|f(x) - 21| = |x^2 + 2x - 15| = |x + 5||x - 3|.$$

To make this small, we need a bound on the size of $|x + 5|$ when x is "close" to 3. For example, if we arbitrarily require that $|x - 3| < 1$, then

$$|x + 5| = |x - 3 + 8| \leqslant |x - 3| + |8| < 1 + 8 = 9.$$

To make $f(x)$ within ε of 21, we shall want to have $|x + 5| < 9$ and $|x - 3| < \varepsilon/9$.

Thus, given any $\varepsilon > 0$, let $\delta = \min\{1, \varepsilon/9\}$. Then for all x satisfying $|x - 3| < \delta$ we have $|x - 3| < 1$ so that $|x + 5| < 9$. It follows that for these x we have

$$|f(x) - 21| = |x + 5||x - 3| < 9|x - 3| < 9\delta \leqslant \varepsilon.$$

20.7 PRACTICE Find a $\delta > 0$ so that $|x - 2| < \delta$ implies that

(a) $|x^2 + x - 6| < 1$.
(b) $|x^2 + x - 6| < 1/n$ for a given $n \in \mathbb{N}$.
(c) $|x^2 + x - 6| < \varepsilon$ for a given $\varepsilon > 0$.

Sequential Criterion for Limits

We now present the key theorem that shows the relationship between limits of functions and limits of sequences.

20.8 THEOREM Let $f\colon D \to \mathbb{R}$ and let c be an accumulation point of D. Then $\lim_{x \to c} f(x) = L$ iff

for every sequence (s_n) in D that converges to c with $s_n \neq c$ for all n, the sequence $(f(s_n))$ converges to L.

Proof: Suppose that $\lim_{x \to c} f(x) = L$ and let (s_n) be a sequence in D that converges to c with $s_n \neq c$, for all n. We must show that $\lim_{n\to\infty} f(s_n) = L$. Now, given any $\varepsilon > 0$, there exists $\delta > 0$ such that $|f(x) - L| < \varepsilon$ whenever $x \in D$ and $0 < |x - c| < \delta$. Furthermore, since $s_n \to c$, there exists N such that $n > N$ implies that $|s_n - c| < \delta$. Thus for $n > N$ we have $0 < |s_n - c| < \delta$ and $s_n \in D$, so that $|f(s_n) - L| < \varepsilon$. Hence $\lim_{n\to\infty} f(s_n) = L$.

Conversely, suppose that L is not a limit of f at c. We must find a sequence (s_n) in D that converges to c with each $s_n \neq c$, and such that $(f(s_n))$ does not converge to L. Since L is not a limit of f at c, there exists $\varepsilon > 0$ such that for every $\delta > 0$ there exists $x \in D$ with $0 < |x - c| < \delta$ such that $|f(x) - L| \geqslant \varepsilon$. In particular, for each $n \in \mathbb{N}$, there exists $s_n \in D$ with $0 < |s_n - c| < 1/n$ such that $|f(s_n) - L| \geqslant \varepsilon$. Now the sequence (s_n) converges to c with $s_n \neq c$ for all n, but $(f(s_n))$ cannot converge to L. ▮

Using Theorem 20.8 and our earlier results on sequences, we conclude that the limit of a function is unique.

20.9 COROLLARY If $f: D \to \mathbb{R}$ and if c is an accumulation point of D, then f can have only one limit at c.

Proof: Exercise 20.6. ▮

One very useful application of the sequential criterion for the limit of a function is to show that a given limit does *not* exist.

20.10 THEOREM Let $f: D \to \mathbb{R}$ and let c be an accumulation point of D. Then the following are equivalent:

(a) f does not have a limit at c.
(b) There exists a sequence (s_n) in D with each $s_n \neq c$ such that (s_n) converges to c, but $(f(s_n))$ is not convergent in $\mathbb{R}$.

Proof: Exercise 20.7. ▮

20.11 EXAMPLE Consider the function $f(x) = \sin(1/x)$ for $x > 0$. (See Figure 20.2.) Using Theorem 20.10, we can show that $\lim_{x\to 0} f(x)$ does not exist. Recall that for all $k \in \mathbb{N}$ we have

$$\sin\left(2\pi k + \frac{n\pi}{2}\right) = \begin{cases} 0, & \text{if } n = 0 \\ 1, & \text{if } n = 1 \\ 0, & \text{if } n = 2 \\ -1, & \text{if } n = 3. \end{cases}$$

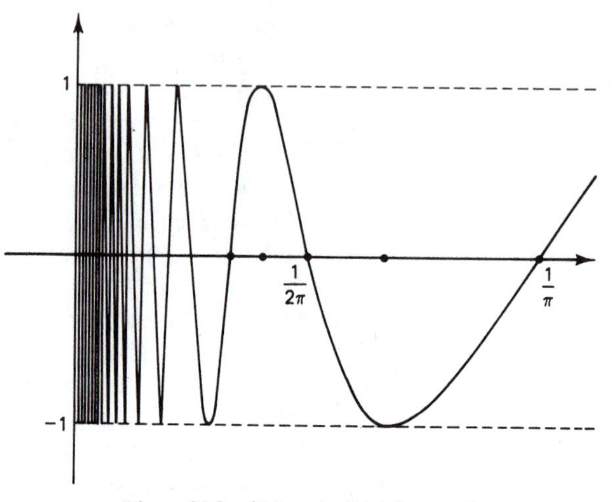

Figure 20.2 $f(x) = \sin(1/x)$ for $x > 0$

Thus, if we let $s_n = 2/(n\pi)$ for all $n \in \mathbb{N}$, then $\lim_{n\to\infty} s_n = 0$. But $(f(s_n))$ is the sequence $(0, 1, 0, -1, 0, 1, 0, -1, \ldots)$, which clearly does not converge.

By combining Theorem 20.8 with the limit theorems for sequences, we can derive corresponding results for limits of functions. Our next theorem is the counterpart of Theorem 17.1, and others are included in the exercises.

20.12 DEFINITION Let $f: D \to \mathbb{R}$ and $g: D \to \mathbb{R}$. We define the **sum** $f + g$ and the **product** fg to be the functions from D to $\mathbb{R}$ given by

$$(f + g)(x) = f(x) + g(x) \quad \text{and} \quad (fg)(x) = f(x)g(x)$$

for all $x \in D$. If $k \in \mathbb{R}$, then the **multiple** $kf: D \to \mathbb{R}$ is the function defined by

$$(kf)(x) = k \cdot f(x), \qquad \text{for all } x \in D.$$

If $g(x) \neq 0$ for all $x \in D$, then the **quotient** $f/g: D \to \mathbb{R}$ is the function defined by

$$\left(\frac{f}{g}\right)(x) = \frac{f(x)}{g(x)}, \qquad \text{for all } x \in D.$$

20.13 THEOREM Let $f: D \to \mathbb{R}$ and $g: D \to \mathbb{R}$, and let c be an accumulation point of D. If $\lim_{x\to c} f(x) = L$, $\lim_{x\to c} g(x) = M$, and $k \in \mathbb{R}$, then

$$\lim_{x\to c} (f + g)(x) = L + M, \qquad \lim_{x\to c} (fg)(x) = LM,$$

$$\text{and} \quad \lim_{x\to c} (kf)(x) = kL.$$

Furthermore, if $g(x) \neq 0$ for all $x \in D$ and $M \neq 0$, then

$$\lim_{x \to c} \left(\frac{f}{g}\right)(x) = \frac{L}{M}.$$

Proof: The proofs of the various formulas are all similar to each other. We shall prove the formula for the sum and leave the others to the reader (Exercise 20.8). Let (s_n) be a sequence in D that converges to c with each $s_n \neq c$. By Theorem 20.8 we have

$$\lim_{n \to \infty} f(s_n) = L \quad \text{and} \quad \lim_{n \to \infty} g(s_n) = M,$$

and it suffices to show that $\lim_{n \to \infty}(f + g)(s_n) = L + M$. Now from Definition 20.12 and Theorem 17.1 we obtain

$$\begin{aligned} \lim_{n \to \infty} (f + g)(s_n) &= \lim_{n \to \infty} [f(s_n) + g(s_n)] \\ &= \lim_{n \to \infty} f(s_n) + \lim_{n \to \infty} g(s_n) \\ &= L + M. \quad \blacksquare \end{aligned}$$

20.14 EXAMPLE Since $\lim_{x \to c} x = c$, it follows from the product formula that $\lim_{x \to c} x^2 = c^2$. By induction we easily obtain $\lim_{x \to c} x^n = c^n$ for all $n \in \mathbb{N}$. Combining this with the sum and multiple formulas, we see that for any polynomial P and for any $c \in \mathbb{R}$ we have $\lim_{x \to c} P(x) = P(c)$.

20.15 PRACTICE Evaluate $\displaystyle\lim_{x \to 1} \frac{x^2 + 2x - 5}{x^2 + 3x - 5}$.

20.16 EXAMPLE Let $f(x) = (2x^2 - 3x + 1)/(x - 1)$ for $x \neq 1$ and $f(1) = 5$, as in Example 20.5. Since $\lim_{x \to 1} (x - 1) = 0$, we cannot use the quotient formula directly, but once we have simplified $f(x)$ to $2x - 1$ for $x \neq 1$, then we can use Theorem 20.13 to obtain

$$\lim_{x \to 1} f(x) = \lim_{x \to 1} (2x - 1) = 2(1) - 1 = 1.$$

One-Sided Limits

If the domain of f is an interval (a, b), then $\lim_{x \to a} f(x)$ will only involve points x that are close to a and greater than a. We sometimes indicate this by writing $\lim_{x \to a+} f(x)$, which is called the **right-hand limit** of f at a. More precisely, $\lim_{x \to a+} f(x) = L$ iff for every $\varepsilon > 0$ there exists $\delta > 0$ such that $|f(x) - L| < \varepsilon$ whenever $x \in (a, b)$ and $a < x < a + \delta$.

Similarly, the **left-hand limit** of f at b is given by $\lim_{x \to b^-} f(x) = L$ iff for every $\varepsilon > 0$ there exists $\delta > 0$ such that $|f(x) - L| < \varepsilon$ whenever $x \in (a, b)$ and $b - \delta < x < b$. Occasionally, we may use the one-sided limit notation when a function is defined throughout a deleted neighborhood of some point c, but we wish to consider values on only one side of c. Of course, $\lim_{x \to c} f(x) = L$ iff both one-sided limits exist and are equal to L. (See Exercise 20.16.)

ANSWERS TO PRACTICE PROBLEMS

20.4 Let $\delta = \varepsilon$. Then, if $0 < |x - c| < \delta$, we have $|f(x) - c| = |x - c| < \delta = \varepsilon$. Thus $\lim_{x \to c} x = c$.

20.7 First note that $|x^2 + x - 6| = |x + 3||x - 2|$. Thus we need to have an upper bound on the size of $|x + 3|$. Now if $|x - 2| < 1$ then $|x + 3| = |x - 2 + 5| \leqslant |x - 2| + |5| < 6$, so that $|x^2 + x - 6| = |x + 3||x - 2| < 6|x - 2|$. Thus in part (a) we take $\delta = 1/6$ and in part (b) we let $\delta = 1/(6n)$. In both cases $\delta \leqslant 1$, so the inequality $|x + 3| < 6$ applies. For part (c), to ensure that $\delta \leqslant 1$, we set $\delta = \min\{1, \varepsilon/6\}$.

20.15 Using Example 20.14 and the quotient formula, we have

$$\lim_{x \to 1} \frac{x^2 + 2x - 5}{x^2 + 3x - 5} = \frac{-2}{-1} = 2.$$

EXERCISES

20.1 Determine the following limits.

(a) $\displaystyle\lim_{x \to 1} \frac{x^3 + 5}{x^2 + 2}$ (b) $\displaystyle\lim_{x \to 1} \frac{x^2 + 2x - 3}{x^2 - 1}$

(c) $\displaystyle\lim_{x \to 1} \frac{\sqrt{x} - 1}{x - 1}$ (d) $\displaystyle\lim_{x \to 0} \frac{x^2 + 4x}{x^2 + 2x}$

(e) $\displaystyle\lim_{x \to 0} \frac{x^2 + 3x}{x^2 + 1}$ (f) $\displaystyle\lim_{x \to 0} \frac{\sqrt{4 + x} - 2}{x}$

(g) $\displaystyle\lim_{x \to 0^-} \frac{4x}{|x|}$ (h) $\displaystyle\lim_{x \to 1^+} \frac{x^2 - 1}{|x - 1|}$

20.2 Use definition 20.1 to prove each limit.

(a) $\lim_{x \to 5} x^2 - 3x + 1 = 11$

(b) $\lim_{x \to -2} x^2 + 2x + 7 = 7$

(c) $\lim_{x \to 2} x^3 = 8$

20.3 Find the following limits and prove your answers.

(a) $\displaystyle\lim_{x \to 0} |x|$

(b) $\lim_{x \to 0} x^2/|x|$

(c) $\lim_{x \to c} \sqrt{x}$, where $c \geq 0$

20.4 Let $f: D \to \mathbb{R}$ and let c be an accumulation point of D. Suppose that $\lim_{x \to c} f(x) = L$.

(a) Prove that $\lim_{x \to c} |f(x)| = |L|$.

(b) If $f(x) \geqslant 0$ for all $x \in D$, prove that $\lim_{x \to c} \sqrt{f(x)} = \sqrt{L}$.

20.5 Determine whether or not the following limits exist. Justify your answers.

(a) $\lim_{x \to 0+} \dfrac{1}{x}$

(b) $\lim_{x \to 0+} \left| \sin \dfrac{1}{x} \right|$

(c) $\lim_{x \to 0+} x \sin \dfrac{1}{x}$

20.6 Prove Corollary 20.9

(a) by using Definition 20.1,

(b) by using Theorems 20.8 and 16.14.

20.7 Prove Theorem 20.10.

20.8 Finish the proof of Theorem 20.13.

20.9 Let f, g and h be functions from D into $\mathbb{R}$, and let c be an accumulation point of D. Suppose that $f(x) \leqslant g(x) \leqslant h(x)$, for all $x \in D$ with $x \neq c$, and suppose that $\lim_{x \to c} f(x) = \lim_{x \to c} h(x) = L$. Prove that $\lim_{x \to c} g(x) = L$.

20.10 Let $f: D \to \mathbb{R}$ and let c be an accumulation point of D. Suppose that $a \leqslant f(x) \leqslant b$ for all $x \in D$ with $x \neq c$, and suppose that $\lim_{x \to c} f(x)$ exists. Prove that $a \leqslant \lim_{x \to c} f(x) \leqslant b$.

20.11 Let f and g be functions from D into $\mathbb{R}$ and let c be an accumulation point of D. Suppose that there exists a neighborhood U of c and a real number M such that $|g(x)| \leqslant M$ for all $x \in U \cap D$. If $\lim_{x \to c} f(x) = 0$, prove that $\lim_{x \to c} (fg)(x) = 0$.

***20.12** Let f: $D \to \mathbb{R}$ and let c be an accumulation point of D. Suppose that $\lim_{x \to c} f(x) > 0$. Prove that there exists a deleted neighborhood U of c such that $f(x) > 0$ for all $x \in U \cap D$.

20.13 Define $f: \mathbb{R} \to \mathbb{R}$ by $f(x) = x$ if x is rational and $f(x) = 0$ if x is irrational. Show that f has a limit at c iff $c = 0$.

***20.14** Let $f: D \to \mathbb{R}$ and let c be an accumulation point of D. Suppose that f has a limit at c. Prove that f is bounded on a neighborhood of c. That is, prove that there exists a neighborhood U of c and a real number M such that $|f(x)| \leqslant M$ for all $x \in U \cap D$.

20.15 Suppose that $f: \mathbb{R} \to \mathbb{R}$ is a function such that $f(x + y) = f(x) + f(y)$ for all $x, y \in \mathbb{R}$. Prove that f has a limit at 0 iff f has a limit at every point c in $\mathbb{R}$.

20.16 Let f be a function defined on a deleted neighborhood of a point c. Then $\lim_{x \to c} f(x) = L$ iff $\lim_{x \to c+} f(x) = L$ and $\lim_{x \to c-} f(x) = L$.

Section 21 CONTINUOUS FUNCTIONS

In the preceding section we saw that the limit of a function f at a point c is independent of the nature of the function *at c*. It may be that $f(c)$ does not exist, and even if it does exist it may differ from the value of the limit. When it happens that the limit of f at c is equal to $f(c)$, the function is said to be continuous at c.

In this section we give a precise "ε–δ" definition of continuity and then show its relationship to limits, sequences, and neighborhoods. We also show that algebraic combinations of continuous functions are continuous. As usual, the functions will all be defined on a nonempty subset D of $\mathbb{R}$.

21.1 DEFINITION Let $f: D \to \mathbb{R}$ and let $c \in D$. We say that f is **continuous at** c if

for every $\varepsilon > 0$ there exists a $\delta > 0$ such that $|f(x) - f(c)| < \varepsilon$ whenever $|x - c| < \delta$ and $x \in D$.

If f is continuous at each point of a subset S of D, then f is said to be **continuous on** S. If f is continuous on its domain D, then f is said to be **continuous.**

Notice that the definition of continuity at a point c requires c to be in D, but it does not require c to be an accumulation point of D. Thus the notion of continuity is slightly more general in application than the notion of a limit. Actually, the difference is not as significant as it might seem, since if c is an isolated point of D, then f is automatically continuous at c. (Recall that a point of D that is not an accumulation point of D is called an isolated point of D.) Indeed, if c is an isolated point of D, then there exists $\delta > 0$ such that, if $|x - c| < \delta$ and $x \in D$, then $x = c$. Thus whenever $|x - c| < \delta$ and $x \in D$, we have

$$|f(x) - f(c)| = 0 < \varepsilon$$

for all $\varepsilon > 0$. Hence f is continuous at c.

21.2 THEOREM Let f: $D \to \mathbb{R}$ and let $c \in D$. Then the following three conditions are equivalent:

(a) f is continuous at c.
(b) If (x_n) is any sequence in D such that (x_n) converges to c, then $\lim f(x_n) = f(c)$.
(c) For every neighborhood V of $f(c)$ there exists a neighborhood U of c such that $f(U \cap D) \subseteq V$.

Furthermore, if c is an accumulation point of D, then the above are all equivalent to

(d) f has a limit at c and $\lim_{x \to c} f(x) = f(c)$.

Proof: Suppose first that c is an isolated point of D. Then there exists a neighborhood U of c such that $U \cap D = \{c\}$. It follows that, for any neighborhood V of $f(c)$, $f(U \cap D) = \{f(c)\} \subseteq V$. Thus (c) always holds. Similarly, by Exercise 16.12, if (x_n) is a sequence in D converging to c, then $x_n \in U$ for all n greater than some M. But this implies that $x_n = c$ for $n > M$, so $\lim f(x_n) = f(c)$. Thus (b) also holds. We have already observed that (a) applies, so (a), (b), and (c) are all equivalent.

Now suppose that c is an accumulation point of D. Then (a)$\Leftrightarrow$(d) is Definition 20.1, (d)$\Leftrightarrow$(c) is Theorem 20.2, and (d)$\Leftrightarrow$(b) is essentially Theorem 20.8. ▮

21.3 EXAMPLE Let p be a polynomial. We observed in Example 20.14 that, for any $c \in \mathbb{R}$, $\lim_{x \to c} p(x) = p(c)$. It follows that p is continuous on $\mathbb{R}$.

21.4 PRACTICE Illustrate the preceding example by using the limit formulas in Theorem 20.13 to show that for any $c \in \mathbb{R}$ the polynomial $p(x) = 3x^2 - 2x + 5$ is continuous at c.

21.5 EXAMPLE Let $f(x) = x \sin(1/x)$ for $x \neq 0$ and $f(0) = 0$. From the graph in Figure 21.1 it appears that f may be continuous at 0. Let us prove that it actually is. Since

$$|f(x) - f(0)| = \left| x \sin \frac{1}{x} \right| \leqslant |x|, \qquad \text{for all } x,$$

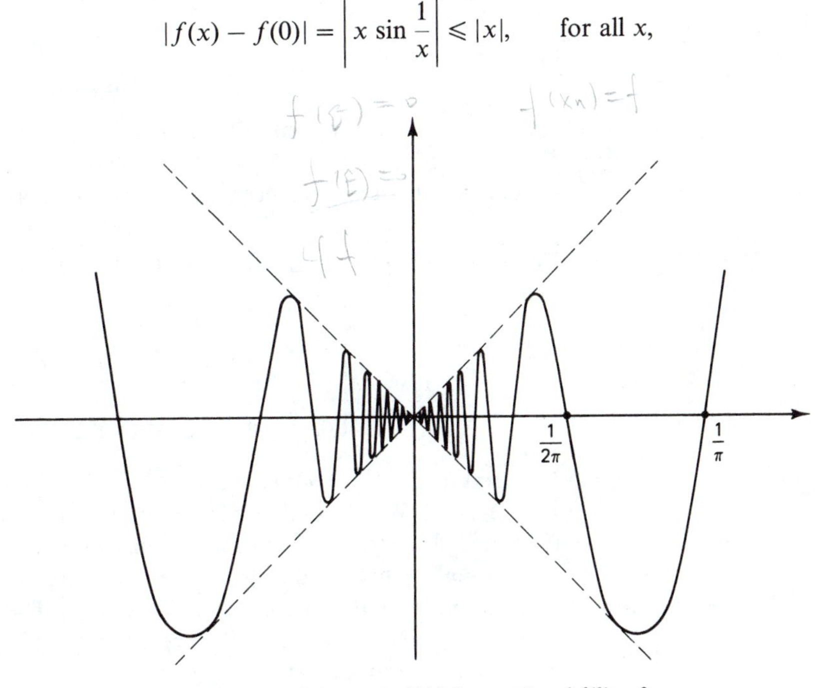

Figure 21.1 $f(x) = x \sin(1/x)$ for $x \neq 0$ and $f(0) = 0$

given $\varepsilon > 0$ we may let $\delta = \varepsilon$. Then when $|x - 0| < \delta$ we have $|f(x) - f(0)| \leqslant |x| < \delta = \varepsilon$. Hence f is continuous at 0. We shall return to this example later to see that f is continuous on $\mathbb{R}$.

By taking the negation of (a) and (b) in Theorem 21.2, we obtain the following useful characterization of discontinuity.

21.6 THEOREM Let $f: D \to \mathbb{R}$ and let $c \in D$. Then f is discontinuous at c iff there exists a sequence (x_n) in D such that (x_n) converges to c but the sequence $(f(x_n))$ does not converge to $f(c)$.

21.7 EXAMPLE Let $f(x) = 1/x$ for $x \in D = (-\infty, 0) \cup (0, \infty)$. Since $\lim_{x \to c} 1/x = 1/c$ for all $c \in D$ by Theorem 20.13, f is continuous on D. But f is not continuous on $\mathbb{R}$ for two reasons. In the first place, f is not defined at 0 so it cannot possibly be continuous there. In the second place, even if we were to define $f(0) = k$ for some $k \in \mathbb{R}$ (as we did in Example 21.5), then f would still not be continuous at 0. Indeed, since $1/n \to 0$ and $\lim f(1/n) = +\infty$, the sequence $(f(1/n))$ is not convergent. Thus there is no way to define f at 0 to make it continuous there.

21.8 EXAMPLE To obtain a function that is discontinuous at every real number, we define $f: \mathbb{R} \to \mathbb{R}$ by

$$f(x) = \begin{cases} 1, & \text{if } x \text{ is rational} \\ 0, & \text{if } x \text{ is irrational.} \end{cases}$$

If $c \in \mathbb{R}$, then every neighborhood of c contains rational points at which $f(x) = 1$ and also irrational points at which $f(x) = 0$. Thus $\lim_{x \to c} f(x)$ cannot possibly exist. Hence f is discontinuous at each $c \in \mathbb{R}$.

21.9 EXAMPLE For a more exotic example, let $f: (0, 1) \to \mathbb{R}$ be the Dirichlet function defined by

$$f(x) = \begin{cases} \dfrac{1}{n}, & \text{if } x = \dfrac{m}{n} \text{ is rational in lowest terms} \\ 0, & \text{if } x \text{ is irrational.} \end{cases}$$

For example, $f(\frac{1}{3}) = f(\frac{2}{3}) = \frac{1}{3}$, $f(\frac{3}{5}) = \frac{1}{5}$, $f(\sqrt{2}/5) = 0$, and so on. (See Figure 21.2.) We claim that f is continuous at each irrational number in $(0, 1)$ and discontinuous at each rational in $(0, 1)$.

Suppose that c is a rational number in $(0, 1)$. Let (x_n) be a sequence of irrationals in $(0, 1)$ that converges to c. Then $f(x_n) = 0$ for all n, so $\lim f(x_n) = 0$. But $f(c) > 0$ since c is rational. Thus $\lim f(x_n) \neq f(c)$ and f is discontinuous at c.

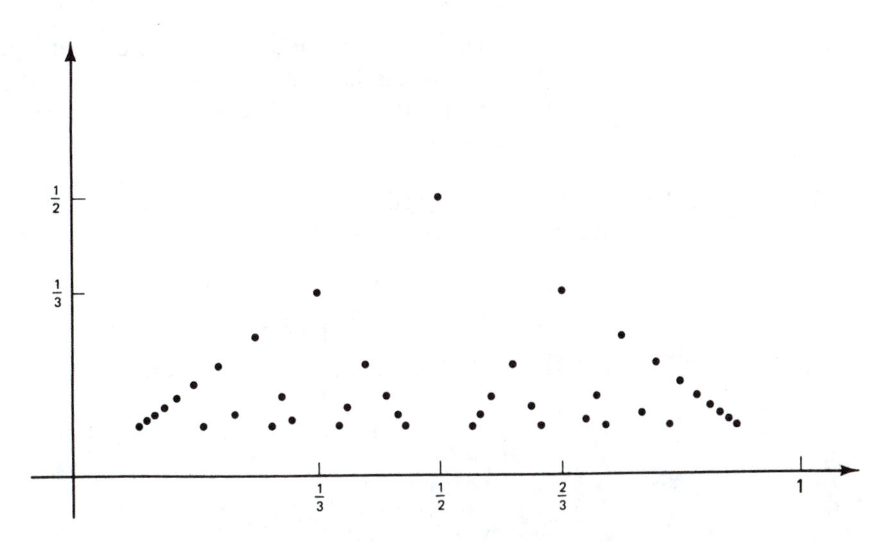

Figure 21.2 Part of the Dirichlet Function

On the other hand, suppose that d is an irrational number in $(0, 1)$. Given any $\varepsilon > 0$, by the Archimedean property (Theorem 12.10) there exists $k \in \mathbb{N}$ such that $1/k < \varepsilon$. Now there are only a finite number of rationals in $(0, 1)$ whose denominators are less than k. Thus there exists $\delta > 0$ such that all the rationals in $(d - \delta, d + \delta)$ have a denominator (in lowest terms) greater than or equal to k. It follows that if $x \in (0, 1)$ and $|x - d| < \delta$ then $|f(x) - f(d)| = |f(x)| \leqslant 1/k < \varepsilon$. Hence f is continuous at d.

Since continuity can be described in terms of limits, and limits are preserved under algebraic operations, our next theorem should come as no surprise.

21.10 THEOREM Let f and g be functions from D to $\mathbb{R}$, and let $c \in D$. Suppose that f and g are continuous at c. Then

(a) $f + g$ and fg are continuous at c,
(b) f/g is continuous at c if $g(c) \neq 0$.

Proof: Let (x_n) be a sequence in D converging to c. To show that $f + g$ is continuous at c it suffices by Theorem 21.2 to show that $\lim (f + g)(x_n) = (f + g)(c)$. From Definition 20.12 and Theorem 17.1, we have

$$\begin{aligned}\lim (f + g)(x_n) &= \lim [f(x_n) + g(x_n)] \\ &= \lim f(x_n) + \lim g(x_n) \\ &= f(c) + g(c) = (f + g)(c).\end{aligned}$$

The proofs for the products and quotient are similar, the only difference being that for f/g we have to choose the sequence (x_n) so that $g(x_n) \neq 0$ for all n. Recall that the quotient is defined only at those points $x \in D$ for which $g(x) \neq 0$. ∎

21.11 EXAMPLE Let f and g be functions from D to $\mathbb{R}$ and define the functions $\max(f, g)$ and $\min(f, g)$ from D to $\mathbb{R}$ by

$$\max(f, g)(x) = \max\{f(x), g(x)\},$$
$$\min(f, g)(x) = \min\{f(x), g(x)\}.$$

In Exercise 21.9 you are asked to show that

$$\max(f, g) = \frac{1}{2}(f + g) + \frac{1}{2}|f - g|,$$

$$\min(f, g) = \frac{1}{2}(f + g) - \frac{1}{2}|f - g|.$$

In Exercise 21.8 you are asked to show that $|f|$ is continuous wherever f is continuous. Using these results and Theorem 21.10, we conclude that $\max(f, g)$ and $\min(f, g)$ are continuous wherever f and g are both continuous.

For our final theorem in this section, we show that the composition of two continuous functions is continuous. The proof is particularly easy when written using the terminology of neighborhoods.

21.12 THEOREM Let $f: D \to \mathbb{R}$ and $g: E \to \mathbb{R}$ be functions such that $f(D) \subseteq E$. If f is continuous at a point $c \in D$ and g is continuous at $f(c)$, then the composition $g \circ f: D \to \mathbb{R}$ is continuous at c.

Proof: Let $e = f(c)$ and let W be any neighborhood of $g(e)$. Since g is continuous at e, there exists a neighborhood V of e such that $g(V \cap E) \subseteq W$. Since $e = f(c)$ and f is continuous at c, there exists a neighborhood U of c such that $f(U \cap D) \subseteq V$. Since $f(D) \subseteq E$, we have $f(U \cap D) \subseteq (V \cap E)$. It follows that $g(f(U \cap D)) \subseteq W$, so $g \circ f$ is continuous at c by Theorem 21.2. ∎

21.13 EXAMPLE Let $g(x) = \sin x$, $h(x) = 1/x$, and $i(x) = x$. Then for all $x \neq 0$ we have

$$x \sin \frac{1}{x} = [(i)(g \circ h)](x).$$

If we assume that $\sin x$ is continuous for all x, then using Theorems 21.10 and 21.12 we conclude that $x \sin(1/x)$ is continuous for all $x \neq 0$. It follows from Example 21.5 that the function f defined by $f(x) = x \sin(1/x)$ for $x \neq 0$ and $f(0) = 0$ is continuous on $\mathbb{R}$.

ANSWERS TO PRACTICE PROBLEMS

21.4 Suppose that $x_n \to c$. Then by Theorem 20.13 we have

$$\begin{aligned}\lim p(x_n) &= \lim\,[3x_n^2 + 2x_n + 5]\\ &= 3[\lim x_n]^2 - 2[\lim x_n] + 5\\ &= 3c^2 - 2c + 5 = p(c).\end{aligned}$$

Thus, by Theorem 21.2, p is continuous at c.

EXERCISES

21.1 Let $f(x) = (x^2 - 4x - 5)/(x - 5)$ for $x \neq 5$. How should $f(5)$ be defined so that f will be continuous at 5?

21.2 Define $f\colon \mathbb{R} \to \mathbb{R}$ by $f(x) = x^2 - 3x + 5$. Use Definition 21.1 to prove that f is continuous at 2.

21.3 Find an example of a function $f\colon \mathbb{R} \to \mathbb{R}$ that is continuous at exactly one point.

21.4 Find an example of a function $f\colon \mathbb{R} \to \mathbb{R}$ that is discontinuous at every real number but such that f^2 is continuous on $\mathbb{R}$.

21.5 Prove or give a counterexample: Every sequence of real numbers is a continuous function.

21.6 Consider the formula

$$f(x) = \lim_{n\to\infty} \frac{x^n}{1 + x^n}.$$

Let $D = \{x\colon f(x) \in \mathbb{R}\}$. Calculate $f(x)$ for all $x \in D$ and determine where $f\colon D \to \mathbb{R}$ is continuous.

21.7 Define $f\colon \mathbb{R} \to \mathbb{R}$ by $f(x) = 5x$ if x is rational and $f(x) = x^2 + 6$ if x is irrational. Prove that f is discontinuous at 1 and continuous at 2. Are there any other points besides 2 at which f is continuous?

***21.8** Let $f\colon D \to \mathbb{R}$ and define $|f|\colon D \to \mathbb{R}$ by $|f|(x) = |f(x)|$. Suppose that f is continuous at $c \in D$. Prove that $|f|$ is continuous at c.

***21.9** Define max (f, g) and min (f, g) as in Example 21.11. Show that

$$\max(f, g) = \frac{1}{2}(f + g) + \frac{1}{2}|f - g| \text{ and } \min(f, g) = \frac{1}{2}(f + g) - \frac{1}{2}|f - g|.$$

21.10 Let $f\colon D \to \mathbb{R}$ and suppose that $f(x) \geqslant 0$ for all $x \in D$. Define $\sqrt{f}\colon D \to \mathbb{R}$ by $\sqrt{f}(x) = \sqrt{f(x)}$. If f is continuous at $c \in D$, prove that $\sqrt{f}$ is continuous at c.

***21.11** Let $f\colon D \to \mathbb{R}$ be continuous at $c \in D$ and suppose that $f(c) > 0$. Prove that there exists $\alpha > 0$ and a neighborhood U of c such that $f(x) > \alpha$ for all $x \in U \cap D$.

21.12 Let $f\colon D \to \mathbb{R}$ be continuous at $c \in D$. Prove that there exists $M > 0$ and a neighborhood U of c such that $|f(x)| \leqslant M$ for all $x \in U \cap D$.

21.13 Let $f: \mathbb{R} \to \mathbb{R}$ be a continuous function and let $k \in \mathbb{R}$. Prove that the set $f^{-1}(\{k\})$ is closed.

21.14 Suppose that $f: (a, b) \to \mathbb{R}$ is continuous and that $f(r) = 0$ for every rational number $r \in (a, b)$. Prove that $f(x) = 0$ for all $x \in (a, b)$.

21.15 Suppose that $f: \mathbb{R} \to \mathbb{R}$ is a continuous function such that $f(x + y) = f(x) + f(y)$ for all $x, y \in \mathbb{R}$. Prove that there exists $k \in \mathbb{R}$ such that $f(x) = kx$, for every $x \in \mathbb{R}$.

Section 22 PROPERTIES OF CONTINUOUS FUNCTIONS

In this section we develop a number of the important properties of continuous functions. One of these, the intermediate value property, is probably familiar to the reader from calculus, where it is usually presented without proof. Once again, we shall consider functions whose domain D is a nonempty subset of $\mathbb{R}$.

A function $f: D \to \mathbb{R}$ is said to be **bounded** if its range $f(D)$ is a bounded subset of $\mathbb{R}$. That is, f is bounded if there exists $M \in \mathbb{R}$ such that $|f(x)| \leqslant M$ for all $x \in D$. Unfortunately, a continuous function may not be bounded even when its domain is bounded.

22.1 PRACTICE Let $D = (0, 1)$. Find a continuous function that is not bounded on D.

If it happens, however, that the domain of a continuous function is both closed and bounded, then the function will be bounded. In fact, we can prove the following stronger result. Recall from the Heine–Borel theorem (14.5) that a subset of $\mathbb{R}$ is compact iff it is closed and bounded.

22.2 THEOREM Let D be a compact subset of $\mathbb{R}$ and suppose that $f: D \to \mathbb{R}$ is continuous. Then $f(D)$ is compact.

Proof: From the Heine–Borel theorem it suffices to show that $f(D)$ is closed and bounded. To this end, suppose that $f(D)$ is not bounded. Then for each $n \in \mathbb{N}$ there exists a point $x_n \in D$ such that $|f(x_n)| > n$. Since D is bounded, the Bolzano–Weierstrass theorem (19.6) implies that the sequence (x_n) has a subsequence (x_{n_k}) converging to some $x_0 \in \mathbb{R}$. Since D is closed, we must have $x_0 \in D$. (See Exercise 16.11.) Since f is therefore continuous at x_0, $f(x_{n_k})$ converges to $f(x_0)$. In particular, $f(x_{n_k})$ must be bounded. But this contradicts the fact that $|f(x_{n_k})| > n_k \geqslant k$ for all $k \in \mathbb{N}$. We conclude that $f(D)$ is bounded.

It remains to show that $f(D)$ is closed. Let (y_n) be a convergent sequence in $f(D)$ and let $y_0 = \lim y_n$. It suffices (Exercise 16.11) to show that $y_0 \in f(D)$. Since $y_n \in f(D)$ for each n, there exists a sequence (x_n) in D such that $f(x_n) = y_n$ for all n. Since D is bounded and closed, it follows as in the first part of the proof that there exists a subsequence (x_{n_k}) converging to some point x_0 in D. Since f is continuous at x_0, we have

$$f(x_0) = \lim_{k \to \infty} f(x_{n_k}) = \lim_{k \to \infty} (y_{n_k}).$$

But (y_{n_k}) is a subsequence of (y_n), so $\lim_{k \to \infty} y_{n_k} = y_0$. Thus $f(x_0) = y_0$ and $y_0 \in f(D)$. ∎

22.3 COROLLARY Let D be a compact subset of $\mathbb{R}$ and suppose that $f: D \to \mathbb{R}$ is continuous. Then f assumes minimum and maximum values on D. That is, there exist points x_1 and x_2 in D such that $f(x_1) \leqslant f(x) \leqslant f(x_2)$ for all $x \in D$.

Proof: We know from Theorem 22.2 that $f(D)$ is compact. Thus (Lemma 14.4) $f(D)$ has both a minimum, say y_1, and a maximum, say y_2. Since $y_1, y_2 \in f(D)$, there exist $x_1, x_2 \in D$ such that $f(x_1) = y_1$ and $f(x_2) = y_2$. It follows that $f(x_1) \leqslant f(x) \leqslant f(x_2)$ for all $x \in D$. ∎

To apply Corollary 22.3, it is necessary that D be both closed and bounded. For example, the identity mapping on the unbounded set $[0, \infty)$ is continuous, but it certainly does not assume a maximum value. If D is not closed, then f may not attain its sup and inf on D even if $f(D)$ is bounded.

22.4 PRACTICE Let $D = (0, 1)$. Find a continuous function $f: D \to \mathbb{R}$ such that f is bounded on D but does not assume max and min values on D.

While the previous results apply to the continuous image of any compact subset of $\mathbb{R}$, if we require D to be a compact interval, then we obtain the following additional properties.

22.5 LEMMA Let $f: [a, b] \to \mathbb{R}$ be continuous and suppose that $f(a) < 0 < f(b)$. Then there exists a point c in (a, b) such that $f(x) = 0$.

Proof: Our strategy is to let c be the largest x for which $f(x) \leqslant 0$. More precisely, let $S = \{x \in [a, b]: f(x) \leqslant 0\}$. Since $a \in S$, S is nonempty. Thus $c = \sup S$ exists as a real number in $[a, b]$. We claim that $f(c) = 0$. Indeed, suppose that $f(c) < 0$. Then there exists a neighborhood U of c such that $f(x) < 0$ for all $x \in U \cap D$. (See Exercise 21.11.) This contradicts c being an upper bound for S. Similarly, if $f(c) > 0$, then there exists a neighborhood U of c such that $f(x) > 0$ for all $x \in U \cap D$. This contradicts c being the *least* upper bound for S. We conclude that $f(c) = 0$, as desired. Finally, since $f(a) < 0 < f(b)$ and $f(c) = 0$, it must be that $c \in (a, b)$. ▮

22.6 THEOREM (Intermediate Value Theorem) Suppose that $f: [a, b] \to \mathbb{R}$ is continuous. Then f has the intermediate value property on $[a, b]$. That is, if k is any value between $f(a)$ and $f(b)$ [i.e., $f(a) < k < f(b)$ or $f(b) < k < f(a)$], then there exists $c \in (a, b)$ such that $f(c) = k$.

Proof: Let k be any number between $f(a)$ and $f(b)$. If $f(a) < f(b)$, then apply Lemma 22.5 to the continuous function $g: [a, b] \to \mathbb{R}$ given by $g(x) = f(x) - k$. Then $g(a) = f(a) - k < 0$ and $g(b) = f(b) - k > 0$. Thus there exists $c \in (a, b)$ such that $g(c) = f(c) - k = 0$. If $f(a) > f(b)$, a similar argument applies to the function $g(x) = k - f(x)$. ▮

The idea behind the intermediate value theorem is very simple. It says that the graph of f must cross any horizontal line between $y = f(a)$ and $y = f(b)$. (See Figure 22.1.) Intuitively, if the graph of f is below $y = k$ at a and above $y = k$ at b, then for f to be continuous on $[a, b]$ it must cross $y = k$ somewhere in between. Thus the graph of a continuous function can have no "jumps."

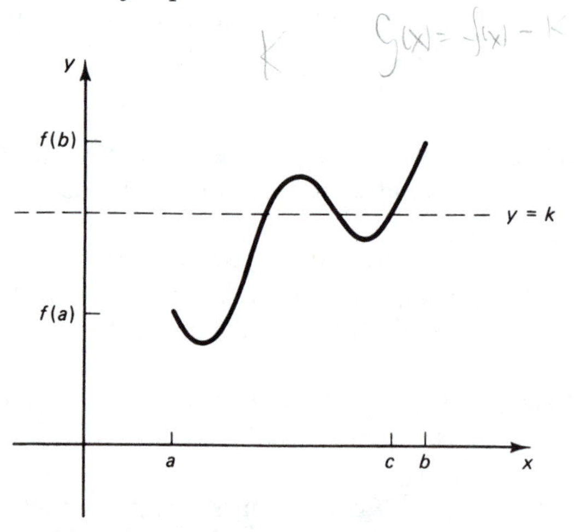

Figure 22.1 The Intermediate Value Theorem

22.7 EXAMPLE Using the intermediate value theorem, we can show that every positive number has a positive nth root. Suppose that $k > 0$ and $n \in \mathbb{N}$. Let $f(x) = x^n$. Then $f(0) = 0 < k$. Furthermore, if $b = k + 1$, then

$$b^n = (1 + k)^n \geqslant 1 + kn > k$$

by Bernoulli's inequality (Exercise 10.10). Thus $f(b) > k$. Since f is continuous, we conclude that there exists $c \in (0, b)$ such that $f(c) = k$. Thus $c^n = k$ and c is an nth root of k.

22.8 EXAMPLE The intermediate value theorem is really a very powerful result that can be useful in a wide variety of settings. To illustrate this diversity, let C be any bounded closed subset of the plane. (For a subset of the plane to be bounded it must be contained in some circle. For it to be closed it must include all the points that are, roughly speaking, on its "edges.") We claim that there is a square S that circumscribes C. That is, $C \subseteq S$ and each side of S intersects C. Although the details of the proof are beyond our reach, the idea of the proof is as follows. Given any $\theta \in [0, 2\pi]$, let r be a ray from the origin having angle θ with the positive x-axis, as in Figure 22.2. This ray determines a unique circumscribing rectangle whose sides are parallel and perpendicular to r. Let $A(\theta)$ be the length of the sides parallel to r and let $B(\theta)$ be the length of the sides perpendicular to r. Define $f: [0, 2\pi] \to \mathbb{R}$ by $f(\theta) = A(\theta) - B(\theta)$.

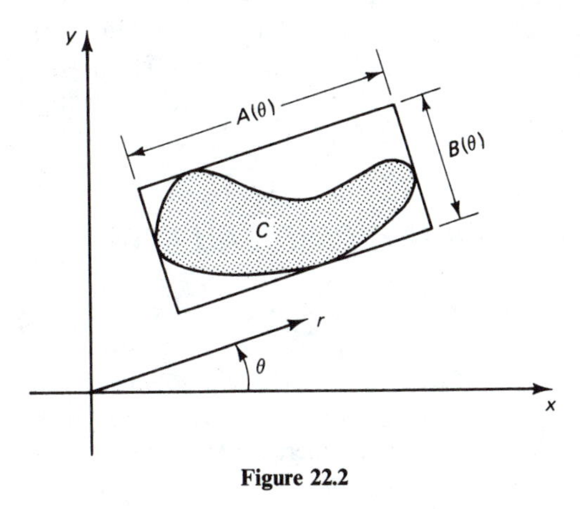

Figure 22.2

If for some particular θ the circumscribing rectangle is not a square, then $A(\theta) \neq B(\theta)$, and let us suppose that we have $f(\theta) = A(\theta) - B(\theta) > 0$. If we replace the angle θ by $\theta + \pi/2$, then the circumscribing rectangle is unchanged except the labeling of its sides is reversed. That is,

$$A\left(\theta + \frac{\pi}{2}\right) = B(\theta) \quad \text{and} \quad B\left(\theta + \frac{\pi}{2}\right) = A(\theta).$$

Thus $f(\theta + \pi/2) < 0$. Now it is reasonable to assume (and not too difficult to verify) that f is a continuous function on $[0, 2\pi]$. (A small change in angle will produce a small change in the lengths of the sides.) Thus by the intermediate value theorem there must be some angle θ_0 between θ and $\theta + \pi/2$ such that $f(\theta_0) = 0$. But then $A(\theta_0) = B(\theta_0)$ and the circumscribing rectangle for this angle is a square.

22.9 PRACTICE Assuming that cos x is a continuous function, prove that $x = \cos x$ for some x in $(0, \pi/2)$.

We conclude this section with a theorem that combines two earlier results. Further properties of continuous functions are included in the exercises.

22.10 THEOREM Let I be a compact interval and suppose that $f: I \to \mathbb{R}$ is a continuous function. Then the set $f(I)$ is a compact interval.

Proof: Corollary 22.3 implies that there exist x_1, x_2 in I such that $f(x_1) \leqslant f(x) \leqslant f(x_2)$ for all $x \in I$. Let $m_1 = f(x_1)$ and $m_2 = f(x_2)$. Then $f(I) \subseteq [m_1, m_2]$. Now if $k \in (m_1, m_2)$, then Theorem 22.6 implies that $k = f(c)$ for some $c \in (x_1, x_2) \subseteq I$. Thus $(m_1, m_2) \subseteq f(I)$. Finally, since $m_1, m_2 \in f(I)$, we have $[m_1, m_2] \subseteq f(I)$. Hence $f(I)$ is the compact interval $[m_1, m_2]$. ▮

ANSWERS TO PRACTICE PROBLEMS

22.1 There are many possibilities. One simple one is $f(x) = 1/x$ for $x \in D = (0, 1)$. Then $f(D) = (1, \infty)$.

22.4 $f(x) = x$ is one possibility.

22.9 Let $f(x) = x - \cos x$. Then $f(0) = -1$ and $f(\pi/2) = \pi/2$. Since f is continuous, there exists x_0 in $(0, \pi/2)$ such that $f(x_0) = 0$. That is, $x_0 = \cos x_0$.

EXERCISES

22.1 Let $f: D \to \mathbb{R}$ be continuous. For each of the following, prove or give a counterexample.

(a) If D is open, then $f(D)$ is open.
(b) If D is closed, then $f(D)$ is closed.
(c) If D is not open, then $f(D)$ is not open.
(d) If D is not closed, then $f(D)$ is not closed.
(e) If D is not compact, then $f(D)$ is not compact.
(f) If D is unbounded, then $f(D)$ is unbounded.

(g) If D is finite, then $f(D)$ is finite.
(h) If D is infinite, then $f(D)$ is infinite.
(i) If D is an interval, then $f(D)$ is an interval.
(j) If D is an interval that is not open, then $f(D)$ is an interval that is not open.

22.2 Show that $2^x = 3x$ for some $x \in (0, 1)$.

22.3 Show that the equation $3^x = x^2$ has at least one real solution.

22.4 Show that any polynomial of odd degree has at least one real root.

22.5 Suppose that $f: [a, b] \to [a, b]$ is continuous. Prove that f has a **fixed point**. That is, prove that there exists $c \in [a, b]$ such that $f(c) = c$.

22.6 Suppose that $f: [a, b] \to \mathbb{R}$ and $g: [a, b] \to \mathbb{R}$ are continuous functions such that $f(a) \leqslant g(a)$ and $f(b) \geqslant g(b)$. Prove that $f(c) = g(c)$ for some $c \in [a, b]$.

22.7 Let $f: [a, b] \to \mathbb{R}$ be continuous and suppose that f has only rational values. Prove that f is constant on $[a, b]$.

22.8 Suppose that $f: [0, 1] \to \mathbb{R}$ is two-to-one. That is, for each $y \in \mathbb{R}$, $f^{-1}(\{y\})$ is either empty or contains exactly two points.
(a) Find an example of such a function.
(b) Prove that no such function can be continuous.

***22.9** Let $f: \mathbb{R} \to \mathbb{R}$. Prove that f is continuous on $\mathbb{R}$ iff $f^{-1}(G)$ is an open set whenever G is an open set.

***22.10** Let $f: \mathbb{R} \to \mathbb{R}$. Prove that f is continuous on $\mathbb{R}$ iff $f^{-1}(H)$ is a closed set whenever H is a closed set.

***22.11** Let f be a function defined on an interval I. We say that f is **strictly increasing** if $x_1 < x_2$ in I implies that $f(x_1) < f(x_2)$. Similarly, f is **strictly decreasing** if $x_1 < x_2$ in I implies that $f(x_1) > f(x_2)$. Prove the following.
(a) If f is continuous and injective on I, then f is strictly increasing or strictly decreasing.
(b) If f is strictly increasing and if $f(I)$ is an interval, then f is continuous. Furthermore, f^{-1} is a strictly increasing continuous function on $f(I)$.

22.12 Define $f: \mathbb{R} \to \mathbb{R}$ by $f(x) = \sin(1/x)$ if $x \neq 0$ and $f(0) = 0$.
(a) Show that f is not continuous at 0.
(b) Show that f has the intermediate value property on $\mathbb{R}$.

22.13 Let $f: D \to \mathbb{R}$ and let $c \in D$. We say that f is **bounded on a neighborhood** of c if there exists a neighborhood U of c and a number M such that $|f(x)| \leqslant M$ for all $x \in U \cap D$.
(a) Suppose that f is bounded on a neighborhood of each x in D and that D is compact. Prove that f is bounded on D.
(b) Suppose that f is bounded on a neighborhood of each x in D, but that D is not compact. Show that f is not necessarily bounded on D, even when f is continuous.
(c) Suppose that $f: [a, b] \to \mathbb{R}$ has a limit at each x in $[a, b]$. Prove that f is bounded on $[a, b]$.

22.14 Let D be a compact subset of $\mathbb{R}$ and suppose that $f: \mathbb{R} \to \mathbb{R}$ is continuous. Use Exercise 22.9 and Definition 14.1 to prove that $f(D)$ is compact.

22.15 A subset S of $\mathbb{R}$ is said to be **disconnected** if there exist disjoint open sets U and V in $\mathbb{R}$ such that $S \subseteq U \cup V$, $S \cap U \neq \emptyset$, and $S \cap V \neq \emptyset$. If S is not disconnected, then it is said to be **connected**. Suppose that S is connected and that $f: \mathbb{R} \to \mathbb{R}$ is continuous. Prove that $f(S)$ is connected.

Section 23 UNIFORM CONTINUITY

Given a function $f: D \to \mathbb{R}$, for f to be continuous on D it is required that

> for every $x_0 \in D$ and for every $\varepsilon > 0$ there exists a $\delta > 0$ such that $|f(x) - f(x_0)| < \varepsilon$ whenever $|x - x_0| < \delta$ and $x \in D$.

By considering the order of the quantifiers, we see that the δ that is chosen may depend both on ε and on the point x_0. If it happens that, given $\varepsilon > 0$, a $\delta > 0$ can be found that will work for all x_0, then the function is said to be uniformly continuous on D. As usual, we require D to be a nonempty subset of $\mathbb{R}$.

23.1 DEFINITION Let $f: D \to \mathbb{R}$. We say that f is **uniformly continuous** on D if

> for every $\varepsilon > 0$ there exists a $\delta > 0$ such that $|f(x) - f(y)| < \varepsilon$ whenever $|x - y| < \delta$ and $x, y \in D$.

It should be clear that if a function is uniformly continuous on a set D then it is certainly continuous on D. Furthermore, while it is proper to speak of a function being continuous at a point, uniform continuity is a property that applies to a function *on a set*. We never speak of a function being uniformly continuous at a point.

23.2 EXAMPLE Consider the function $f(x) = 2x$ for all $x \in \mathbb{R}$. Given any $\varepsilon > 0$, we want to make $|f(x) - f(y)| < \varepsilon$ by making x sufficiently close to y. Now

$$|f(x) - f(y)| = |2x - 2y| = 2|x - y|.$$

Thus we take $\delta = \varepsilon/2$. Then whenever $|x - y| < \delta$ we have

$$|f(x) - f(y)| = 2|x - y| < 2\delta = \varepsilon.$$

We conclude that f is uniformly continuous on $\mathbb{R}$.

23.3 EXAMPLE Suppose that we let $f(x) = x^2$. Then

$$|f(x) - f(y)| = |x^2 - y^2| = |x + y| \cdot |x - y|.$$

This time, to make $|f(x) - f(y)| < \varepsilon$, we need to know something about the size of $|x + y|$. In Figure 23.1 we see that for a given $\varepsilon > 0$, as x and y

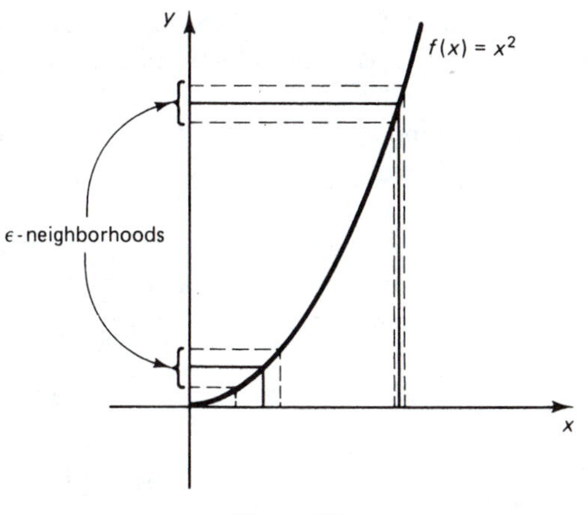

Figure 23.1

increase, the value of δ must decrease. Let us prove that f is *not* uniformly continuous on $\mathbb{R}$.

Suppose that we take $\varepsilon = 1$. (It turns out in this case that any $\varepsilon > 0$ would work out as well.) We must show that, given any $\delta > 0$, there exist $x, y \in \mathbb{R}$ such that $|x - y| < \delta$ and $|f(x) - f(y)| \geqslant 1$. For any x, if we let $y = x + \delta/2$, then $|x - y| = \delta/2 < \delta$. Thus to make

$$1 \leqslant |x + y| \cdot |x - y| = |x + y| \cdot \frac{\delta}{2}$$

we need to have $|x + y| \geqslant 2/\delta$. This prompts us to let $x = 1/\delta$. Here is the formal proof:

> Let $\varepsilon = 1$. Then, given any $\delta > 0$, let $x = 1/\delta$ and $y = 1/\delta + \delta/2$. Then $|x - y| = \delta/2 < \delta$, but
>
> $$\begin{aligned} |f(x) - f(y)| &= |x + y| \cdot |x - y| \\ &= \left|\frac{1}{\delta} + \frac{1}{\delta} + \frac{\delta}{2}\right| \cdot \frac{\delta}{2} > \frac{2}{\delta} \cdot \frac{\delta}{2} = 1. \end{aligned}$$
>
> Thus f is not uniformly continuous on $\mathbb{R}$. ∎

23.4 EXAMPLE The difficulty with $f(x) = x^2$ in Example 23.13 is that we allowed $|x + y|$ to be arbitrarily large. If we restrict the domain to a compact set, then this cannot happen. For example, if $D = [-5, 5]$, then $|x + y| \leqslant 10$. Thus given $\varepsilon > 0$, if $\delta = \varepsilon/10$ and $|x - y| < \delta$, we have

$$\begin{aligned} |f(x) - f(y)| &= |x^2 - y^2| = |x + y| \cdot |x - y| \\ &\leqslant 10|x - y| < 10\delta = \varepsilon. \end{aligned}$$

Thus $f(x) = x^2$ is uniformly continuous on $[-5, 5]$. What we have just shown is a special case of the following theorem.

23.5 THEOREM Suppose that $f: D \to \mathbb{R}$ is continuous on a compact set D. Then f is uniformly continuous on D.

Proof: Let $\varepsilon > 0$ be given. Since f is continuous on D, f is continuous at each $x \in D$. Thus for each $x \in D$ there exists $\delta_x > 0$ such that

$$|f(x) - f(y)| < \frac{\varepsilon}{2} \qquad \text{whenever } |x - y| < \delta_x$$

and $y \in D$. Now the family of neighborhoods

$$\mathscr{F} = \left\{N\left(x; \frac{\delta_x}{2}\right): x \in D\right\}$$

is an open cover of D. Since D is compact, $\mathscr{F}$ contains a finite subcover. That is, there exists $x_1, \ldots, x_n$ in D such that

$$D \subseteq N\left(x_1; \frac{\delta_{x_1}}{2}\right) \cup \cdots \cup N\left(x_n; \frac{\delta_{x_n}}{2}\right).$$

Now let $\delta = \min \{\delta_{x_1}/2, \ldots, \delta_{x_n}/2\}$. We claim that this δ works in the definition of uniform continuity.

Indeed, suppose that $x, y \in D$ and $|x - y| < \delta$. Then $x \in N(x_i; \delta_{x_i}/2)$ for some i among $1, \ldots, n$. (See Figure 23.2.) Since $|x - y| < \delta \leqslant \delta_{x_i}/2$, we have

$$|y - x_i| \leqslant |y - x| + |x - x_i| < \frac{\delta_{x_i}}{2} + \frac{\delta_{x_i}}{2} = \delta_{x_i}.$$

Thus $|f(y) - f(x_i)| < \varepsilon/2$. But we also have $|x - x_i| < \delta_{x_i}/2 < \delta_{x_i}$, so $|f(x) - f(x_i)| < \varepsilon/2$. It follows that

$$|f(x) - f(y)| \leqslant |f(x) - f(x_i)| + |f(x_i) - f(y)|$$

$$< \frac{\varepsilon}{2} + \frac{\varepsilon}{2} = \varepsilon. \blacksquare$$

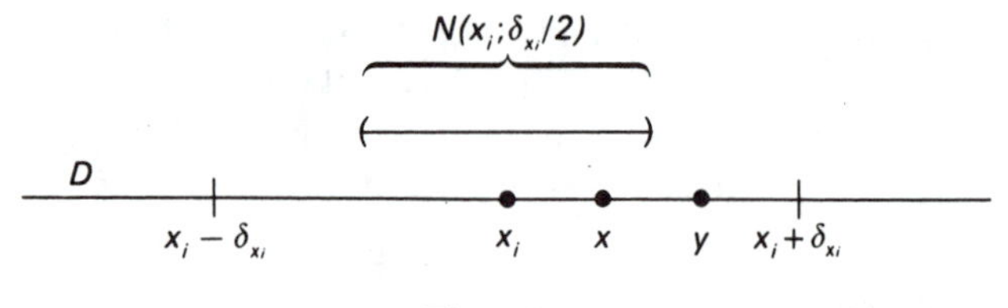

Figure 23.2

The proof of Theorem 23.5 was intentionally given in terms of neighborhoods because this illustrates the usefulness of the open cover property of compactness. For an alternative proof using the closed-bounded property and sequences, see Exercise 23.11.

The impact of uniform continuity on sequences is given in the next theorem. Recall that a continuous function does not necessarily preserve convergence. That is, the continuous image of a convergent sequence need not be convergent if the limit of the sequence is not in the domain of the function.

23.6 PRACTICE Find a continuous function $f\colon D \to \mathbb{R}$ and a Cauchy sequence (x_n) in D such that $(f(x_n))$ is divergent.

23.7 THEOREM Let $f\colon D \to \mathbb{R}$ be uniformly continuous on D and suppose that (x_n) is a Cauchy sequence in D. Then $(f(x_n))$ is a Cauchy sequence.

Proof: Given any $\varepsilon > 0$, since f is uniformly continuous on D there exists $\delta > 0$ such that

$$|f(x) - f(y)| < \varepsilon \qquad \text{whenever } |x - y| < \delta$$

and $x, y \in D$. Since (x_n) is Cauchy, there exists a number N such that

$$|x_n - x_m| < \delta \qquad \text{whenever } m, n > N.$$

Thus for $m, n > N$ we have $|f(x_n) - f(x_m)| < \varepsilon$, so $(f(x_n))$ is a Cauchy sequence. ▮

Using Theorem 23.7, we can derive a useful test to determine if a function is uniformly continuous on a bounded open interval. We say that a function $\tilde{f}\colon E \to \mathbb{R}$ is an **extension of a function** $f\colon D \to \mathbb{R}$ if $D \subseteq E$ and $f(x) = \tilde{f}(x)$ for all $x \in D$.

23.8 THEOREM A function $f\colon (a, b) \to \mathbb{R}$ is uniformly continuous on (a, b) iff it can be extended to a function $\tilde{f}$ that is continuous on $[a, b]$.

Proof: If f can be extended to a function $\tilde{f}$ that is continuous on the compact set $[a, b]$, then f is uniformly continuous on $[a, b]$ by Theorem 23.5. It follows that $\tilde{f}$ (and hence f) is also uniformly continuous on the subset (a, b).

Conversely, suppose that f is uniformly continuous on (a, b). We claim that $\lim_{x \to a} f(x)$ and $\lim_{x \to b} f(x)$ both exist as real numbers. To see this, let (s_n) be a sequence in (a, b) that converges to a. Then (s_n) is a Cauchy sequence, so Theorem 23.7 implies that $(f(s_n))$ is also Cauchy. Theorem 18.12 then implies that $(f(s_n))$

converges to some real number, say p. It follows from the contrapositive of Theorem 20.10 that we have $\lim_{x \to a} f(x) = p$. Similarly, we conclude that $\lim_{x \to b} f(x) = q$, for some $q \in \mathbb{R}$.

If we define $\tilde{f}: [a, b] \to \mathbb{R}$ by

$$\tilde{f}(x) = \begin{cases} f(x), & \text{if } a < x < b \\ p, & \text{if } x = a \\ q, & \text{if } x = b, \end{cases}$$

then $\tilde{f}$ will be an extension of f. Since f is continuous on (a, b), so is $\tilde{f}$. But $\tilde{f}$ is also continuous at a and b by Theorem 21.2(d), so $\tilde{f}$ is continuous on $[a, b]$. ▮

23.9 PRACTICE Use Theorem 23.8 to determine whether or not the function $f(x) = \sin(1/x)$ is uniformly continuous on the interval $(0, 1/\pi)$.

ANSWERS TO PRACTICE PROBLEMS

23.6 We may define $f: (0, \infty) \to \mathbb{R}$ by $f(x) = 1/x$ and let $x_n = 1/n$. Then (x_n) is Cauchy, but since $f(x_n) = n$ the sequence $(f(x_n))$ is divergent.

23.9 Since $\lim_{x \to 0} f(x)$ does not exist (Example 20.11), f cannot be extended to a function that is continuous on $[0, 1/\pi]$. Thus f is not uniformly continuous on $(0, 1/\pi)$.

EXERCISES

23.1 Determine which of the following continuous functions are uniformly continuous on the given set. Justify your answers.

(a) $f(x) = \dfrac{e^x}{x}$ on $[2, 5]$

(b) $f(x) = \dfrac{e^x}{x}$ on $(0, 2)$

(c) $f(x) = x^2 + 3x - 5$ on $[0, 4]$

(d) $f(x) = x^2 + 3x - 5$ on $(1, 3)$

(e) $f(x) = \dfrac{1}{x^2}$ on $(0, 1)$

(f) $f(x) = \dfrac{1}{x^2}$ on $(0, \infty)$

(g) $f(x) = x \sin \dfrac{1}{x}$ on $(0, 1)$

23.2 Prove that each function is uniformly continuous on the given set by directly verifying the ε–δ property in Definition 23.1.

(a) $f(x) = x^3$ on $[0, 2]$

(b) $f(x) = \dfrac{1}{x}$ on $[2, \infty)$

(c) $f(x) = \dfrac{x-1}{x+1}$ on $[0, \infty)$

23.3 Prove that $f(x) = \sqrt{x}$ is uniformly continuous on $[0, \infty)$.

23.4 Let f and g be real-valued functions that are uniformly continuous on D. Prove that $f + g$ is uniformly continuous on D.

23.5 Let $f: D \to \mathbb{R}$ be uniformly continuous on D and let $k \in \mathbb{R}$. Prove that the function kf is uniformly continuous on D.

23.6 Let f and g be real-valued functions that are uniformly continuous on D, and suppose that $g(x) \neq 0$ for all $x \in D$.

(a) Find an example to show that the function f/g need not be uniformly continuous on D.

(b) Prove that if D is compact then f/g must be uniformly continuous on D.

23.7 Prove or give a counterexample: If $f: A \to B$ is uniformly continuous on A and $g: B \to C$ is uniformly continuous on B, then $g \circ f: A \to C$ is uniformly continuous on A.

23.8 Find two real-valued functions f and g that are uniformly continuous on a set D, but such that their product fg is not uniformly continuous on D.

23.9 Let $f: D \to \mathbb{R}$ be uniformly continuous on the bounded set D. Prove that f is bounded on D.

23.10 (a) Let f and g be real-valued functions that are bounded and uniformly continuous on D. Prove that their product fg is uniformly continuous on D.

(b) Let f and g be real-valued functions that are uniformly continuous on a bounded set D. Prove that their product fg is uniformly continuous on D.

23.11 Prove Theorem 23.5 by justifying the following steps.

(a) Suppose that f is not uniformly continuous on D. Then there exists $\varepsilon > 0$ such that, for every $n \in \mathbb{N}$, there exist x_n and y_n in D with $|x_n - y_n| < 1/n$ and $|f(x_n) - f(y_n)| \geqslant \varepsilon$.

(b) Apply Theorem 19.6 to obtain a convergent subsequence (x_{n_k}) whose limit, say x, belongs to D.

(c) Show that $\lim_{k \to \infty} y_{n_k} = x$.

(d) Show that $(f(x_{n_k}))$ and $(f(y_{n_k}))$ both converge to $f(x)$, to obtain a contradiction.

23.12 A function $f: \mathbb{R} \to \mathbb{R}$ is said to be **periodic** if there exists a number $k > 0$ such that $f(x + k) = f(x)$ for all $x \in \mathbb{R}$. Suppose that $f: \mathbb{R} \to \mathbb{R}$ is continuous and periodic. Prove that f is bounded and uniformly continuous on $\mathbb{R}$.

Section 24 CONTINUITY IN METRIC SPACES†

In this section we apply some of the important topics of the last two chapters to the more general setting of a metric space. As we would expect from our earlier contact with metric spaces in Section 15, most of the definitions are quite similar, with the notion of absolute value being replaced by the metric function d.

24.1 DEFINITION A sequence (s_n) in a metric space (S, d) is said to **converge** if there exists a point $s \in S$ such that

> for each $\varepsilon > 0$ there exists a real number N such that $n > N$ implies that $d(s_n, s) < \varepsilon$.

In this case we say that s_n converges to s, and we write $s_n \to s$ or $\lim s_n = s$.

For a fixed point s in S, we can think of the real numbers $d(s_n, s)$ as a sequence in $\mathbb{R}$. Thus, to show that a sequence (s_n) converges to s in the metric space (S, d), it suffices to show that the real sequence $d(s_n, s)$ converges to 0 in $\mathbb{R}$. Furthermore, since $d(s_n, s) \geqslant 0$ for all n, we can do this by finding a positive real sequence (a_n) such that $d(s_n, s) \leqslant a_n$ for all n and $a_n \to 0$. (See Theorem 16.8.)

24.2 EXAMPLE In $\mathbb{R}^2$ with the usual Euclidean metric, define a sequence (s_n) by

$$s_n = \left(\frac{1}{n}, \frac{n}{n+1}\right).$$

Then $s_n \to (0, 1)$. One way to see this is to note that

$$\begin{aligned}
d(s_n, (0, 1)) &\leq d\left(s_n, \left(\frac{1}{n}, 1\right)\right) + d\left(\left(\frac{1}{n}, 1\right), (0, 1)\right) \\
&= \sqrt{\left(\frac{1}{n} - \frac{1}{n}\right)^2 + \left(\frac{n}{n+1} - 1\right)^2} + \sqrt{\left(\frac{1}{n} - 0\right)^2 + (1-1)^2} \\
&= \sqrt{\left(\frac{-1}{n+1}\right)^2} + \sqrt{\left(\frac{1}{n}\right)^2} \\
&= \frac{1}{n+1} + \frac{1}{n} < \frac{2}{n}
\end{aligned}$$

Since $(2/n) \to 0$ in $\mathbb{R}$, we know that $s_n \to (0, 1)$ in $\mathbb{R}^2$.

† This section may be skipped, if desired, since it is not used in later sections. It does depend, however, on the earlier optional Section 15 on metric spaces.

24.3 DEFINITION Let (S, d) and (S^*, d^*) be metric spaces. A function $f: S \to S^*$ is **continuous at a point** c in S if for every $\varepsilon > 0$ there exists a $\delta > 0$ such that

$$d^*[f(x), f(c)] < \varepsilon \qquad \text{whenever } d(x, c) < \delta.$$

If f is continuous at each point of a set D, then f is said to be **continuous** on D.

24.4 EXAMPLE Let (S, d) be a metric space and let $p \in S$. Then the function $f: S \to \mathbb{R}$ defined by $f(x) = d(x, p)$ is continuous on S. To see this, let $x, y \in S$. From the triangle inequality we have

$$d(y, p) \leqslant d(y, x) + d(x, p),$$

so

$$d(y, p) - d(x, p) \leqslant d(x, y).$$

Likewise,

$$d(x, p) \leq d(x, y) + d(y, p),$$

so

$$d(x, p) - d(y, p) \leqslant d(x, y).$$

It follows that

$$|d(x, p) - d(y, p)| \leqslant d(x, y).$$

Thus, given any $\varepsilon > 0$, if $d(x, y) < \varepsilon$ we have

$$|f(x) - f(y)| = |d(x, p) - d(y, p)| \leqslant d(x, y) < \varepsilon.$$

That is, f is continuous at the arbitrary point y and hence continuous on S.

If $S = S^* = \mathbb{R}$ and d and d^* are the usual absolute value metric, then Definition 24.3 is consistent with our earlier Definition 21.1 for continuity in $\mathbb{R}$. As before, we would expect to be able to characterize continuity in terms of sequences and neighborhoods (compare Theorem 21.2).

24.5 THEOREM Let (S, d) and (S^*, d^*) be metric spaces, let $f: S \to S^*$, and let $c \in S$. Then the following three conditions are equivalent:

(a) f is continuous at c.
(b) If (x_n) is any sequence in S such that x_n converges to c, then $f(x_n)$ converges to $f(c)$ in S^*.
(c) For every neighborhood V of $f(c)$ in S^* there exists a neighborhood U of c in S such that $f(U) \subseteq V$.

Proof: (a) $\Rightarrow$ (b) Suppose f is continuous at c and let (x_n) be a sequence in S such that $x_n \to c$. Given $\varepsilon > 0$, since f is continuous at c there exists a $\delta > 0$ such that $d^*(f(x), f(c)) < \varepsilon$ whenever $d(x, c) < \delta$. Since $x_n \to c$, there exists a real number N such that $n > N$ implies that $d(x_n, c) < \delta$. Thus for $n > N$ we have $d^*(f(x_n), f(c)) < \varepsilon$ and so $f(x_n) \to f(c)$.

(b) $\Rightarrow$ (c) We prove this by establishing the contrapositive. That is, suppose there exists a neighborhood $V = N(f(c); \beta)$ of $f(c)$ such that, for all neighborhoods U of c, $f(U) \nsubseteq V$. We must find a sequence (x_n) in S such that $x_n \to c$ but $(f(x_n))$ does not converge to $f(c)$. For each $n \in \mathbb{N}$, let $U_n = N(c; 1/n)$. Then, since $f(U_n) \nsubseteq V$, there exists a point x_n in U_n such that $f(x_n) \notin V$. Clearly $x_n \to c$, but $f(x_n)$ cannot converge to $f(c)$ because none of the $f(x_n)$ are in V. That is, $d^*(f(x_n), f(c)) \geqslant \beta > 0$ for all n.

(c) $\Rightarrow$ (a) Given any $\varepsilon > 0$, let $V = N(f(c); \varepsilon)$. By our hypotheses in (a), there exists a neighborhood $U = N(c; \delta)$ such that $f(U) \subseteq V$. But then whenever $d(x, c) < \delta$ we have $x \in U$, so $f(x) \in V$ and $d^*(f(x), f(c)) < \varepsilon$. Thus f is continuous at c. ▮

24.6 EXAMPLE Let (S, d) and (S^*, d^*) be metric spaces and let $p \in S^*$. Define a function $f: S \to S^*$ by $f(x) = p$ for all $x \in S$. Then f is continuous on S. Indeed, let $x \in S$ and let V be a neighborhood of p. Given any neighborhood U of x, we have $f(U) = \{p\} \subseteq V$, so f is continuous at x by Theorem 24.5. Since x was an arbitrary point in S, f is continuous on S.

Since the singleton set $\{p\}$ is not generally an open set, this example also illustrates that the continuous image of an open set need not be open. It is true, however, that the *pre*-image of an open set will always be open (see Exercise 22.9). It turns out that this latter characterization of continuity is often the most useful. We state this result as our next theorem.

24.7 THEOREM Let (S, d) and (S^*, d^*) be metric spaces and let $f: S \to S^*$. Then f is continuous on S iff $f^{-1}(G)$ is an open set in S whenever G is an open set in S^*.

Proof: Suppose f is continuous on S and let G be an open subset of S^*. We must show that every point in $f^{-1}(G)$ is an interior point of $f^{-1}(G)$. To this end, let $x \in f^{-1}(G)$ so that $f(x) \in G$. Since G is open there exists a neighborhood V of $f(x)$ such that $V \subseteq G$. Since f is continuous at x, Theorem 24.5 implies the existence of a neighborhood U of x such that $f(U) \subseteq V$. But $f(U) \subseteq V \subseteq G$ implies that $U \subseteq f^{-1}(G)$. Hence x is an interior point of $f^{-1}(G)$.

Conversely, suppose $f^{-1}(G)$ is open in S whenever G is open in S^*. Let $x \in S$ and let V be a neighborhood of $f(x)$. To show that f is continuous at x, it suffices by Theorem 24.5 to show that there exists a neighborhood U of x such that $f(U) \subseteq V$. Now since V is a neighborhood, V is open (Theorem 15.6). Hence $f^{-1}(V)$ is an open set containing x. Thus there exists a neighborhood U of x such that $U \subseteq f^{-1}(V)$. But then $f(U) \subseteq V$, so f is continuous at x. Since x was an arbitrary point in S, f is continuous on S. ∎

24.8 PRACTICE Let S be a nonempty set, let d be the "discrete" metric described in Example 15.2(c), and let (S^*, d^*) be any metric space. Prove that *every* function $f: S \to S^*$ is continuous on S with this metric.

In Section 22, when we proved that the continuous image of a compact set in $\mathbb{R}$ is compact (Theorem 22.2), our proof used the Heine–Borel theorem. Even though the same approach cannot be used in metric spaces, the result still holds true. In fact, the proof in this more general context is even easier than it was in $\mathbb{R}$.

24.9 THEOREM Let (S, d) and (S^*, d^*) be metric spaces, let $f: S \to S^*$ be continuous on S, and let D be a compact subset of S. Then $f(D)$ is a compact subset of S^*.

Proof: Let $\mathscr{F}$ be an open cover of $f(D)$. Since f is continuous on S, Theorem 24.7 implies that, for each $U \in \mathscr{F}$, $f^{-1}(U)$ is open in S. Furthermore, since $f(D) \subseteq \bigcup \mathscr{F}$, it follows that $D \subseteq \bigcup \{f^{-1}(U): U \in \mathscr{F}\}$. That is, $\{f^{-1}(U): U \in \mathscr{F}\}$ is an open cover for D. Since D is compact, there exist finitely many sets $U_1, U_2, \ldots, U_n$ in $\mathscr{F}$ such that

$$D \subseteq f^{-1}(U_1) \cup f^{-1}(U_2) \cup \cdots \cup f^{-1}(U_n).$$

Since $f(f^{-1}(B)) \subseteq B$ for every $B \subseteq S^*$, we must have

$$f(D) \subseteq U_1 \cup U_2 \cup \cdots \cup U_n.$$

Thus $\{U_1, U_2, \ldots, U_n\}$ is a finite subcover of $\mathscr{F}$ for $f(D)$, and $f(D)$ is compact. ∎

As an important corollary to Theorem 24.9, we state the following generalization of Corollary 22.3.

24.10 COROLLARY Let f be a continuous real-valued function defined on a metric space (S, d), and let D be a compact subset of S. Then f assumes maximum and minimum values on D.

Proof: Exercise 24.4. ∎

We conclude this section by relating uniform continuity to metric spaces.

24.11 DEFINITION Let (S, d) and (S^*, d^*) be metric spaces, let D be a subset of S, and suppose $f: D \to S^*$. We say that f is **uniformly continuous** on D if for every $\varepsilon > 0$ there exists a $\delta > 0$ such that

$$d^*(f(x), f(y)) < \varepsilon \qquad \text{whenever } d(x, y) < \delta \text{ and } x, y \in D.$$

24.12 PRACTICE Is the function described in Example 24.4 uniformly continuous on S?

24.13 PRACTICE Is the function f in Practice 24.8 always uniformly continuous on S?

24.14 THEOREM Let (S, d) and (S^*, d^*) be metric spaces, let $f: S \to S^*$ be continuous on S, and let D be a compact subset of S. Then f is uniformly continuous on D.

Proof: The argument is the same as the proof of Theorem 23.5, except that the absolute value notation must be replaced by the appropriate metric d or d^*. ∎

ANSWERS TO PRACTICE PROBLEMS

24.8 With the discrete metric, *every* subset of S is open. (See the answer to Practice 15.8.) So, in particular, $f^{-1}(G)$ will be open in S whenever G is open in S^*. It follows from Theorem 24.7 that f is continuous on S.

24.12 Yes, because we had $|f(x) - f(y)| \leqslant d(x, y)$ for all $x, y \in S$.

24.13 Yes. Given any $\varepsilon > 0$, let $\delta = 1$. Then if $d(x, y) < \delta$ we must have $x = y$, so that $d^*(f(x), f(y)) = d^*(f(x), f(x)) = 0 < \varepsilon$.

EXERCISES

24.1 Find the limit of each sequence in $\mathbb{R}^2$. Justify your answers as in Example 24.2.

(a) $s_n = \left(\dfrac{1}{3n}, 5\right)$ (b) $s_n = \left(\dfrac{1}{n}, \dfrac{1}{n^2}\right)$

(c) $s_n = \left(\dfrac{-1}{n+2}, \dfrac{2n+1}{n+1}\right)$ (d) $s_n = \left(\dfrac{(-1)^n}{n}, \dfrac{4n}{2n-1}\right)$

24.2 Let (S, d) and (S^*, d^*) be metric spaces and suppose $f: S \to S^*$ and $g: S \to S^*$ are both continuous on S. If $D \subseteq S$ and $f(x) = g(x)$ for all $x \in D$, prove that $f(x) = g(x)$ for all $x \in \text{cl } D$.

24.3 Use Example 24.4 and Theorem 24.7 to prove that in a metric space, neighborhoods are always open sets.

24.4 Prove Corollary 24.10.

24.5 (a) Let S and T be metric spaces and suppose $f: S \to T$ is bijective. If S is compact and f is continuous on S, prove that $f^{-1}: T \to S$ is continuous on T.

(b) Show that the compactness of S is necessary in part (a) by finding a continuous bijection f from $[0, 2\pi)$ onto the unit circle T in $\mathbb{R}^2$ such that f^{-1} is not continuous on T.

24.6 Let R, S, and T be metric spaces. Suppose $f: R \to S$ is continuous on R and $g: S \to T$ is continuous on S. Use Theorem 24.7 to prove that $g \circ f$ is continuous on R.

24.7 Let (R, d'), (S, d), and (T, d^*) be metric spaces and suppose $f: R \to S$ and $g: S \to T$. If f is uniformly continuous on R and g is uniformly continuous on S, prove that $g \circ f$ is uniformly continuous on R.

24.8 Let R, S, and T be metric spaces with S compact. Suppose $f: R \to S$ and $g: S \to T$, with g being continuous on S and bijective. Define $h = g \circ f$.

(a) Prove that f is continuous if h is continuous.

(b) Prove that f is uniformly continuous if h is uniformly continuous.

24.9 Let D be a nonempty subset of a metric space (S, d). If $x \in S$, we define the **distance** from x to D by

$$d(x, D) = \inf \{d(x, p): p \in D\}.$$

(a) Prove that $d(x, D) = 0$ iff $x \in \text{cl } D$.

(b) If D is compact, prove that there exists a point $p_0 \in D$ such that $d(x, D) = d(x, p_0)$.

(c) Prove that $f(x) = d(x, D)$ is a uniformly continuous function on S (even when D is not compact).

24.10 Let A and B be nonempty subsets of a metric space (S, d). We define the **distance** from A to B by

$$d(A, B) = \inf \{d(a, b): a \in A \text{ and } b \in B\}.$$

(a) Give an example to show that it is possible for two disjoint closed sets A and B to satisfy $d(A, B) = 0$.

(b) Prove that if A is closed and B is compact and $A \cap B = \varnothing$, then $d(A, B) > 0$.

24.11 Let B be a subset of a metric space (S, d). The set B is said to be **disconnected** if there exist disjoint open subsets U and V in S such that

$$B \subseteq U \cup V, \qquad B \cap U \neq \varnothing, \quad \text{and} \quad B \cap V \neq \varnothing.$$

A set B is **connected** if it is not disconnected. Let (S^*, d^*) be a metric space and suppose $f: S \to S^*$ is continuous. Prove that if B is a connected subset of S, then $f(B)$ is a connected subset of S^*. That is, the continuous image of a connected set is connected.

Exercises 24.12 to 24.17 use the following concepts: In the metric space $\mathbb{R}^n$ with the usual Euclidean metric, we can define a linear structure by setting

$$(x_1, \ldots, x_n) + (y_1, \ldots, y_n) = (x_1 + y_1, \ldots, x_n + y_n)$$

and

$$\lambda(x_1, \ldots, x_n) = (\lambda x_1, \ldots, \lambda x_n)$$

for arbitrary points $\mathbf{x} = (x_1, \ldots, x_n)$ and $\mathbf{y} = (y_1, \ldots, y_n)$ in $\mathbb{R}^n$ and for $\lambda \in \mathbb{R}$.

24.12 Let $\mathbf{p} \in \mathbb{R}^n$. Prove that $d(\mathbf{x} + \mathbf{p}, \mathbf{y} + \mathbf{p}) = d(\mathbf{x}, \mathbf{y})$ for all $\mathbf{x}, \mathbf{y} \in \mathbb{R}^n$.

24.13 Let $\mathbf{p} \in \mathbb{R}^n$. Prove that $f\colon \mathbb{R}^n \to \mathbb{R}^n$ defined by $f(\mathbf{x}) = \mathbf{p} + \mathbf{x}$ is uniformly continuous on $\mathbb{R}^n$.

24.14 Let $\lambda \in \mathbb{R}$. Prove that $d(\lambda\mathbf{x}, \lambda\mathbf{y}) = |\lambda| d(\mathbf{x}, \mathbf{y})$ for all $\mathbf{x}, \mathbf{y} \in \mathbb{R}^n$.

24.15 Let $\lambda \in \mathbb{R}$. Prove that $f\colon \mathbb{R}^n \to \mathbb{R}^n$ defined by $f(\mathbf{x}) = \lambda\mathbf{x}$ is continuous on $\mathbb{R}^n$.

24.16 Let $\mathbf{p} \in \mathbb{R}^n$ and $\lambda \in \mathbb{R}$. Prove that $f\colon \mathbb{R}^n \to \mathbb{R}^n$ defined by $f(\mathbf{x}) = \mathbf{p} + \lambda\mathbf{x}$ is continuous on $\mathbb{R}^n$.

24.17 If $A, B \subseteq \mathbb{R}^n$ and $\lambda \in \mathbb{R}$, we define $A + B = \{\mathbf{a} + \mathbf{b}\colon \mathbf{a} \in A \text{ and } \mathbf{b} \in B\}$ and $\lambda B = \{\lambda\mathbf{b}\colon \mathbf{b} \in B\}$. If A consists of a single point, say $\mathbf{p}$, then we often write $\mathbf{p} + B$ instead of $A + B$. The set $\mathbf{p} + B$ is called a **translate** of B. The set λB is called a **scalar multiple** of B. More generally, if $\lambda \neq 0$, the set $\mathbf{p} + \lambda B$ is said to be **homothetic** to B.

(a) Prove that each set homothetic to an open set is open.
(b) Prove that each set homothetic to a closed set is closed.
(c) Prove that $A + B = \bigcup_{\mathbf{a} \in A} (\mathbf{a} + B) = \bigcup_{\mathbf{b} \in B} (A + \mathbf{b})$.
(d) Prove or give a counterexample: If A is open, then for any set B, $A + B$ is open.
(e) Prove or give a counterexample: If A and B are both closed, then $A + B$ is closed.

6

Differentiation

Having developed our skill at working with limits, we now apply this understanding to the important process of differentiation. Most of the topics covered here will be at least somewhat familiar to the reader from the standard calculus course. In that earlier course a good deal of time was spent on the applications of derivatives in physics, geometry, economics, and the like. By way of contrast, the focus of this chapter will be on the theoretical aspects of differentiation that are often treated more superficially in the introductory course.

After establishing the basic properties of the derivative in Section 25, we prove the mean value theorem and develop some of its consequences in Section 26. In Section 27, we examine indeterminate forms and derive L'Hospital's rule for evaluating them. Finally, in Section 27 we give a brief discussion of Taylor's theorem.

Section 25 THE DERIVATIVE

25.1 DEFINITION Let f be a real-valued function defined on an interval I containing the point c. We say that f is **differentiable** at c (or has a **derivative** at c) if the limit

$$\lim_{x \to c} \frac{f(x) - f(c)}{x - c}$$

exists and is finite. We denote the derivative of f at c by $f'(c)$ so that

$$f'(c) = \lim_{x \to c} \frac{f(x) - f(c)}{x - c}$$

whenever the limit exists and is finite. If the function f is differentiable at each point of the set $S \subseteq I$, then f is said to be differentiable on S, and the function $f': S \to \mathbb{R}$ is called the derivative of f on S.

25.2 EXAMPLE Let $f(x) = x^2$ for each $x \in \mathbb{R}$. Then for any $c \in \mathbb{R}$ we have

$$f'(c) = \lim_{x \to c} \frac{f(x) - f(c)}{x - c} = \lim_{x \to c} \frac{x^2 - c^2}{x - c}$$
$$= \lim_{x \to c} (x + c) = 2c.$$

It is customary to regard f' as a function of x when f is a function of x, so we have $f'(x) = 2x$ for all $x \in \mathbb{R}$.

Applying the sequential criterion for limits (Theorem 20.8), we obtain the following sequential condition for derivatives. The sequential approach is often useful when trying to show that a given function is *not* differentiable at a particular point.

25.3 THEOREM Let I be an interval containing the point c and suppose that $f: I \to \mathbb{R}$. Then f is differentiable at c iff, for every sequence (x_n) in $I \setminus \{c\}$ that converges to c, the sequence

$$\left(\frac{f(x_n) - f(c)}{x_n - c} \right)$$

converges. Furthermore, if f is differentiable at c, then the sequence of quotients above will converge to $f'(c)$.

25.4 EXAMPLE Let $f(x) = |x|$ for each $x \in \mathbb{R}$, and let $x_n = (-1)^n/n$ for $n \in \mathbb{N}$. Then the sequence (x_n) converges to 0, but the corresponding sequence of quotients does not converge. Indeed, when n is even, $x_n = 1/n$ so that

$$\frac{f(x_n) - f(0)}{x_n - 0} = \frac{1/n - 0}{1/n - 0} = 1.$$

But when n is odd we have $x_n = -1/n$, so that

$$\frac{f(x_n) - f(0)}{x_n - 0} = \frac{1/n - 0}{-1/n - 0} = -1.$$

Since the two subsequences have different limits, the sequence

$$\left(\frac{f(x_n) - f(0)}{x_n - 0}\right)$$

does not converge. Thus f is not differentiable at zero.

25.5 PRACTICE Define $f\colon \mathbb{R} \to \mathbb{R}$ by $f(x) = x \sin(1/x)$ if $x \neq 0$ and $f(0) = 0$. Determine whether or not f is differentiable at $x = 0$.

We see from Example 25.4 and Practice 25.5 that it is possible for a function to be continuous at a point without being differentiable at the point. On the other hand, it is easy to prove that if f is differentiable at a point, then it must also be continuous there.

25.6 THEOREM If $f\colon I \to \mathbb{R}$ is differentiable at a point $c \in I$, then f is continuous at c.

Proof: For every $x \in I$ with $x \neq c$, we have

$$f(x) = (x - c)\,\frac{f(x) - f(c)}{x - c} + f(c).$$

Since $f'(c)$ exists, we know that

$$\lim_{x \to c} \frac{f(x) - f(c)}{x - c} = f'(c) \in \mathbb{R}.$$

Thus by Theorem 20.13 we obtain

$$\begin{aligned}\lim_{x \to c} f(x) &= \left[\lim_{x \to c} (x - c)\right]\left[\lim_{x \to c} \frac{f(x) - f(c)}{x - c}\right] + \lim_{x \to c} f(c) \\ &= 0 \cdot f'(c) + f(c) = f(c).\end{aligned}$$

Hence Theorem 21.2 implies that f is continuous at c. ∎

We now present the useful (and familiar) rules for taking the derivative of sums, products, and quotients of functions.

25.7 THEOREM Suppose that $f\colon I \to \mathbb{R}$ and $g\colon I \to \mathbb{R}$ are differentiable at $c \in I$. Then

(a) If $k \in \mathbb{R}$, then the function kf is differentiable at c and

$$(kf)'(c) = k \cdot f'(c).$$

(b) The function $f + g$ is differentiable at c and

$$(f + g)'(c) = f'(c) + g'(c).$$

(c) (Product Rule) The function fg is differentiable at c and

$$(fg)'(c) = f(c)g'(c) + g(c)f'(c).$$

(d) (Quotient Rule) If $g(c) \neq 0$, then the function f/g is differentiable at c and

$$\left(\frac{f}{g}\right)'(c) = \frac{g(c)f'(c) - f(c)g'(c)}{[g(c)]^2}.$$

Proof: Parts (a) and (b) are left as exercises.

(c) For every $x \in I$ with $x \neq c$, we have

$$\frac{(fg)(x) - (fg)(c)}{x - c} = f(x)\frac{g(x) - g(c)}{x - c} + g(c)\frac{f(x) - f(c)}{x - c}.$$

Theorem 25.6 implies that f is continuous at c, so $\lim_{x \to c} f(x) = f(c)$. Since f and g are differentiable at c, we conclude (using Theorem 20.13) that

$$(fg)'(c) = \lim_{x \to c} \frac{(fg)(x) - (fg)(c)}{x - c} = f(c)g'(c) + g(c)f'(c).$$

(d) Since $g(c) \neq 0$ and g is continuous at c, there exists an interval $J \subseteq I$ with $c \in J$ such that $g(x) \neq 0$ for all $x \in J$. (See Exercise 21.11.) For all $x \in J$ with $x \neq c$, we have

$$\frac{f}{g}(x) - \frac{f}{g}(c) = \frac{f(x)}{g(x)} - \frac{f(c)}{g(c)} = \frac{g(c)f(x) - f(c)g(x)}{g(x)g(c)}$$
$$= \frac{g(c)f(x) - g(c)f(c) + g(c)f(c) - f(c)g(x)}{g(x)g(c)}$$

It follows that

$$\left(\frac{f}{g}\right)'(c) = \lim_{x \to c} \frac{(f/g)(x) - (f/g)(c)}{x - c}$$
$$= \lim_{x \to c} \left\{\left[g(c)\frac{f(x) - f(c)}{x - c} - f(c)\frac{g(x) - g(c)}{x - c}\right]\left[\frac{1}{g(x)g(c)}\right]\right\}$$
$$= \frac{g(c)f'(c) - f(c)g'(c)}{[g(c)]^2}.$$ ∎

25.8 EXAMPLE To illustrate the use of Theorem 25.7, let us show that for any $n \in \mathbb{N}$, if $f(x) = x^n$ for all $x \in \mathbb{R}$, then $f'(x) = nx^{n-1}$ for all $x \in \mathbb{R}$. Our proof is by induction.

When $n = 1$ we have $f(x) = x$, so that

$$f'(x) = \lim_{t \to x} \frac{f(t) - f(x)}{t - x} = \lim_{t \to x} \frac{t - x}{t - x} = \lim_{t - x} 1 = 1 = 1x^0,$$

and the formula holds. Now suppose that the formula holds for $n = k$. That is, if $f(x) = x^k$, then $f'(x) = kx^{k-1}$. The function $g(x) = x^{k+1}$ we write as the product of two functions and use the product rule of Theorem 25.7. Let $f(x) = x^k$ and $h(x) = x$. Then $g(x) = f(x)h(x)$, so that

$$\begin{aligned} g'(x) &= f(x)h'(x) + h(x)f'(x) \\ &= (x^k)(1) + (x)(kx^{k-1}) = (k+1)x^k. \end{aligned}$$

Thus the formula holds for $n = k + 1$, and by induction we conclude that it holds for all $n \in \mathbb{N}$.

We also note that the formula holds for $n = 0$. That is, if $f(x) = 1$ for all x, then f is differentiable for all x and $f'(x) = 0$.

25.9 PRACTICE Let n be a negative integer and let $f(x) = x^n$ for $x \neq 0$. Note that $-n$ is positive and if $g(x) = x^{-n}$, then $f(x) = (1/g)(x)$. Use Example 25.8 and the quotient rule of Theorem 25.7 to show that $f'(x) = nx^{n-1}$.

In Section 21 we proved that the composition of two continuous functions is continuous. A similar result holds for the composition of differentiable functions, and it is known as the chain rule.

25.10 THEOREM (Chain Rule) Let I and J be intervals in $\mathbb{R}$, let $f: I \to \mathbb{R}$ and $g: J \to \mathbb{R}$, where $f(I) \subseteq J$, and let $c \in I$. If f is differentiable at c and g is differentiable at $f(c)$, then the composite function $g \circ f$ is differentiable at c and

$$(g \circ f)'(c) = g'(f(c)) \cdot f'(c).$$

Proof: Following our usual approach, we write

$$\frac{g \circ f(x) - g \circ f(c)}{x - c} = \frac{g(f(x)) - g(f(c))}{f(x) - f(c)} \cdot \frac{f(x) - f(c)}{x - c}.$$

It would seem that by taking the limit of both sides as $x \to c$ we would obtain the desired result. The only problem is that $f(x) - f(c)$ may be zero even when $x - c \neq 0$. Thus the first factor in the right-hand side may have a zero denominator. To circumvent this problem, we note that

$$\lim_{y \to f(c)} \frac{g(y) - g(f(c))}{y - f(c)} = g'(f(c))$$

since g is differentiable at $f(c)$. Thus we define a new function h: $J \to \mathbb{R}$ by

$$h(y) = \begin{cases} \dfrac{g(y) - g(f(c))}{y - f(c)}, & \text{if } y \neq f(c) \\ g'(f(c)), & \text{if } y = f(c) \end{cases}$$

and see that h is continuous at $f(c)$.

Now since f is differentiable at c, Theorem 25.6 implies that f is continuous at c. Hence $h \circ f$ is continuous at c by Theorem 21.12, so that

$$\lim_{x \to c} h \circ f(x) = h(f(c)) = g'(f(c)).$$

It follows from our definition of h that

$$g(y) - g(f(c)) = h(y)[y - f(c)], \qquad \text{for all } y \in J.$$

Thus, if $x \in I$, then $f(x) \in J$, so that

$$g \circ f(x) - g \circ f(c) = [h \circ f(x)][f(x) - f(c)].$$

But then for $x \in I$ with $x \neq c$, we have

$$\frac{g \circ f(x) - g \circ f(c)}{x - c} = [h \circ f(x)]\left[\frac{f(x) - f(c)}{x - c}\right].$$

Now we can take limits as $x \to c$ to obtain

$$\begin{aligned}(g \circ f)'(c) &= \lim_{x \to c} \frac{g \circ f(x) - g \circ f(c)}{x - c} \\ &= \lim_{x \to c} [h \circ f(x)] \cdot \lim_{x \to c} \frac{f(x) - f(c)}{x - c} \\ &= g'(f(c)) \cdot f'(c).\end{aligned}$$

∎

25.11 EXAMPLE Let us return to the function defined in Practice 25.5. That is, $f(x) = x \sin(1/x)$ if $x \neq 0$ and $f(0) = 0$. Using the fact that the derivative of $\sin x$ is $\cos x$ for all $x \in \mathbb{R}$, we compute the derivative of f at any point $x \neq 0$. Let $h(x) = \sin(1/x)$, so that $f(x) = xh(x)$. Then for $x \neq 0$ the chain rule gives us

$$h'(x) = \left[\cos \frac{1}{x}\right][-x^{-2}],$$

where we have differentiated $1/x = x^{-1}$ using the formula from Practice 25.9. Thus by the product rule of Theorem 25.7 we have

$$\begin{aligned} f'(x) &= xh'(x) + h(x) \\ &= -\frac{1}{x}\cos\frac{1}{x} + \sin\frac{1}{x} \end{aligned}$$

for all $x \neq 0$. We saw in Practice 25.5 that $f'(0)$ does not exist, so f is a (continuous) function that has a derivative at every real x except $x = 0$.†

ANSWERS TO PRACTICE PROBLEMS

25.5 For $x \neq 0$ we have

$$\frac{f(x) - f(0)}{x - 0} = \frac{x \sin(1/x)}{x} = \sin\frac{1}{x}.$$

In Example 20.11 we showed that $\lim_{x\to 0} \sin(1/x)$ does not exist. Thus f is not differentiable at $x = 0$.

25.9 Since $-n \in \mathbb{N}$, the function $g(x) = x^{-n}$ is differentiable by Example 25.8 and $g'(x) = -nx^{-n-1}$. Using the quotient rule of Theorem 25.7, we have

$$\begin{aligned} f'(x) = \left(\frac{1}{g}\right)'(x) &= \frac{[g(x)][0] - [1][g'(x)]}{[g(x)]^2} \\ &= \frac{-(-n)x^{-n-1}}{(x^{-n})^2} = nx^{n-1}. \end{aligned}$$

EXERCISES

25.1 Use Definition 25.1 to find the derivative of each function.

(a) $f(x) = 2x + 7$ for $x \in \mathbb{R}$

(b) $f(x) = x^3$ for $x \in \mathbb{R}$

(c) $f(x) = \dfrac{1}{x}$ for $x \neq 0$

(d) $f(x) = \sqrt{x}$ for $x > 0$

(e) $f(x) = \dfrac{1}{\sqrt{x}}$ for $x > 0$

25.2 Complete the proof of parts (a) and (b) of Theorem 25.7.

† We remark in passing that there exist functions that are continuous for all real x but have a derivative at *no* points. We shall prove this in Chapter 9 (Theorem 36.9) as an application of uniform convergence.

25.3 Let $f(x) = x^{1/3}$ for $x \in \mathbb{R}$.

(a) Use Definition 25.1 to prove that $f'(x) = \frac{1}{3}x^{-2/3}$ for $x \neq 0$.

(b) Show that f is not differentiable at $x = 0$.

***25.4** Let $f(x) = x^2 \sin(1/x)$ for $x \neq 0$ and $f(0) = 0$.

(a) Use the chain rule and the product rule to show that f is differentiable at each $c \neq 0$ and find $f'(c)$. (You may assume that the derivative of $\sin x$ is $\cos x$ for all $x \in \mathbb{R}$.)

(b) Use Definition 25.1 to show that f is differentiable at $x = 0$ and find $f'(0)$.

(c) Show that f' is not continuous at $x = 0$.

25.5 Determine for which values of x each function from $\mathbb{R}$ to $\mathbb{R}$ is differentiable and find the derivative.

(a) $f(x) = |x - 1|$ (b) $f(x) = |x^2 - 1|$

(c) $f(x) = |x|$ (d) $f(x) = x|x|$

***25.6** Let $f(x) = x^2 \sin(1/x^2)$ for $x \neq 0$ and $f(0) = 0$.

(a) Show that f is differentiable on $\mathbb{R}$.

(b) Show that f' is not bounded on the interval $[-1, 1]$.

25.7 Let $f(x) = x^2$ if $x \geqslant 0$ and $f(x) = 0$ if $x < 0$.

(a) Show that f is differentiable at $x = 0$.

(b) Find $f'(x)$ for all real x and sketch the graph of f'.

(c) Is f' continuous on $\mathbb{R}$? Is f' differentiable on $\mathbb{R}$?

25.8 Let $f(x) = x^2$ if x is rational and $f(x) = 0$ if x is irrational.

(a) Prove that f is continuous at exactly one point, namely at $x = 0$.

(b) Prove that f is differentiable at exactly one point, namely at $x = 0$.

25.9 Prove: If a polynomial $p(x)$ is divisible by $(x - a)^2$, then $p'(x)$ is divisible by $(x - a)$.

25.10 Let $f: I \to J$, $g: J \to K$, and $h: K \to \mathbb{R}$, where I, J, and K are intervals. Suppose that f is differentiable at $c \in I$, g is differentiable at $f(c)$, and h is differentiable at $g(f(c))$. Prove that $h \circ (g \circ f)$ is differentiable at c and find the derivative.

25.11 Let f, g, and h be real-valued functions that are differentiable on an interval I. Prove that the product function $fgh: I \to \mathbb{R}$ is differentiable on I and find $(fgh)'$.

25.12 Suppose that $f: I \to \mathbb{R}$ and $g: I \to \mathbb{R}$ are differentiable at $c \in I$ and that $g(c) \neq 0$.

(a) Use Exercise 25.1(c) and the chain rule (Theorem 25.10) to show that $(1/g)'(c) = -g'(c)/[g(c)]^2$.

(b) Use part (a) and the product rule [Theorem 25.7(c)] to derive the quotient rule [Theorem 25.7(d)].

25.13 Let I and J be intervals and suppose that the function $f: I \to J$ is twice differentiable on I. That is, the derivative f' exists and is itself differentiable on I. (We denote the derivative of f' by f''.) Suppose also that the function $g: J \to \mathbb{R}$ is twice differentiable on J. Prove that $g \circ f$ is twice differentiable on I and find $(g \circ f)''$.

25.14 Let $f: I \to \mathbb{R}$ where I is an open interval containing the point c, and let $k \in \mathbb{R}$. Prove the following.

(a) f is differentiable at c with $f'(c) = k$ iff $\lim_{h\to 0}(f(c+h) - f(c))/h = k$.

*(b) If f is differentiable at c with $f'(c) = k$, then $\lim_{h\to 0}(f(c+h) - f(c-h))/2h = k$.

(c) If f is differentiable at c with $f'(c) = k$, then $\lim_{n\to\infty} n[f(c + 1/n) - f(c)] = k$.

(d) Find counterexamples to show that the converses of parts (b) and (c) are not true.

25.15 A function $f: \mathbb{R} \to \mathbb{R}$ is called an **even** function if $f(-x) = f(x)$ for all $x \in \mathbb{R}$. If $f(-x) = -f(x)$ for all $x \in \mathbb{R}$, then f is called an **odd** function.

(a) Prove that if f is a differentiable even function, then f' is an odd function.

(b) Prove that if f is a differentiable odd function, then f' is an even function.

Section 26 THE MEAN VALUE THEOREM

The mean value theorem (also called the law of the mean) is one of the most important theoretical results in differential calculus. Its proof depends on the fact that a continuous function defined on a closed interval assumes its maximum and minimum values (Corollary 22.3). In this section we establish the theorem and derive several of its corollaries. We begin with a preliminary result about maxima and minima that is also of interest in its own right.

26.1 THEOREM If f is differentiable on an open interval (a, b) and if f assumes its maximum or minimum at a point $c \in (a, b)$, then $f'(c) = 0$.

Proof: Suppose that f assumes its maximum at c. That is, $f(x) \leqslant f(c)$ for all $x \in (a, b)$. Let (x_n) be a sequence converging to c such that $a < x_n < c$ for all n. (Since $a < c$ we may, for example, take $x_n = c - 1/n$ for n sufficiently large.) Then, since f is differentiable at c, Theorem 25.3 implies that the sequence

$$\left(\frac{f(x_n) - f(c)}{x_n - c}\right)$$

converges to $f'(c)$. But each term in this sequence of quotients is nonnegative since $f(x_n) \leqslant f(c)$ and $x_n < c$. Thus $f'(c) \geqslant 0$ by Corollary 17.5.

Similarly, let (y_n) be a sequence converging to c such that $c < y_n < b$ for all n. Then the terms of the sequence

$$\left(\frac{f(y_n) - f(c)}{y_n - c}\right)$$

are all nonpositive, since $f(y_n) \leqslant f(c)$ and $y_n > c$. Since the sequence of quotients again converges to $f'(c)$, we must have $f'(c) \leqslant 0$. Therefore, we conclude that $f'(c) = 0$.

The case in which f has a minimum at c is handled in a similar manner and is left to the reader. ▮

In beginning calculus when you wished to find the maximum (or minimum) of a continuous function f on a closed interval $[a, b]$, you were told to consider the following three types of points:

1. The points x where $f'(x) = 0$
2. The endpoints a and b
3. The points where f is not differentiable

In most, but not all, applications the extreme value will actually occur at a point of the first type where the derivative is zero. It is Theorem 26.1 that justifies this approach.

For our present purposes, we shall use Theorem 26.1 to prove Rolle's theorem, which is a special case of the mean value theorem.

26.2 THEOREM (Rolle's Theorem) Let f be a continuous function on $[a, b]$ that is differentiable on (a, b) and such that $f(a) = f(b) = 0$. Then there exists at least one point $c \in (a, b)$ such that $f'(c) = 0$.

Proof: Since f is continuous and $[a, b]$ is compact, Corollary 22.3 implies that there exist points x_1 and x_2 in $[a, b]$ such that $f(x_1) \leqslant f(x) \leqslant f(x_2)$ for all $x \in [a, b]$. If x_1 and x_2 are both endpoints of $[a, b]$, then $f(x) = f(a) = f(b) = 0$ for all $x \in [a, b]$. In this case f is a constant function and $f'(x) = 0$ for *all* $x \in (a, b)$. Otherwise, f assumes either a maximum or a minimum at some point $c \in (a, b)$. But then, by Theorem 26.1, $f'(c) = 0$. ▮

The geometric interpretation of Rolle's theorem is that, if the graph of a differentiable function touches the x-axis at a and b, then for some point c between a and b there is a horizontal tangent. (See Figure 26.1.) If we allow the function to have different values at the endpoints, then we cannot be assured of a horizontal tangent, but there will be a point c in (a, b) such that the tangent to the graph at $x = c$ will be parallel to the

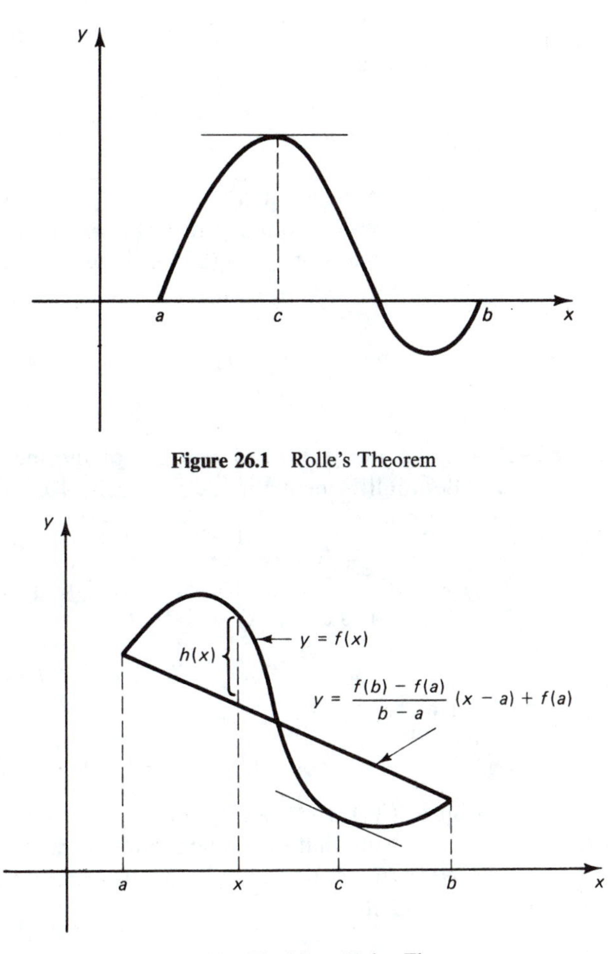

Figure 26.1 Rolle's Theorem

Figure 26.2 The Mean Value Theorem

chord between the endpoints of the graph. (See Figure 26.2.) This is the essence of the mean value theorem.

26.3 THEOREM (Mean Value Theorem) Let f be a continuous function on $[a, b]$ that is differentiable on (a, b). Then there exists at least one point $c \in (a, b)$ such that

$$f'(c) = \frac{f(b) - f(a)}{b - a}.$$

Proof: Figure 26.2 suggests that we might try to apply Rolle's theorem to the function h representing the difference between the graph of f and the chord. To this end, we let g be the function whose graph is the chord between the endpoints $(a, f(a))$ and $(b, f(b))$. That is,

$$g(x) = \frac{f(b) - f(a)}{b - a}(x - a) + f(a), \qquad \text{for all } x \in [a, b].$$

Then the function $h = f - g$ is continuous on $[a, b]$ and differentiable on (a, b). Since $h(a) = h(b) = 0$, h satisfies the hypotheses of Rolle's theorem (26.2). Thus there exists a point $c \in (a, b)$ such that $h'(c) = 0$. But then

$$0 = h'(c) = f'(c) - g'(c) = f'(c) - \frac{f(b) - f(a)}{b - a}. \qquad \blacksquare$$

26.4 EXAMPLE As an illustration of one use of the mean value theorem, let us derive Bernoulli's inequality (see Exercise 10.10) for $x > 0$:

$$(1 + x)^n \geqslant 1 + nx \qquad \text{for all } n \in \mathbb{N}.$$

Let $f(t) = (1 + t)^n$ on the interval $[0, x]$. Then by the mean value theorem there exists $c \in (0, x)$ such that

$$f(x) - f(0) = f'(c)(x - 0).$$

Thus we have

$$(1 + x)^n - 1 = n(1 + c)^{n-1}(x) \geqslant nx,$$

since $f'(c) = n(1 + c)^{n-1}$, $1 + c > 1$, and $n - 1 \geqslant 0$.

Note that this same argument can be used for any *real* number $n \geqslant 1$ as soon as we know that the derivative formula holds for all real exponents.

26.5 PRACTICE Use the mean value theorem to establish Bernoulli's inequality for $x \in (-1, 0)$ by letting $f(t) = (1 + t)^n$ on the interval $[x, 0]$.

In our next several results we show how the mean value theorem can be used to relate the properties of a function f and its derivative f'.

26.6 THEOREM Let f be continuous on $[a, b]$ and differentiable on (a, b). If $f'(x) = 0$ for all $x \in (a, b)$, then f is constant on $[a, b]$.

Proof: Suppose that f were not constant on $[a, b]$. Then there would exist x_1 and x_2 such that $a \leqslant x_1 < x_2 \leqslant b$ and $f(x_1) \neq f(x_2)$. But then by the mean value theorem, for some $c \in (x_1, x_2)$ we have $f'(c) = [f(x_2) - f(x_1)]/(x_2 - x_1) \neq 0$, a contradiction. $\blacksquare$

26.7 COROLLARY Let f and g be continuous on $[a, b]$ and differentiable on (a, b). Suppose that $f'(x) = g'(x)$ for all $x \in (a, b)$. Then there exists a constant C such that $f = g + C$ on $[a, b]$.

Proof: This follows directly by applying Theorem 26.6 to the function $f - g$. ▮

Recall from Exercise 22.11 that a function f is said to be **strictly increasing** on an interval I if $x_1 < x_2$ in I implies that $f(x_1) < f(x_2)$. [For **strictly decreasing** we have $x_1 < x_2$ in I implies that $f(x_1) > f(x_2)$.]

26.8 THEOREM Let f be differentiable on an interval I. Then

(a) if $f'(x) > 0$ for all $x \in I$, then f is strictly increasing on I, and
(b) if $f'(x) < 0$ for all $x \in I$, then f is strictly decreasing on I.

Proof: (a) Suppose that $f'(x) > 0$ for all $x \in I$, and let $x_1, x_2 \in I$ with $x_1 < x_2$. Since f is continuous on $[x_1, x_2]$ and differentiable on (x_1, x_2), the mean value theorem implies that there exists a point $c \in (x_1, x_2)$ such that

$$f(x_2) - f(x_1) = f'(c)(x_2 - x_1).$$

Since $f'(c) > 0$ and $x_2 - x_1 > 0$, we must have $f(x_2) > f(x_1)$. Thus f is strictly increasing on I.

The proof of (b) is similar. ▮

In Theorem 22.6 we saw that continuous functions satisfy the intermediate value property. In our next theorem we show that derivatives also have this property, even though they are not necessarily continuous (see Exercise 25.4).

26.9 THEOREM (Intermediate Value Theorem for Derivatives) Let f be differentiable on $[a, b]$ and suppose that k is a number between $f'(a)$ and $f'(b)$. Then there exists a point $c \in (a, b)$ such that $f'(c) = k$.

Proof: We assume that $f'(a) < k < f'(b)$, the proof of the other case being similar. For $x \in [a, b]$, let $g(x) = f(x) - kx$. Then g is differentiable on $[a, b]$ and $g'(a) < 0 < g'(b)$. Since g is continuous on $[a, b]$, Corollary 22.3 implies that g assumes its minimum at some point $c \in [a, b]$. We claim that $c \in (a, b)$.

Since

$$g'(b) = \lim_{x \to b} \frac{g(x) - g(b)}{x - b} > 0,$$

the ratio $[g(x) - g(b)]/(x - b)$ is positive for all $x \in U \cap [a, b]$, where U is some deleted neighborhood of b. (See Exercise 20.12.) Thus, for $x < b$ with $x \in U$, we must have $g(x) < g(b)$. Hence $g(b)$ is not the minimum of g on $[a, b]$, so $c \neq b$. Since $g'(a) > 0$, a similar argument shows that $c \neq a$.

We conclude that $c \in (a, b)$, so Theorem 26.1 implies that $g'(c) = 0$. But then $f'(c) = g'(c) + k = k$. ∎

We conclude this section by looking at the derivative of an inverse function.

26.10 THEOREM (Inverse Function Theorem) Suppose that f is differentiable on an interval I and $f'(x) \neq 0$ for all $x \in I$. Then f is injective, f^{-1} is differentiable on $f(I)$, and

$$(f^{-1})'(y) = \frac{1}{f'(x)},$$

where $y = f(x)$.

Proof: Since $f'(x) \neq 0$ for all $x \in I$, it follows from Theorem 26.9 that either $f'(x) > 0$ for all $x \in I$ or $f'(x) < 0$ for all $x \in I$. We shall assume that $f'(x) > 0$ for all $x \in I$, the other case being similar. Theorem 26.8 says that f is strictly increasing on I, so that f is injective. Since f is continuous, $f(I)$ is an interval and in Exercise 22.11 you were asked to show that f^{-1} is continuous on $f(I)$.

To see that f^{-1} is differentiable on $f(I)$, let $y \in f(I)$ and let (y_n) be any sequence in $f(I)\backslash\{y\}$ that converges to y. Let $x = f^{-1}(y)$ and for each $n \in \mathbb{N}$ let $x_n = f^{-1}(y_n)$. Then, since f^{-1} is continuous, the sequence (x_n) converges to x. Furthermore, each $x_n \neq x$ because f^{-1} is injective. Since f has a nonzero derivative at x, we have

$$\lim_{n \to \infty} \frac{f(x_n) - f(x)}{x_n - x} = f'(x) \neq 0,$$

so

$$\lim_{n \to \infty} \frac{f^{-1}(y_n) - f^{-1}(y)}{y_n - y} = \lim_{n \to \infty} \frac{x_n - x}{f(x_n) - f(x)} = \frac{1}{f'(x)}.$$

Thus by Theorem 25.3, f^{-1} is differentiable at y and $(f^{-1})'(y) = 1/f'(x)$. ∎

26.11. PRACTICE Let $n \in \mathbb{N}$ and let $y = f(x) = x^{1/n}$ for $x > 0$. Then f is the inverse of the function $g(y) = y^n$. Use Theorem 26.10 to verify the familiar derivative formula for f: $f'(x) = (1/n)x^{1/n - 1}$.

ANSWERS TO PRACTICE PROBLEMS

26.5 We obtain $f(0) - f(x) = f'(c)(0 - x)$ for some $c \in (x, 0)$. Thus we have

$$1 - (1 + x)^n = n(1 + c)^{n-1}(-x) \leqslant -nx,$$

since $0 < 1 + c < 1$ and $n - 1 \geqslant 0$. It follows that $(1 + x)^n \geqslant 1 + nx$.

26.11 $f'(x) = \dfrac{1}{g'(y)} = \dfrac{1}{ny^{n-1}} = \dfrac{1}{n(x^{1/n})^{n-1}} = \dfrac{1}{n} x^{1/n - 1}$

EXERCISES

26.1 Let $f(x) = x^2 - 3x + 4$ for $x \in [0, 2]$.

(a) Find where f is strictly increasing and where it is strictly decreasing.

(b) Find the maximum and minimum of f on $[0, 2]$.

26.2 Repeat Exercise 26.1 for $f(x) = |x^2 - 1|$ on $[0, 2]$.

26.3 Use the mean value theorem to establish the following inequalities. (You may assume any relevant derivative formulas from calculus.)

(a) $e^x > 1 + x$ for $x > 0$

(b) $\dfrac{x - 1}{x} < \log x < x - 1$ for $x > 1$

(c) $\dfrac{1}{8} < \sqrt{51} - 7 < \dfrac{1}{7}$

(d) $\sqrt{1 + x} < 1 + \dfrac{1}{2}x$ for $x > 0$

(e) $\sqrt{1 + x} < 5 + \dfrac{x - 24}{10}$ for $x > 24$

(f) $\sin x \leqslant x$ for $x \geqslant 0$

(g) $|\cos x - \cos y| \leqslant |x - y|$ for $x, y \in \mathbb{R}$

(h) $x < \tan x$ for $0 < x < \dfrac{\pi}{2}$

(i) $\arctan x < \dfrac{\pi}{4} + \dfrac{x - 1}{2}$ for $x > 1$

(j) $\left|\dfrac{\sin ax - \sin bx}{x}\right| \leqslant |a - b|$ for $x \neq 0$

***26.4** A function f is said to be **increasing** on an interval I if $x_1 < x_2$ in I implies that $f(x_1) \leqslant f(x_2)$. [For **decreasing**, replace $f(x_1) \leqslant f(x_2)$ by $f(x_1) \geqslant f(x_2)$.] Suppose that f is differentiable on an interval I. Prove the following:

(a) f is increasing on I iff $f'(x) \geqslant 0$ for all $x \in I$.

(b) f is decreasing on I iff $f'(x) \leqslant 0$ for all $x \in I$.

26.5 Show that the converses of parts (a) and (b) of Theorem 26.8 are false by finding counterexamples.

26.6 Let f be differentiable on $(0, 1)$ and continuous on $[0, 1]$. Suppose that $f(0) = 0$ and that f' is increasing on $(0, 1)$. (See Exercise 26.4.) Let $g(x) = f(x)/x$ for $x \in (0, 1)$. Prove that g is increasing on $(0, 1)$.

***26.7** Let f be differentiable on $[a, b]$. Suppose that $f'(x) \geqslant 0$ for all $x \in [a, b]$ and that f' is not identically zero on any subinterval of $[a, b]$. Prove that f is strictly increasing on $[a, b]$.

26.8 Let f be differentiable on $\mathbb{R}$. Suppose that $f(0) = 0$ and that $1 \leqslant f'(x) \leqslant 2$ for all $x \geqslant 0$. Prove that $x \leqslant f(x) \leqslant 2x$ for all $x \geqslant 0$.

26.9 Suppose that f is differentiable on $\mathbb{R}$ and that $f(0) = 0$, $f(1) = 2$, and $f(2) = 2$.

(a) Show that there exists $c_1 \in (0, 1)$ such that $f'(c_1) = 2$.
(b) Show that there exists $c_2 \in (1, 2)$ such that $f'(c_2) = 0$.
(c) Show that there exists $c_3 \in (0, 2)$ such that $f'(c_3) = \frac{5}{4}$.

26.10 Let $f(x) = x + 2x^2 \sin(1/x)$ for $x \neq 0$ and $f(0) = 0$.

(a) Show that $f'(0) = 1$.
(b) Show that f is not strictly increasing on any neighborhood of 0.
(c) Why doesn't part (b) contradict Theorem 26.8?

26.11 (a) Let f be differentiable on (a, b) and suppose that there exists $m \in \mathbb{R}$ such that $|f'(x)| \leqslant m$ for all $x \in (a, b)$. Prove that f is uniformly continuous on (a, b).

(b) Find an example of a function f that is differentiable and uniformly continuous on $(0, 1)$, but such that f' is unbounded on $(0, 1)$.

26.12. Let $f(x) = 2$ for $x < 0$ and $f(x) = 2x$ for $x \geqslant 0$.

(a) Prove that there does not exist a function F such that $F'(x) = f(x)$ for all $x \in \mathbb{R}$.

(b) Find examples of two functions g and h, not differing by a constant, such that $g'(x) = h'(x) = f(x)$ for all $x \neq 0$.

26.13 Let f and g be differentiable on $\mathbb{R}$. Suppose that $f(0) = g(0)$ and that $f'(x) \leqslant g'(x)$ for all $x \geqslant 0$. Show that $f(x) \leqslant g(x)$ for all $x \geqslant 0$.

26.14 Let f be differentiable on (a, b) and continuous on $[a, b]$. Suppose that $f(a) = f(b) = 0$. Prove that for each $k \in \mathbb{R}$ there exists $c \in (a, b)$ such that $f'(c) = kf(c)$. *Hint*: Apply Rolle's theorem to the function $g(x) = e^{-kx}f(x)$.

26.15 Let f be differentiable on an open interval I containing the point c. Suppose that $\lim_{x \to c} f'(x)$ exists (and is finite).

(a) Prove that f' is continuous at c.
(b) How does this differ from the situation in Exercise 25.4?

26.16 Let $y = f(x) = \tan x$ for $x \in (-\pi/2, \pi/2)$. Use Theorem 26.10 to obtain the derivative of $f^{-1}(y) = \arctan y$.

26.17 Let $r = m/n$ be a nonzero rational number in lowest terms, Let $f(x) = x^r$ for $x > 0$. Use Practice 26.11 and the chain rule to show that $f'(x) = rx^{r-1}$.

26.18 Let f be differentiable on (a, b) and suppose that $f'(c) \neq 0$ for some point $c \in (a, b)$. Suppose further that f^{-1} exists on some interval containing $f(c)$

and that f^{-1} is differentiable at $f(c)$. Use the chain rule (Theorem 25.10) to find $(f^{-1})'(f(c))$.

26.19 Let f be defined on an interval I. Suppose that there exists $M > 0$ and $\alpha > 0$ such that

$$|f(x) - f(y)| \leqslant M|x - y|^{\alpha}$$

for all $x, y \in I$. (Such a function is said to satisfy a **Lipschitz condition** of order α on I.)

(a) Prove that f is uniformly continuous on I.

(b) If $\alpha > 1$, prove that f is constant on I. *Hint*: First show that f is differentiable on I.

(c) Show by an example that if $\alpha = 1$, then f is not necessarily differentiable on I.

(d) Prove that if g is differentiable on an interval I and if g' is bounded on I, then g satisfies a Lipschitz condition of order 1 on I.

Section 27 L'HOSPITAL'S RULE

There are many situations in analysis where one wants to evaluate the limit of a quotient of functions:

$$\lim_{x \to c} \frac{f(x)}{g(x)}.$$

Suppose that $\lim_{x \to c} f(x) = L$ and $\lim_{x \to c} g(x) = M$. If $g(x) \neq 0$ for x close to c and $M \neq 0$, then Theorem 20.13 tells us that the limit of the quotient f/g is equal to L/M. If both L and M are zero, we can sometimes evaluate the limit of f/g by canceling a common factor in the quotient (as in Example 20.16). In this section we derive another technique that is often easier to use than factoring and has wider application.

In general, when $L = M = 0$, the limit of the quotient f/g is called an "indeterminate," because different values may be obtained for the limit, depending on the particular f and g. (The notation "0/0" is sometimes used to refer to this ambiguous situation.) For example, given any real number k, let $f(x) = kx$ and $g(x) = x$. Then

$$\lim_{x \to 0} \frac{f(x)}{g(x)} = \lim_{x \to 0} \frac{kx}{x} = \lim_{x \to 0} k = k.$$

Thus we see that the indeterminate form 0/0 can lead to any real number as a limit. In fact, it may also have no limit at all (Exercise 27.9).

As a prelude to our main theorem on evaluating the indeterminate 0/0, we derive the following generalization of the mean value theorem that is due to Cauchy.

27.1 THEOREM (Cauchy Mean Value Theorem) Let f and g be functions that are continuous on $[a, b]$ and differentiable on (a, b). Then there exists at least one point $c \in (a, b)$ such that

$$[f(b) - f(a)]g'(c) = [g(b) - g(a)]f'(c).$$

Proof: Let $h(x) = [f(b) - f(a)]g(x) - [g(b) - g(a)]f(x)$ for each $x \in [a, b]$. Then h is continuous on $[a, b]$ and differentiable on (a, b). Furthermore,

$$h(a) = f(b)g(a) - g(b)f(a) = h(b).$$

Thus, by the mean value theorem (26.3), there exists $c \in (a, b)$ such that $h'(c) = 0$. That is,

$$[f(b) - f(a)]g'(c) - [g(b) - g(a)]f'(c) = 0. \qquad \blacksquare$$

27.2 THEOREM (L'Hospital's Rule) Let f and g be continuous on $[a, b]$ and differentiable on (a, b). Suppose that $c \in [a, b]$ and that $f(c) = g(c) = 0$. Suppose also that $g'(x) \neq 0$ for $x \in U$, where U is the intersection of (a, b) and some deleted neighborhood of c. If

$$\lim_{x \to c} \frac{f'(x)}{g'(x)} = L, \qquad \text{with } L \in \mathbb{R},$$

then

$$\lim_{x \to c} \frac{f(x)}{g(x)} = L.$$

Proof: Let (x_n) be a sequence in U that converges to c. By the Cauchy mean value theorem there is a sequence (c_n) such that c_n is between x_n and c for each n, and such that

$$[f(x_n) - f(c)]g'(c_n) = [g(x_n) - g(c)]f'(c_n).$$

Since $g'(x) \neq 0$ for all $x \in U$, g is injective on $U \cup \{c\}$ (see Exercise 26.7). Thus $g(x_n) \neq 0$ for all n. Since $f(c) = g(c) = 0$, we have

$$\frac{f(x_n)}{g(x_n)} = \frac{f(x_n) - f(c)}{g(x_n) - g(c)} = \frac{f'(c_n)}{g'(c_n)}, \qquad \text{for all } n.$$

Furthermore, since $x_n \to c$ and c_n is between x_n and c, it follows that $c_n \to c$. Thus by Theorem 20.8, $\lim_{n\to\infty} [f'(c_n)/g'(c_n)] = L$. But then $\lim_{n\to\infty} [f(x_n)/g(x_n)] = L$, so $\lim_{x\to c} [f(x)/g(x)] = L$ also. $\blacksquare$

27.3 EXAMPLE Let $f(x) = 2x^2 - 3x + 1$ and $g(x) = x - 1$. (See Example 20.5.) Then $f(1) = g(1) = 0$. Now $f'(x) = 4x - 3$ and $g'(x) = 1$, so that

$$\lim_{x\to 1} \frac{2x^2 - 3x + 1}{x - 1} = \lim_{x\to 1} \frac{4x - 3}{1} = 1.$$

27.4 EXAMPLE If we assume the familiar properties of the trigonometric, exponential, and logarithmic functions, we can evaluate limits that we would not be able to handle by factoring. For example, let $f(x) = 1 - \cos x$ and $g(x) = x^2$. Then $f(0) = g(0) = 0$, $f'(x) = \sin x$, and $g'(x) = 2x$. Thus we have

$$\lim_{x \to 0} \frac{1 - \cos x}{x^2} = \lim_{x \to 0} \frac{\sin x}{2x}$$

provided that the second limit exists. Since $\sin x \to 0$ and $2x \to 0$ as $x \to 0$, we again have the indeterminate form $0/0$. Since the hypotheses of L'Hospital's rule are still satisfied for this new quotient, we apply the rule again to obtain

$$\lim_{x \to 0} \frac{1 - \cos x}{x^2} = \lim_{x \to 0} \frac{\sin x}{2x} = \lim_{x \to 0} \frac{\cos x}{2} = \frac{1}{2}.$$

Note that $g'(x)$ must be nonzero in a deleted neighborhood of 0, but that it is permitted that $g'(0) = 0$.

27.5 PRACTICE Use L'Hospital's rule to evaluate $\lim_{x \to 0} \frac{e^{2x} - 1}{x}$.

Limits at Infinity

In some situations we wish to evaluate the limit of a function (or a quotient of functions) for larger and larger values of the variable. We make this precise in the following definition, which is patterned after the notion of a limit of a sequence (Definition 16.2).

27.6 DEFINITION Let $f: (b, \infty) \to \mathbb{R}$, where $b \in \mathbb{R}$. We say that $L \in \mathbb{R}$ is the limit of f as $x \to \infty$, and we write

$$\lim_{x \to \infty} f(x) = L,$$

provided that for each $\varepsilon > 0$ there exists a real number $N > b$ such that $x > N$ implies that $|f(x) - L| < \varepsilon$.

The limit of a function f as $x \to \infty$ is sometimes referred to loosely as "the limit of f at ∞." There are many similarities between this concept and the limit of f at a real number. In particular, it is easy to see (Exercise 27.6) that the limit of f as $x \to \infty$ is unique if it exists.

Very often as $x \to \infty$ the values of a given function also get large. This leads to the following definition.

27.7 DEFINITION Let $f\colon (b, \infty) \to \mathbb{R}$. We say that f **tends to** ∞ as $x \to \infty$ and we write

$$\lim_{x \to \infty} f(x) = \infty,$$

provided that given any $\alpha \in \mathbb{R}$ there exists $N > b$ such that $x > N$ implies that $f(x) > \alpha$.

Using these concepts of infinite limits, we are now in a position to prove L'Hospital's rule for indeterminates of the form ∞/∞. In our proof we shall need to use some algebraic properties of limits at ∞ (analogous to Theorem 20.13) and the fact that for any $k \in \mathbb{R}$, $\lim_{x \to \infty} k/f(x) = 0$ whenever $\lim_{x \to \infty} f(x) = \infty$. (See Exercises 27.7 and 27.8.)

27.8 THEOREM (L'Hospital's Rule) Let f and g be differentiable on (b, ∞). Suppose that $\lim_{x \to \infty} f(x) = \lim_{x \to \infty} g(x) = \infty$, and that $g'(x) \neq 0$ for $x \in (b, \infty)$. If

$$\lim_{x \to \infty} \frac{f'(x)}{g'(x)} = L, \qquad \text{where } L \in \mathbb{R},$$

then

$$\lim_{x \to \infty} \frac{f(x)}{g(x)} = L.$$

Proof: Given $\varepsilon > 0$, there exists $N_1 > b$ such that $x > N_1$ implies that

$$\left| \frac{f'(x)}{g'(x)} - L \right| < \frac{\varepsilon}{2}.$$

Since $\lim_{x \to \infty} f(x) = \lim_{x \to \infty} g(x) = \infty$, there exists $N_2 > N_1$ such that $x > N_2$ implies that $f(x) > 0$ and $g(x) > 0$. Furthermore, there exists $N_3 > N_2$ such that $x > N_3$ implies that $f(x) > f(N_2)$ and $g(x) > g(N_2)$. For any $x > N_3$ the Cauchy mean value theorem implies that there exists a point c in (N_2, x) such that

$$\frac{f'(c)}{g'(c)} = \frac{f(x) - f(N_2)}{g(x) - g(N_2)} = \frac{f(x)}{g(x)} \cdot \frac{1 - f(N_2)/f(x)}{1 - g(N_2)/g(x)}.$$

But then for $x > N_3$ we have

$$\frac{f(x)}{g(x)} = \frac{f'(c)}{g'(c)} \cdot F(x), \qquad \text{where } F(x) = \frac{1 - g(N_2)/g(x)}{1 - f(N_2)/f(x)}.$$

Now from our hypotheses on f and g we see that $\lim_{x \to \infty} F(x) = 1$.

To show that $f(x)/g(x)$ is close to L, we write

$$\begin{aligned}\left|\frac{f(x)}{g(x)} - L\right| &= \left|\frac{f'(c)}{g'(c)}F(x) - L\right| \\ &\leqslant \left|\frac{f'(c)}{g'(c)}F(x) - \frac{f'(c)}{g'(c)}\right| + \left|\frac{f'(c)}{g'(c)} - L\right| \\ &= \left|\frac{f'(c)}{g'(c)}\right|\left|F(x) - 1\right| + \left|\frac{f'(c)}{g'(c)} - L\right|\end{aligned}$$

for $x > N_3$. Now the point c depends on x, but we always have $c > N_2 > N_1$, so

$$\left|\frac{f'(c)}{g'(c)} - L\right| < \frac{\varepsilon}{2}.$$

This in turn implies that $|f'(c)/g'(c)| < |L| + \varepsilon/2$. Finally, since $\lim_{x\to\infty} F(x) = 1$, there exists $N_4 > N_3$ such that for $x > N_4$ we have

$$|F(x) - 1| < \frac{\varepsilon}{2(|L| + \varepsilon/2)}.$$

It follows that $x > N_4$ implies that $|f(x)/g(x) - L| < \varepsilon$, so that $\lim_{x\to\infty} f(x)/g(x) = L$. ∎

27.9 PRACTICE Evaluate $\lim_{x\to\infty} (\log x)/x$.

There are other limiting situations involving two functions that can give rise to ambiguous values. These indeterminates are indicated by the symbols $0\cdot\infty$, 0^0, 1^∞, ∞^0, and $\infty - \infty$, and are evaluated by using algebraic manipulations, logarithms, or exponentials to change them into one of the forms $0/0$ or ∞/∞.

27.10 EXAMPLE For $x > 0$, let $f(x) = x$ and $g(x) = -\log x$. Then $\lim_{x\to 0} f(x)g(x)$ is indeterminate of the form $0\cdot\infty$. To evaluate the limit, we write

$$\lim_{x\to 0} (x)(-\log x) = \lim_{x\to 0} \frac{-\log x}{1/x} = \lim_{x\to 0} \frac{-1/x}{-1/x^2} = \lim_{x\to 0} x = 0.$$

27.11 EXAMPLE To evaluate $\lim_{x\to 0} x^x$, where $x > 0$, we let $y = x^x$. Then $\log y = x \log x$, and by Example 27.10, $\log y \to 0$ as $x \to 0$. Thus, since e^x is a continuous function, we have $y = e^{\log y} \to e^0 = 1$ as $x \to 0$.

ANSWERS TO PRACTICE PROBLEMS

27.5 $\lim_{x \to 0} \frac{e^{2x} - 1}{x} = \lim_{x \to 0} \frac{2e^{2x}}{1} = 2$

27.9 $\lim_{x \to \infty} \frac{\log x}{x} = \lim_{x \to \infty} \frac{1/x}{1} = 0$

EXERCISES

27.1 Evaluate the following limits.

(a) $\lim_{x \to 0} \frac{\sin x}{x}$

(b) $\lim_{x \to 0+} \frac{\sin x}{\sqrt{x}}$

(c) $\lim_{x \to 0} \frac{\sin x - x}{x^3}$

(d) $\lim_{x \to 1} \frac{\cos(\pi x/2)}{x - 1}$

(e) $\lim_{x \to 1} \frac{\log x}{x - 1}$

(f) $\lim_{x \to 0+} (1 + 3x)^{1/x}$

(g) $\lim_{x \to \infty} \left(1 + \frac{1}{x}\right)^x$

(h) $\lim_{x \to 0} \left(\frac{1}{x} - \frac{1}{\sin x}\right)$

27.2 Evaluate the following limits.

(a) $\lim_{x \to 0} \frac{\tan x - x}{x^3}$

(b) $\lim_{x \to 0} \frac{\sin x - x}{e^x - 1}$

(c) $\lim_{x \to \infty} \frac{x^2}{e^x}$

(d) $\lim_{x \to 0+} \frac{\log \sin x}{\log x}$

(e) $\lim_{x \to \infty} \frac{(\log x)^2}{x}$

(f) $\lim_{x \to 1} \frac{\log x}{x^2 + x - 2}$

(g) $\lim_{x \to \infty} (1 + 3x)^{1/x}$

(h) $\lim_{x \to 0+} x^{2x}$

27.3 Indicate what is wrong with the following result:

$$\lim_{x \to 1} \frac{2x^2 - x - 1}{3x^2 - 5x + 2} = \lim_{x \to 1} \frac{4x - 1}{6x - 5} = \lim_{x \to 1} \frac{4}{6} = \frac{2}{3}.$$

27.4 Suppose that $f(x) = g(1/x)$ for $x > 0$ and let $L \in \mathbb{R}$. Prove that $\lim_{x \to \infty} f(x) = L$ iff $\lim_{x \to 0} g(x) = L$.

27.5 Let $f: (b, \infty) \to \mathbb{R}$. Prove that $\lim_{x \to \infty} f(x) = L$ iff for every sequence (x_n) in (b, ∞) with $\lim_{n \to \infty} x_n = \infty$, the sequence $(f(x_n))$ converges to L.

27.6 Let $f: (a, \infty) \to \mathbb{R}$. Prove: If the limit of f as $x \to \infty$ exists, then it is unique.

27.7 Let f and g be real-valued functions defined on (b, ∞). Suppose that $\lim_{x\to\infty} f(x) = L$ and $\lim_{x\to\infty} g(x) = M$, where $L, M \in \mathbb{R}$. Prove the following.

(a) $\lim_{x\to\infty} (f + g)(x) = L + M$.
(b) $\lim_{x\to\infty} (fg)(x) = LM$.
(c) If $k \in \mathbb{R}$, then $\lim_{x\to\infty} (kf) = kL$.
(d) If $g(x) \neq 0$ for $x > b$ and $M \neq 0$, then $\lim_{x\to\infty} (f/g)(x) = L/M$.

27.8 Let $f\colon (b, \infty) \to \mathbb{R}$ and let $k \in \mathbb{R}$. Prove that $\lim_{x\to\infty} [k/f(x)] = 0$ whenever $\lim_{x\to\infty} f(x) = \infty$.

27.9 Find examples of functions f and g such that $f \to 0$ and $g \to 0$ as $x \to 0$, $\lim_{x\to 0} f/g$ does not exist, and that satisfy the following conditions.

(a) f and g are nonzero in a deleted neighborhood of 0.
(b) $\lim_{x\to 0^-} f/g$ and $\lim_{x\to 0^+} f/g$ exist and are finite.
(c) g is not identically zero in any neighborhood of 0, but the one-sided limits in part (b) do not exist (finite or infinite).

27.10 Suppose that a curve is described parametrically by $x = f(t)$ and $y = g(t)$, where $a \leqslant t \leqslant b$. If f and g are continuous on $[a, b]$ and differentiable on (a, b), we may apply the Cauchy mean value theorem to f and g to obtain a point $c \in (a, b)$ such that

$$\frac{g'(c)}{f'(c)} = \frac{g(b) - g(a)}{f(b) - f(a)},$$

as long as $f'(c) \neq 0$. Interpret this result geometrically.

27.11 (a) Let $f\colon D \to \mathbb{R}$ and let c be an accumulation point of D. Give a reasonable definition of $\lim_{x\to c} f(x) = \infty$.

(b) Let $f\colon D \to \mathbb{R}$ and let c be an accumulation point of D. Prove that $\lim_{x\to c} f(x) = \infty$ iff for every sequence (s_n) in D that converges to c with $s_n \neq c$ for all n, $\lim_{n\to\infty} f(s_n) = \infty$.

(c) Suppose that f and g are continuous on $[a, b]$ and differentiable on (a, b). Suppose that $c \in [a, b]$ and that $f(c) = g(c) = 0$. Suppose also that $g'(x) \neq 0$ for $x \in U$, where U is the intersection of (a, b) and some deleted neighborhood of c. If $\lim_{x\to c} [f'(x)/g'(x)] = \infty$, prove that $\lim_{x\to c} [f(x)/g(x)] = \infty$.

*(d) Suppose that f and g are differentiable on (a, b), that $\lim_{x\to a} (f(x) = \lim_{x\to a} g(x) = \infty$, and that $g'(x) \neq 0$ for $x \in (a, b)$. If $\lim_{x\to a} [f'(x)/g'(x)] = L$, where $L \in \mathbb{R}$, prove that $\lim_{x\to a} [f(x)/g(x)] = L$.

27.12 Let $x \in \mathbb{R}$. Recall (Definition 13.1) that a neighborhood of x is an open interval of the sort $(x - \varepsilon, x + \varepsilon)$, where $\varepsilon > 0$. We now define a **neighborhood of** ∞ to be an open interval of the sort (b, ∞), where $b \in \mathbb{R}$. Suppose that $f\colon D \to \mathbb{R}$ and let c and L be real numbers or ∞. If $c \in \mathbb{R}$, we require c to be an accumulation point of D, and if $c = \infty$, we require D to be unbounded above. Define $\lim_{x\to c} f(x) = L$ if for every neighborhood V of L there exists a neighborhood U of c such that $f(x) \in V$ for all $x \in U \cap D$ with $x \neq c$. Show that this definition is consistent with Definition 20.1 in the finite case where $c, L \in \mathbb{R}$ and consistent with Definitions 27.6 and 27.7

in the cases where $c = \infty$. (It should also agree with Exercise 27.11 when $c \in \mathbb{R}$ and $L = \infty$.)

27.13 Define a neighborhood of $-\infty$ to be an open interval of the sort $(-\infty, b)$, where $b \in \mathbb{R}$. Let c be a real number or ∞.

(a) Use Exercise 27.12 to give reasonable definitions of $\lim_{x \to c} f(x) = -\infty$ and $\lim_{x \to -\infty} f(x) = c$ in terms of neighborhoods.

(b) Reformulate your answer to part (a) without referring to neighborhoods (similar to the style of Definitions 27.6 and 27.7).

(c) State and prove a sequential criterion for limits involving $-\infty$.

(d) State and prove results similar to L'Hospital's rule for limits involving $-\infty$.

Section 28 TAYLOR'S THEOREM

In this section we derive an important generalization of the mean value theorem that is useful in approximating functions by polynomials. It will also be helpful to us when discussing series of functions in Chapter 9. We begin by describing the higher derivatives of a function.

If f is differentiable on an interval I, then $f': I \to \mathbb{R}$. If $c \in I$ and f' is differentiable at c, then the derivative of f' at c [denoted by $f''(c)$ or $f^{(2)}(c)$] is the **second derivative** of f at c, and f is said to be **twice differentiable** at c. For many functions this process can be repeated. In general, if a function f can be differentiated n times (where $n \in \mathbb{N}$), then f is said to be n-times differentiable and we denote the nth derivative of f by $f^{(n)}$. Thus to say that $f^{(n)}$ exists at a point c is to say that the $(n-1)$st derivative $f^{(n-1)}$ exists in an interval containing c and that $f^{(n-1)}$ is differentiable at c.

28.1 EXAMPLE Let $f(x) = |x^3|$ for $x \in \mathbb{R}$. That is,

$$f(x) = \begin{cases} x^3, & \text{if } x \geqslant 0 \\ -x^3, & \text{if } x < 0. \end{cases}$$

Then $f'(x) = 3x^2$ if $x > 0$ and $f'(x) = -3x^2$ if $x < 0$. It is easy to show (Exercise 28.8) that $f'(0) = 0$, so f is differentiable on $\mathbb{R}$ and we have

$$f'(x) = \begin{cases} 3x^2, & \text{if } x \geqslant 0 \\ -3x^2, & \text{if } x < 0. \end{cases}$$

Differentiating again, we find that

$$f''(x) = \begin{cases} 6x, & \text{if } x \geqslant 0 \\ -6x, & \text{if } x < 0, \end{cases}$$

so that $f''(x) = 6|x|$ for all $x \in \mathbb{R}$. (Again, the only case that requires much effort is when $x = 0$.) Thus f is twice differentiable on $\mathbb{R}$, but the third derivative does not exist at $x = 0$. (Why not?)

If a function is $(n + 1)$-times differentiable on an open interval, then there is an extension of the mean value theorem that applies to the higher derivatives. This result is often referred to as Taylor's theorem because of its relation to the Taylor polynomials, named for Brook Taylor (1685–1731). (See the discussion following the theorem.) Actually, the form of the theorem as it appears here is due to J. L. Lagrange (1736–1813).

28.2 THEOREM (Taylor's Theorem) Let f and its first n derivatives be continuous on $[a, b]$ and differentiable on (a, b), and let $x_0 \in [a, b]$. Then for each $x \in [a, b]$ with $x \neq x_0$ there exists a point c between x and x_0 such that

$$f(x) = f(x_0) + f'(x_0)(x - x_0) + \frac{f''(x_0)}{2!}(x - x_0)^2 + \cdots$$
$$+ \frac{f^{(n)}(x_0)}{n!}(x - x_0)^n + \frac{f^{(n+1)}(c)}{(n+1)!}(x - x_0)^{n+1}.$$

Proof: Fix x in $[a, b]$ with $x \neq x_0$, and let M be the unique solution of

$$f(x) = f(x_0) + f'(x_0)(x - x_0) + \cdots + \frac{f^{(n)}(x_0)}{n!}(x - x_0)^n$$
$$+ M(x - x_0)^{n+1}.$$

(Since $x \neq x_0$, it will always be possible to solve algebraically for M.) Our proof will consist of showing that $M = f^{(n+1)}(c)/(n + 1)!$ for some c between x and x_0.

To this end, we define

$$F(t) = f(t) + f'(t)(x - t) + \cdots + \frac{f^{(n)}(t)}{n!}(x - t)^n + M(x - t)^{n+1}.$$

By our hypotheses on f and its derivatives, we see that F is continuous on $[a, b]$ and differentiable on (a, b). Whence the same properties hold on the interval between x and x_0. Now $F(x) = f(x)$ since all the terms after the first are zero, and $F(x_0) = f(x)$ by our choice of M. Thus, by the mean value theorem (26.3), there exists a point c between x and x_0 such that

$$F'(c) = \frac{F(x) - F(x_0)}{x - x_0} = 0.$$

When we compute the derivative of F, we find that all the terms except the last two cancel in pairs. [Notice that $f'(t)(x-t)$, for example, must be differentiated by the product rule to obtain $f''(t)(x-t) - f'(t)$.] Thus we obtain

$$0 = F'(c) = \frac{f^{(n+1)}(c)}{n!}(x-c)^n - M(n+1)(x-c)^n,$$

so that $M = f^{(n+1)}(c)/(n+1)!$, as desired. ▌

Taylor's theorem can be viewed as an extension of the mean value theorem in the sense that taking $x = b$, $x_0 = a$, and $n = 1$ in Taylor's theorem yields the earlier result.

Another way to look at Taylor's theorem is to consider the problem of approximating a function f near a point x_0 by a sequence of polynomials of the form

$$p_n(x) = a_0 + a_1(x - x_0) + a_2(x - x_0)^2 + \cdots + a_n(x - x_0)^n.$$

If we take $a_0 = f(x_0)$, then the polynomial

$$p_0(x) = a_0$$

(a polynomial of degree zero) will agree with f at x_0. If we wish $p_1(x) = a_0 + a_1(x - x_0)$ to satisfy

$$p_1(x_0) = f(x_0) \quad \text{and} \quad p_1'(x_0) = f'(x_0),$$

then since $p_1'(x) = a_1$ we have $a_1 = f'(x_0)$ and $a_0 = f(x_0)$. (See Figure 28.1.)

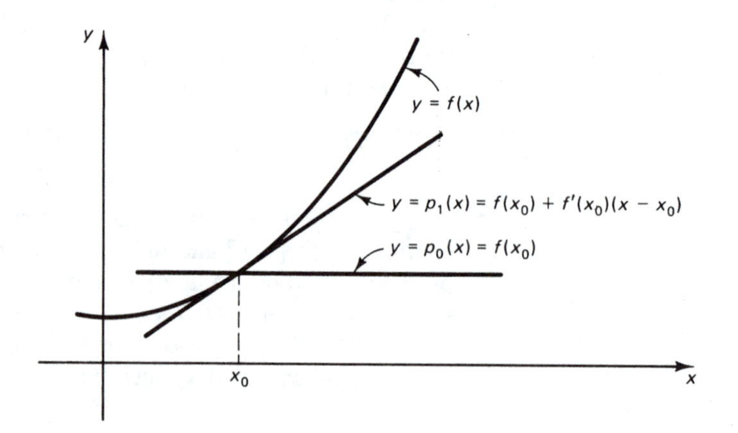

Figure 28.1 The Taylor Polynomials

In general, we want to have p_n and its first n derivatives agree with f and its first n derivatives at $x = x_0$. Now

$$p_n'(x) = a_1 + 2a_2(x - x_0) + 3a_3(x - x_0)^2 + \cdots + na_n(x - x_0)^{n-1}$$

$$p_n''(x) = 2a_2 + 3 \cdot 2a_3(x - x_0) + \cdots + n(n-1)a_n(x - x_0)^{n-2}$$

$$\vdots$$

$$p_n^{(n)}(x) = n!a_n.$$

Evaluating at $x = x_0$, we obtain $a_0 = p_n(x_0) = f(x_0)$, $a_1 = p_n'(x_0) = f'(x_0)$, $a_2 = \frac{1}{2}p_n''(x_0) = \frac{1}{2}f''(x_0), \ldots$, and in general

$$a_k = \frac{p_n^{(k)}(x_0)}{k!} = \frac{f^{(k)}(x_0)}{k!}, \qquad \text{for } k = 0, 1, \ldots, n.$$

[It is customary to let $f^{(0)}$ denote f and let $0! = 1$.] The polynomials we obtain in this manner are called the **Taylor polynomials** for f at $x = x_0$:

$$p_0(x) = f(x_0)$$

$$p_1(x) = f(x_0) + f'(x_0)(x - x_0)$$

$$p_2(x) = f(x_0) + f'(x_0)(x - x_0) + \frac{f''(x_0)}{2!}(x - x_0)^2$$

$$\vdots$$

$$p_n(x) = f(x_0) + f'(x_0)(x - x_0) + \cdots + \frac{f^{(n)}(x_0)}{n!}(x - x_0)^n.$$

In this context we see that Taylor's theorem gives us the remainder (or error)

$$R_n(x) = f(x) - p_n(x) = \frac{f^{(n+1)}(c)}{(n+1)!}(x - x_0)^{n+1}$$

obtained in approximating f by p_n, and it expresses this remainder in terms of the $(n + 1)$st derivative of f. Unfortunately, the exact location of c is not specified, so its usefulness depends on having an estimate on the size of $f^{(n+1)}$ between x and x_0. Other formulas for the remainder are possible, as we shall see in Exercise 31.16. One hopes that as n gets large the remainder term will approach zero, at least for values of x close to x_0.

If f has derivatives of all orders in a neighborhood of x_0, then the limit of the Taylor polynomials is an infinite series known as the **Taylor series** for the function f at the point x_0:

$$\sum_{n=0}^{\infty} \frac{f^{(n)}(x_0)}{n!}(x - x_0)^n = f(x_0) + \frac{f'(x_0)}{1!}(x - x_0) + \frac{f''(x_0)}{2!}(x - x_0)^2 + \cdots.$$

We shall discuss series of functions more fully in Chapter 9.

28.3 EXAMPLE As an illustration of the usefulness of Taylor's theorem in approximations, let us consider $f(x) = e^x$ for $x \in \mathbb{R}$. To find the nth Taylor polynomial for f at $x = 0$, we recall from beginning calculus that $f^{(n)}(x) = e^x$ for all $n \in \mathbb{N}$ so that $f(0) = f'(0) = \cdots = f^{(n)}(0) = e^0 = 1$. Thus we have

$$p_n(x) = 1 + x + \frac{x^2}{2!} + \frac{x^3}{3!} + \cdots + \frac{x^n}{n!}.$$

To find the error involved in approximating e^x by $p_n(x)$, we use the remainder term given by Taylor's theorem. That is,

$$R_n(x) = \frac{f^{(n+1)}(c)}{(n+1)!}x^{n+1} = \frac{e^c x^{n+1}}{(n+1)!},$$

where c is some number between 0 and x.

For example, suppose that we take $n = 5$ and compute the error when $x \in [-1, 1]$. Since c is also in $[-1, 1]$, a simple calculation shows that $|R_5(x)| \leqslant e/6! < 0.0038$. Thus for all $x \in [-1, 1]$ the polynomial

$$1 + x + \frac{x^2}{2} + \frac{x^3}{6} + \frac{x^4}{24} + \frac{x^5}{120}$$

differs from e^x by less than 0.0038.

Furthermore, since $\lim_{n \to \infty} e^k k^{n+1}/(n+1)! = 0$ for all $k \in \mathbb{R}$ (see Exercise 17.15), we see that given any bounded interval $I = [-k, k]$ and any $\varepsilon > 0$, we can find a Taylor polynomial p_n that will approximate e^x to within ε on I if we take n sufficiently large.

28.4 PRACTICE Find the Taylor polynomial p_5 for the function $f(x) = x^3 + 4x^2 - 3x + 5$ at the point $x = 1$. What is the error?

28.5 EXAMPLE Suppose that $f(x) = \sqrt{1+x}$ for $x > 0$. Let us find p_2 and p_3 for f at $x = 0$. We have

$$f(x) = (1+x)^{1/2} \qquad f(0) = 1$$

$$f'(x) = \frac{1}{2}(1+x)^{-1/2} \qquad f'(0) = \frac{1}{2}$$

$$f''(x) = -\frac{1}{4}(1+x)^{-3/2} \qquad f''(0) = -\frac{1}{4}$$

$$f^{(3)}(x) = \frac{3}{8}(1+x)^{-5/2} \qquad f^{(3)}(0) = \frac{3}{8}$$

$$f^{(4)}(x) = -\frac{15}{16}(1+x)^{-7/2} \qquad f^{(4)}(0) = -\frac{15}{16}.$$

Thus $p_2(x) = 1 + x/2 - x^2/8$ and $p_3 = 1 + x/2 - x^2/8 + x^3/16$. Now since $R_2(x) = \frac{3}{8}(1 + c)^{-5/2}x^3/3! > 0$ for $x > 0$, we have $f(x) > p_2(x)$. On the other hand, $R_3(x) = -\frac{15}{16}(1 + c)^{-7/2}x^4/4! < 0$ for $x > 0$, so $f(x) < p_3(x)$. That is, if $x > 0$, then

$$1 + \frac{x}{2} - \frac{x^2}{8} < \sqrt{1 + x} < 1 + \frac{x}{2} - \frac{x^2}{8} + \frac{x^2}{16}.$$

ANSWERS TO PRACTICE PROBLEMS

28.4 $f(x) = x^3 + 4x^2 - 3x + 5 \qquad f(1) = 7$
$f'(x) = 3x^2 + 8x - 3 \qquad f'(1) = 8$
$f''(x) = 6x + 8 \qquad f''(1) = 14$
$f^{(3)}(x) = 6 \qquad f^{(3)}(1) = 6$
$f^{(4)}(x) = f^{(5)}(x) = f^{(6)}(x) = 0$ for all x

Thus $p_5(x) = 7 + 8(x - 1) + 7(x - 1)^2 + 1(x - 1)^3$. Notice that this is the same as $p_4(x)$ and $p_3(x)$ and that the error for all three of these Taylor polynomials is zero. That is, $p_5(x) = p_4(x) = p_3(x) = f(x)$ for all $x \in \mathbb{R}$. In general, if f is a kth-degree polynomial, then $p_n = f$ for all $n \geqslant k$.

EXERCISES

Note: In these exercises you may assume the familiar derivative formulas for the trigonometric, logarithmic, and exponential functions.

28.1 Use Example 28.5 to approximate the following:
(a) $\sqrt{3}$; (b) $\sqrt{2}$; (c) $\sqrt{1.2}$.

28.2 Find a Taylor polynomial that approximates e^x to within 0.2 on the interval $[-2, 2]$.

28.3 Let $f(x) = \sin x$.
(a) Find p_6 for f at $x = 0$.
(b) How accurate is this on the interval $[-1, 1]$?

28.4 Show that if $x \in [0, 1]$, then

$$x - \frac{x^2}{2} + \frac{x^3}{3} - \frac{x^4}{4} \leqslant \log(1 + x) \leqslant x - \frac{x^2}{2} + \frac{x^3}{3}.$$

28.5 Let $f(x) = \cos x$. Use p_5 for f at $x = 0$ to estimate $\cos 1$. What is the error?

28.6 Let $f(x) = \sqrt{x}$. Use p_2 for f at $x = 9$ to estimate $\sqrt{8.8}$. What is the error?

28.7 Suppose that f is defined in a neighborhood of c and suppose that $f''(c)$ exists.

(a) Show that

$$f''(c) = \lim_{h \to 0} \frac{f(c+h) + f(c-h) - 2f(c)}{h^2}$$

(b) Give an example where the limit in part (a) exists but $f''(c)$ does not exist.

28.8 Show that $f'(0) = f''(0) = 0$ in Example 28.1.

28.9 A function $f: D \to \mathbb{R}$ is said to have a **local maximum** (respectively, **minimum**) at a point $x_0 \in D$ if there exists a neighborhood U of x_0 such that $f(x) \leqslant f(x_0)$ [respectively, $f(x) \geqslant f(x_0)$] for all $x \in U \cap D$. Suppose for some integer $n \geqslant 2$ that the derivatives $f', f'', \ldots, f^{(n)}$ exist and are continuous on an open interval I containing x_0 and that $f'(x_0) = \cdots = f^{(n-1)}(x_0) = 0$, but $f^{(n)}(x_0) \neq 0$. Use Taylor's theorem to prove the following:

(a) If n is even and $f^{(n)}(x_0) < 0$, then f has a local maximum at x_0.
(b) If n is even and $f^{(n)}(x_0) > 0$, then f has a local minimum at x_0.
(c) If n is odd, then f has neither a local maximum nor a local minimum at x_0.

28.10 Let $f: I \to \mathbb{R}$, where I is an open interval containing the point x_0. Let $n \in \mathbb{N}$, suppose that f is n-times differentiable at x_0, and let p_n be the nth Taylor polynomial for f at $x = x_0$.

(a) Apply L'Hospital's rule $n - 1$ times and then use the definition of $f^{(n)}(x_0)$ to show that

$$\lim_{x \to x_0} \frac{f(x) - p_n(x)}{(x - x_0)^n} = 0.$$

(b) Why can't we just apply L'Hospital's rule n times in establishing the limit in part (a)?

28.11 The following is an outline of a proof that e is irrational.

(a) Use Theorem 26.8 to show that e^x is strictly increasing on $\mathbb{R}$.
(b) Use Taylor's theorem at $x_0 = 0$ and the estimate $e < 3$ to show that, for all $n \in \mathbb{N}$,

$$0 < e - \left(1 + 1 + \frac{1}{2!} + \frac{1}{3!} + \cdots + \frac{1}{n!}\right) < \frac{3}{(n+1)!}.$$

(c) Suppose that e were rational. That is, suppose there existed $a, b \in \mathbb{N}$ with $b > 0$ such that $e = a/b$. Choose $n > \max\{b, 3\}$. Substitute $e = a/b$ in the inequalities in part (b) and multiply the inequalities by $(n!)$. Show that this leads to the existence of an integer between 0 and $\frac{3}{4}$.

7

Integration

In this chapter we present the theory of Riemann integration. While the process of integration had been developed much earlier in the seventeenth century by Isaac Newton (1642–1727) and Gottfried Leibniz (1646–1716), it was Bernhard Riemann (1826–1866) who formulated the modern definition of the definite integral that is commonly used today. (We note in passing that some of the ideas of integral calculus can even be traced back to Archimedes in the third century B.C.) Subsequent to Riemann's work, other more general approaches to integration were developed, most notably by T. J. Stieltjes (1856–1894) and H. Lebesgue (1875–1941), but we shall not cover them in this book.

Since the reader has already seen many of the important applications on integration, we shall concentrate on a rigorous development of the theory. We begin by defining the Riemann integral in terms of upper and lower sums. In Section 30 we identify two classes of functions that are integrable and then derive several related algebraic properties. The fundamental theorem of calculus is included in Section 31, as is a brief discussion of improper integrals.

Section 29 THE RIEMANN INTEGRAL

29.1 DEFINITION Let $[a, b]$ be an interval in $\mathbb{R}$. A **partition** P of $[a, b]$ is a finite set of points $[x_0, x_1, \ldots, x_n\}$ in $[a, b]$ such that

$$a = x_0 < x_1 < x_2 < \cdots < x_n = b.$$

If P and Q are two partitions of $[a, b]$ with $P \subseteq Q$, then Q is called a **refinement** of P.

29.2 DEFINITION Suppose that f is a bounded function defined on $[a, b]$ and that $P = \{x_0, \ldots, x_n\}$ is a partition of $[a, b]$. For each $i = 1, \ldots, n$ we let

$$M_i(f) = \sup\{f(x): x \in [x_{i-1}, x_i]\}$$

and

$$m_i(f) = \inf\{f(x): x \in [x_{i-1}, x_i]\}.$$

When only one function is under consideration, we may abbreviate these to M_i and m_i, respectively. Letting $\Delta x_i = x_i - x_{i-1}$ $(i = 1, \ldots, n)$, we define the **upper sum** of f with respect to P to be

$$U(f, P) = \sum_{i=1}^{n} M_i \Delta x_i,$$

and the **lower sum** of f with respect to P to be

$$L(f, P) = \sum_{i=1}^{n} m_i \Delta x_i.$$

[Sometimes $U(f, P)$ and $L(f, P)$ are called the upper and lower Darboux sums in honor of Gaston Darboux (1842–1917), who first developed this approach to the Riemann integral.]

Since we are assuming that f is a bounded function on $[a, b]$, there exist numbers m and M such that $m \leqslant f(x) \leqslant M$ for all $x \in [a, b]$. Thus for any partition P of $[a, b]$ we have

$$m(b - a) \leqslant L(f, P) \leqslant U(f, P) \leqslant M(b - a).$$

This implies that the upper and lower sums for f form a bounded set, and it guarantees the existence of the following upper and lower integrals of f.

29.3 DEFINITION Let f be a bounded function defined on $[a, b]$. Then

$$U(f) = \inf\{U(f, P): P \text{ is a partition of } [a, b]\}$$

is called the **upper integral** of f on $[a, b]$. Similarly,

$$L(f) = \sup\{L(f, P): P \text{ is a partition of } [a, b]\}$$

is called the **lower integral** of f on $[a, b]$. If these upper and lower integrals are equal, then we say that f is **Riemann integrable** on $[a, b]$, and we denote their common value by $\int_a^b f$ or by $\int_a^b f(x)\,dx$. That is, if $L(f) = U(f)$, then

$$\int_a^b f = \int_a^b f(x)\,dx = L(f) = U(f)$$

is the **Riemann integral** of f on $[a, b]$.

Since the Riemann integral is the only kind of integral we shall deal with in this book, it will often be convenient to drop the reference to Riemann and simply refer to a function f as being **integrable** on $[a, b]$ and call $\int_a^b f$ the **integral** of f on $[a, b]$.

When the function f is nonnegative on $[a, b]$, we may interpret $\int_a^b f$ intuitively as the area under the graph of f between a and b. (We say "intuitively" since we shall not give a precise definition of "area" under the graph.) Each lower sum $L(f, P)$ represents the area of a union of rectangles with base Δx_i and height m_i. (See Figure 29.1.) Similarly, each upper sum $U(f, P)$ represents the area of a union of rectangles with base Δx_i and height M_i. For any partition P, the area A under the graph of f is seen to satisfy $L(f, P) \leqslant A \leqslant U(f, P)$. But when $\int_a^b f$ exists, it is the *unique* number that lies between $L(f, P)$ and $U(f, P)$ for all partitions P. Thus $\int_a^b f$ corresponds to our intuitive notion of the area under the graph of f between a and b.

29.4 THEOREM Let f be a bounded function on $[a, b]$. If P and Q are partitions of $[a, b]$ and Q is a refinement of P, then

$$L(f, P) \leqslant L(f, Q) \leqslant U(f, Q) \leqslant U(f, P).$$

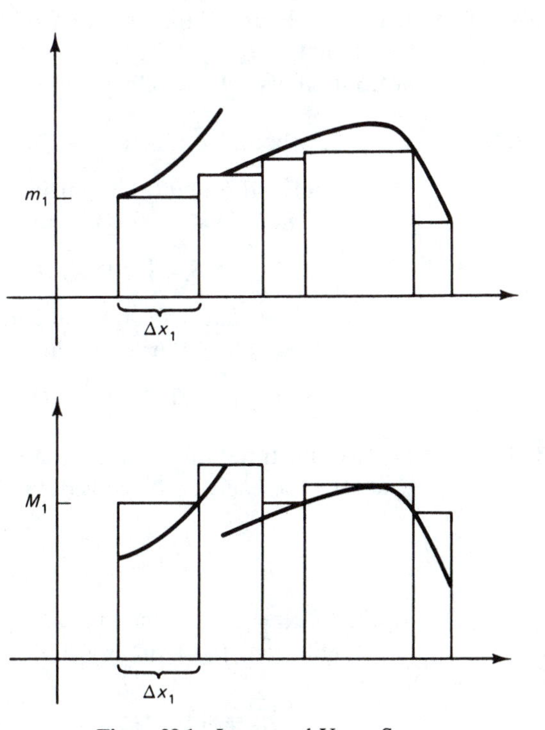

Figure 29.1 Lower and Upper Sums

Proof: The middle inequality follows directly from the definitions of $L(f, Q)$ and $U(f, Q)$. To prove $L(f, P) \leqslant L(f, Q)$, we suppose that $P = \{x_0, x_1, \ldots, x_n\}$ and consider the partition P^* formed by joining some point, say x^*, to P, where $x_{k-1} < x^* < x_k$ for some $k = 1, \ldots, n$. Let

$$t_1 = \inf\{f(x): x \in [x_{k-1}, x^*]\},$$

$$t_2 = \inf\{f(x): x \in [x^*, x_k]\}.$$

Then $t_1 \geqslant m_k$ and $t_2 \geqslant m_k$, where $m_k = \inf\{f(x): x \in [x_{k-1}, x_k]\}$, as usual.

Now the terms in $L(f, P^*)$ and $L(f, P)$ are all the same except those over the subinterval $[x_{k-1}, x_k]$. Thus we have

$$\begin{aligned} &L(f, P^*) - L(f, P) \\ &\quad = [t_1(x^* - x_{k-1}) + t_2(x_k - x^*)] - [m_k(x_k - x_{k-1})] \\ &\quad = (t_1 - m_k)(x^* - x_{k-1}) + (t_2 - m_k)(x_k - x^*) \geqslant 0. \end{aligned}$$

Finally, if the partition Q contains r more points than P, we apply the argument above r times to obtain $L(f, P) \leqslant L(f, Q)$.

The proof that $U(f, Q) \leqslant U(f, P)$ is similar. ▮

29.5 PRACTICE Let f be a bounded function on $[a, b]$. If P and Q are partitions of $[a, b]$, prove that $L(f, P) \leqslant U(f, Q)$. *Hint*: Consider the partition $P \cup Q$ and notice that $P \cup Q$ is a refinement of both P and Q.

29.6 THEOREM Let f be a bounded function on $[a, b]$. Then $L(f) \leqslant U(f)$.

Proof: If P and Q are partitions of $[a, b]$, then by Practice 29.5 we have $L(f, P) \leqslant U(f, Q)$. Thus $U(f, Q)$ is an upper bound for the set

$$S = \{L(f, P): P \text{ is a partition of } [a, b]\}.$$

It follows that $U(f, Q)$ is at least as large as $\sup S = L(f)$. That is, $L(f) \leqslant U(f, Q)$ for each partition Q of $[a, b]$. But then

$$L(f) \leqslant \inf\{U(f, Q): Q \text{ is a partition of } [a, b]\} = U(f). \qquad ▮$$

29.7 EXAMPLE Let us illustrate upper and lower sums by using them to evaluate $\int_0^1 x^2\,dx$. For each $n \in \mathbb{N}$, consider the partition

$$P_n = \left\{0, \frac{1}{n}, \frac{2}{n}, \ldots, \frac{n-1}{n}, 1\right\}$$

in which $\Delta x_i = 1/n$ for each $i = 1, 2, \ldots, n$. Since $f(x) = x^2$ is an increasing function on $[0, 1]$, on any subinterval $[(i-1)/n, i/n]$ we have

$$M_i = \sup\left\{f(x): x \in \left[\frac{i-1}{n}, \frac{i}{n}\right]\right\} = f\left(\frac{i}{n}\right) = \left(\frac{i}{n}\right)^2$$

and

$$m_i = \inf\left\{f(x): x \in \left[\frac{i-1}{n}, \frac{i}{n}\right]\right\} = f\left(\frac{i-1}{n}\right) = \left(\frac{i-1}{n}\right)^2.$$

Thus

$$U(f, P_n) = \sum_{i=1}^{n} \left(\frac{i}{n}\right)^2\left(\frac{1}{n}\right) = \frac{1}{n^3}(1^2 + 2^2 + \cdots + n^2)$$

$$= \frac{1}{n^3}\left[\frac{1}{6}(n)(n+1)(2n+1)\right] = \frac{1}{3}\left(\frac{n+1}{n}\right)\left(\frac{2n+1}{2n}\right),$$

where we have used the formula $1^2 + 2^2 + \cdots + n^2 = n(n+1)(2n+1)/6$ from Exercise 10.1. Similarly,

$$L(f, P_n) = \sum_{i=1}^{n} \left(\frac{i-1}{n}\right)^2\left(\frac{1}{n}\right) = \frac{1}{n^3}[0^2 + 1^2 + \cdots + (n-1)^2]$$

$$= \frac{1}{n^3}\left[\frac{1}{6}(n-1)(n)(2n-1)\right] = \frac{1}{3}\left(\frac{n-1}{n}\right)\left(\frac{2n-1}{2n}\right).$$

Since $\lim_{n\to\infty} U(f, P_n) = 1/3$ and $\lim_{n\to\infty} L(f, P_n) = 1/3$, we must have $U(f) \leqslant 1/3$ and $L(f) \geqslant 1/3$. But since $L(f) \leqslant U(f)$ by Theorem 29.6, this means that $L(f) = U(f) = 1/3$, so that $\int_0^1 x^2\,dx = 1/3$.

29.8 EXAMPLE As a contrast to Example 29.7, we consider the function g: $[0, 2] \to \mathbb{R}$ defined by

$$g(x) = \begin{cases} 1, & \text{if } x \text{ is rational} \\ 0, & \text{if } x \text{ is irrational.} \end{cases}$$

Let $P = \{x_0, x_1, \ldots, x_n\}$ be any partition of $[0, 2]$. Since each subinterval $[x_{i-1}, x_i]$ contains both rational and irrational numbers, we have $M_i = 1$ and $m_i = 0$ for all $i = 1, \ldots, n$. Thus

$$U(g, P) = \sum_{i=1}^{n} M_i \Delta x_i = \sum_{i=1}^{n} \Delta x_i = 2 \quad\text{and}\quad L(g, P) = \sum_{i=1}^{n} m_i \Delta x_i = 0.$$

It follows that $U(g) = 2$ and $L(g) = 0$. Since the upper and lower integrals of g on $[0, 2]$ are not equal, we conclude that g is *not* integrable on $[0, 2]$.

Since not every function is integrable, we are faced with the problem of determining when the integral of a function exists. In Section 30 we identify two large classes of functions that are integrable. Our next theorem will be very useful to us in that task.

29.9 THEOREM Let f be a bounded function on $[a, b]$. Then f is integrable iff for each $\varepsilon > 0$ there exists a partition P of $[a, b]$ such that

$$U(f, P) - L(f, P) < \varepsilon.$$

Proof: Suppose that f is integrable, so that $L(f) = U(f)$. Given $\varepsilon > 0$, there exists a partition P_1 of $[a, b]$ such that

$$L(f, P_1) > L(f) - \frac{\varepsilon}{2}.$$

[This follows from the definition of $L(f)$ as a supremum.] Similarly, there exists a partition P_2 of $[a, b]$ such that

$$U(f, P_2) < U(f) + \frac{\varepsilon}{2}.$$

Let $P = P_1 \cup P_2$ and apply Theorem 29.4 to obtain

$$\begin{aligned} U(f, P) - L(f, P) &\leqslant U(f, P_2) - L(f, P_1) \\ &< \left[U(f) + \frac{\varepsilon}{2}\right] - \left[L(f) - \frac{\varepsilon}{2}\right] \\ &= U(f) - L(f) + \varepsilon = \varepsilon. \end{aligned}$$

Conversely, given $\varepsilon > 0$, suppose that there exists a partition P of $[a, b]$ such that $U(f, P) < L(f, P) + \varepsilon$. Then we have

$$U(f) \leqslant U(f, P) < L(f, P) + \varepsilon \leqslant L(f) + \varepsilon.$$

Since $\varepsilon > 0$ is arbitrary, we must have $U(f) \leqslant L(f)$. But then Theorem 29.6 implies that $L(f) = U(f)$, so that f is integrable. ▌

ANSWERS TO PRACTICE PROBLEMS

29.5 Since $P \cup Q$ is a refinement of both P and Q, Theorem 29.4 implies that $L(f, P) \leqslant L(f, P \cup Q) \leqslant U(f, P \cup Q) \leqslant U(f, Q)$.

EXERCISES

29.1 Suppose that $f(x) = x$ for all $x \in [0, b]$. Show that f is integrable and that $\int_0^b f(x)\,dx = b^2/2$.

29.2 Suppose that $f(x) = c$ for all $x \in [a, b]$. Show that f is integrable and that $\int_a^b f(x)\,dx = c(b - a)$.

29.3 Let $f(x) = x^3$ for $x \in [0, 1]$. Given $n \in \mathbb{N}$, let P_n be the partition of $[0, 1]$ defined in Example 29.7.

(a) Find $L(f, P_n)$ and $U(f, P_n)$.

(b) Find $L(f)$ and $U(f)$.

29.4 Define $f\colon [0,1] \to \mathbb{R}$ by $f(x) = x$ if x is rational and $f(x) = 0$ if x is irrational.

(a) Given any partition $P = \{x_0, x_1, \ldots, x_n\}$ of $[0,1]$, show that $U(f, P) > \frac{1}{2}$. *Hint*: Note that $M_i = x_i$ and that $x_i > (x_i + x_{i-1})/2$ since $x_i > x_{i-1}$.

(b) For $n \in \mathbb{N}$, let P_n be the partition of $[0,1]$ defined in Example 29.7. Show that $\lim_{n\to\infty} U(f, P_n) = \frac{1}{2}$.

(c) Show that $U(f) = \frac{1}{2}$ and $L(f) = 0$, so that f is not integrable on $[0,1]$.

29.5 Let f be a bounded function on $[a, b]$. Suppose that there exists a sequence (P_n) of partitions of $[a, b]$ such that

$$\lim_{n\to\infty} [U(f, P_n) - L(f, P_n)] = 0.$$

(a) Prove that f is integrable.

(b) Prove that $\int_a^b f = \lim_{n\to\infty} U(f, P_n) = \lim_{n\to\infty} L(f, P_n)$.

29.6 Let f be a bounded function on $[a, b]$ and suppose that $f(x) \geqslant 0$ for all $x \in [a, b]$. Prove that $L(f) \geqslant 0$.

***29.7** Let f be continuous on $[a, b]$ and suppose that $f(x) \geqslant 0$ for all $x \in [a, b]$. Prove that if $L(f) = 0$ then $f(x) = 0$ for all $x \in [a, b]$.

29.8 Let f and g be bounded functions on $[a, b]$.

(a) Prove that $U(f + g) \leqslant U(f) + U(g)$.

(b) Find an example to show that a strict inequality may hold in part (a).

29.9 Suppose that f is integrable on $[a, b]$ and that $[c, d] \subseteq [a, b]$. Prove that f is integrable on $[c, d]$.

***29.10** Let $S = \{s_1, s_2, \ldots, s_k\}$ be a finite subset of $[a, b]$. Suppose that f is a bounded function on $[a, b]$ such that $f(x) = 0$ if $x \notin S$. Show that f is integrable and that $\int_a^b f = 0$.

29.11 Suppose that f is integrable on $[a, b]$. Use Theorem 29.9 to prove that f^2 is integrable on $[a, b]$.

29.12 Let f be a bounded function on $[a, b]$. In this exercise we identify an alternative way of obtaining the Riemann integral of f in terms of Riemann sums. (This approach is often used in beginning calculus texts.) Given a partition $P = \{x_0, \ldots, x_n\}$ of $[a, b]$, we define the **mesh** of P by

$$\text{mesh}\,(P) = \max\,\{\Delta x_i : i = 1, \ldots, n\}.$$

A **Riemann sum** of f associated with P is a sum of the form

$$\sum_{i=1}^{n} f(t_i)\Delta x_i,$$

where $t_i \in [x_{i-1}, x_i]$ for $i = 1, \ldots, n$. Notice that the choice of t_i in $[x_{i-1}, x_i]$ is arbitrary, so that there are infinitely many Riemann sums associated with each partition.

(a) Prove that f is integrable on $[a, b]$ iff for every $\varepsilon > 0$ there exists a $\delta > 0$ such that $U(f, P) - L(f, P) < \varepsilon$ whenever P is a partition of $[a, b]$ with mesh $(P) < \delta$.

(b) Prove that f is integrable on $[a, b]$ iff there exists a number r such that for every $\varepsilon > 0$ there exists a $\delta > 0$ such that $|S - r| < \varepsilon$ for every Riemann sum S of f associated with a partition P such that mesh $(P) < \delta$.

Section 30 PROPERTIES OF THE RIEMANN INTEGRAL

In this section we establish some of the basic properties of the Riemann integral and show that monotone functions and continuous functions are integrable. Recall that a function is said to be **monotone** on an interval if it is either increasing or decreasing on the interval. (See Exercise 26.4.)

30.1 THEOREM Let f be a monotone function on $[a, b]$. Then f is integrable.

Proof: We suppose that f is increasing on $[a, b]$; the decreasing case is similar. Since $f(a) \leqslant f(x) \leqslant f(b)$ for all $x \in [a, b]$, f is bounded on $[a, b]$. Now given $\varepsilon > 0$ there exists $k > 0$ such that

$$k[f(b) - f(a)] < \varepsilon.$$

Let $P = \{x_0, \ldots, x_n\}$ be a partition of $[a, b]$ such that $\Delta x_i \leqslant k$ for all i. Since f is increasing, we have

$$m_i = f(x_{i-1}) \quad \text{and} \quad M_i = f(x_i)$$

for $i = 1, \ldots, n$. Thus

$$U(f, P) - L(f, P) = \sum_{i=1}^{n} [f(x_i) - f(x_{i-1})](\Delta x_i)$$

$$\leqslant k \sum_{i=1}^{n} [f(x_i) - f(x_{i-1})] = k[f(b) - f(a)] < \varepsilon.$$

It follows from Theorem 29.9 that f is integrable on $[a, b]$. ∎

30.2 THEOREM Let f be a continuous function on $[a, b]$. Then f is integrable on $[a, b]$.

Proof: Once again we shall use Theorem 29.9. Since f is continuous on the compact set $[a, b]$, it is uniformly continuous on $[a, b]$ by Theorem 23.5. Thus given $\varepsilon > 0$ there exists $\delta > 0$ such that

$$|f(x) - f(y)| < \frac{\varepsilon}{b - a}$$

whenever $x, y \in [a, b]$ and $|x - y| < \delta$. Let $P = \{x_0, \ldots, x_n\}$ be a partition of $[a, b]$ such that $\Delta x_i < \delta$ for all i. Now f assumes its maximum and minimum on each subinterval $[x_{i-1}, x_i]$ by Corollary 22.3. That is, there exist points s_i and t_i in $[x_{i-1}, x_i]$ such that $m_i = f(s_i)$ and $M_i = f(t_i)$. Since $x_i - x_{i-1} < \delta$, we have $|s_i - t_i| < \delta$ and

$$0 \leqslant M_i - m_i = f(t_i) - f(s_i) < \frac{\varepsilon}{b-a}$$

for all i. It follows that

$$U(f, P) - L(f, P) = \sum_{i=1}^{n} (M_i - m_i)(\Delta x_i) < \frac{\varepsilon}{b-a} \sum_{i=1}^{n} \Delta x_i = \varepsilon.$$

Thus by Theorem 29.9, f is integrable on $[a, b]$. ▮

While Theorems 30.1 and 30.2 establish the integrability of two large classes of functions, other functions besides these are also integrable. In Example 30.3 we describe an integrable function that is neither monotone nor continuous on any interval.

30.3 EXAMPLE Let f be the Dirichlet function of Example 21.9 defined on the interval $[0, 1]$. That is,

$$f(x) = \begin{cases} \dfrac{1}{n}, & \text{if } x = \dfrac{m}{n} \text{ is rational in lowest terms} \\ 0, & \text{if } x \text{ is irrational.} \end{cases}$$

We shall show that f is integrable using Theorem 29.9. (See Figure 30.1.) Given any $\varepsilon > 0$, choose an integer N such that $N > 2/\varepsilon$. Then let

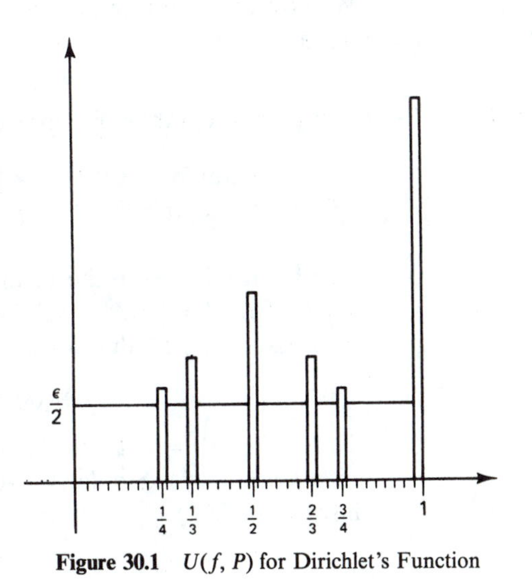

Figure 30.1 $U(f, P)$ for Dirichlet's Function

$Y = \{y_1, y_2, \ldots, y_m\}$ denote the (finite) set of all rational numbers in $[0, 1]$ having denominator less than N (when expressed in lowest terms). Choose a prime number n with $n > 2m/\varepsilon$ and $n \geqslant N$, and consider the partition $P = \{x_0, x_1, \ldots, x_n\}$, where $x_i = i/n$ for all i, so that each $\Delta x_i = 1/n$. (Making n prime with $n \geqslant N$ implies that $P \cap Y = \{0, 1\}$.)

We now separate the set of indices of the partition P into two disjoint subsets depending on whether or not the corresponding subinterval contains a point of Y. Let

$$A = \{i\colon [x_{i-1}, x_i] \cap Y = \varnothing\} \quad \text{and} \quad B = \{i\colon [x_{i-1}, x_i] \cap Y \neq \varnothing\}.$$

Note that if $i \in A$ then all the rationals in $[x_{i-1}, x_i]$ have denominator $\geqslant N$, so that $M_i \leqslant 1/N \leqslant \varepsilon/2$. Of course, $M_i \leqslant 1$ for all i. Also note that

$$\sum_{i \in B} \Delta x_i \leqslant m\left(\frac{1}{n}\right) < m\left(\frac{\varepsilon}{2m}\right) = \frac{\varepsilon}{2},$$

since there are only m points in Y. Thus we have

$$U(f, P) = \sum_{i=1}^{n} M_i \Delta x_i = \sum_{i \in A} M_i \Delta x_i + \sum_{i \in B} M_i \Delta x_i$$

$$< \frac{\varepsilon}{2} \sum_{i \in A} \Delta x_i + \sum_{i \in B} (1)(\Delta x_i) < \frac{\varepsilon}{2} + \frac{\varepsilon}{2} = \varepsilon.$$

Clearly, we have $L(f, P) = 0$, so $U(f, P) - L(f, P) < \varepsilon$ and f is integrable by Theorem 29.9.

We now turn our attention to proving several algebraic properties of the integral.

30.4 THEOREM Let f and g be integrable functions on $[a, b]$ and let $k \in \mathbb{R}$. Then

(a) kf is integrable and $\int_a^b kf = k \int_a^b f$, and
(b) $f + g$ is integrable and $\int_a^b (f + g) = \int_a^b f + \int_a^b g$.

Proof: (a) If $k = 0$, the result is trivial. Consider the case where $k > 0$. Let $P = \{x_0, \ldots, x_n\}$ be any partition of $[a, b]$. It follows from Exercise 12.5 that for all i we have

$$M_i(kf) = k \cdot M_i(f).$$

Thus $U(kf, P) = k \cdot U(f, P)$, so again using Exercise 12.5 we have $U(kf) = k \cdot U(f)$. Similarly, we obtain $L(kf) = k \cdot L(f)$. Since f is integrable, $L(f) = U(f)$, so

$$L(kf) = k \cdot L(f) = k \cdot U(f) = U(kf).$$

We conclude that kf is integrable and that

$$\int_a^b kf = U(kf) = k \cdot U(f) = k \int_a^b f.$$

The case where $k < 0$ is left to the reader (Exercise 30.1).

(b) Recall from Exercise 12.10 that

$$\sup [(f+g)(D)] \leqslant \sup f(D) + \sup g(D).$$

Thus for any partition P we have $M_i(f+g) \leqslant M_i(f) + M_i(g)$ for each i, so

$$U(f+g, P) \leqslant U(f, P) + U(g, P).$$

Similarly, since $\inf [(f+g)(D)] \geqslant \inf f(D) + \inf g(D)$, we obtain for any partition P,

$$L(f+g, P) \geqslant L(f, P) + L(g, P).$$

We now appeal to Theorem 29.9. Given $\varepsilon > 0$, there exist partitions P_1 and P_2 of $[a, b]$ such that

$$U(f, P_1) < L(f, P_1) + \frac{\varepsilon}{2} \quad \text{and} \quad U(g, P_2) < L(g, P_2) + \frac{\varepsilon}{2}.$$

Let $P = P_1 \cup P_2$. Then by Theorem 29.4,

$$U(f, P) < L(f, P) + \frac{\varepsilon}{2} \quad \text{and} \quad U(g, P) < L(g, P) + \frac{\varepsilon}{2}.$$

Combining inequalities, we obtain

$$\begin{aligned} U(f+g, P) &\leqslant U(f, P) + U(g, P) \\ &< L(f, P) + L(g, P) + \varepsilon \leqslant L(f+g, P) + \varepsilon. \end{aligned}$$

It follows from Theorem 29.9 that $f + g$ is integrable.

Furthermore, since

$$\begin{aligned} \int_a^b (f+g) &= U(f+g) \leqslant U(f+g, P) \\ &< L(f, P) + L(g, P) + \varepsilon \\ &\leqslant L(f) + L(g) + \varepsilon = \int_a^b f + \int_a^b g + \varepsilon \end{aligned}$$

and

$$\begin{aligned} \int_a^b (f+g) &= L(f+g) \geqslant L(f+g, P) \\ &> U(f, P) + U(g, P) - \varepsilon \\ &\geqslant U(f) + U(g) - \varepsilon = \int_a^b f + \int_a^b g - \varepsilon, \end{aligned}$$

we must have

$$\int_a^b (f+g) = \int_a^b f + \int_a^b g. \quad \blacksquare$$

30.5 PRACTICE Suppose that f and g are integrable on $[a, b]$ and that $f(x) \leqslant g(x)$ for all $x \in [a, b]$. Prove that $\int_a^b f \leqslant \int_a^b g$.

30.6 THEOREM Suppose that f is integrable on both $[a, c]$ and $[c, b]$. Then f is integrable on $[a, b]$. Furthermore, $\int_a^b f = \int_a^c f + \int_c^b f$.

Proof: Given $\varepsilon > 0$, there exist partitions P_1 of $[a, c]$ and P_2 of $[c, b]$ such that

$$U(f, P_1) - L(f, P_1) < \frac{\varepsilon}{2} \quad \text{and} \quad U(f, P_2) - L(f, P_2) < \frac{\varepsilon}{2}.$$

Let $P = P_1 \cup P_2$. Then P is a partition of $[a, b]$ and we have

$$\begin{aligned} U(f, P) - L(f, P) &= U(f, P_1) + U(f, P_2) - L(f, P_1) - L(f, P_2) \\ &= [U(f, P_1) - L(f, P_1)] \\ &\quad + [U(f, P_2) - L(f, P_2)] \\ &< \frac{\varepsilon}{2} + \frac{\varepsilon}{2} = \varepsilon. \end{aligned}$$

Thus f is integrable on $[a, b]$ by Theorem 29.9. Furthermore,

$$\begin{aligned} \int_a^b f &\leqslant U(f, P) = U(f, P_1) + U(f, P_2) \\ &< L(f, P_1) + L(f, P_2) + \varepsilon \leqslant \int_a^c f + \int_c^b f + \varepsilon, \end{aligned}$$

and also

$$\begin{aligned} \int_a^b f &\geqslant L(f, P) = L(f, P_1) + L(f, P_2) \\ &> U(f, P_1) + U(f, P_2) - \varepsilon \geqslant \int_a^c f + \int_c^b f - \varepsilon, \end{aligned}$$

so that $\int_a^b f = \int_a^c f + \int_c^b f$, as desired. $\blacksquare$

Because of our experience with continuous functions and differentiable functions, we might hope that the composition of two integrable functions would be integrable. Unfortunately, this is not the case. (See Exercise 30.5.) We do, however, have the following very useful result.

30.7 THEOREM Suppose that f is integrable on $[a, b]$ and g is continuous on $[c, d]$, where $f([a, b]) \subseteq [c, d]$. Then $g \circ f$ is integrable on $[a, b]$.

Proof: Given any $\varepsilon > 0$, let $K = \sup \{|g(t)| : t \in [c, d]\}$ and choose $\varepsilon' > 0$ such that $\varepsilon'(b - a + 2K) < \varepsilon$. Since g is continuous on $[c, d]$, it is uniformly continuous on $[c, d]$ by Theorem 23.5. Thus there exists $\delta > 0$ such that $\delta < \varepsilon'$ and such that $|g(s) - g(t)| < \varepsilon'$ whenever $|s - t| < \delta$ and $s, t \in [c, d]$. Since f is integrable on $[a, b]$, there exists a partition $P = \{x_0, \ldots, x_n\}$ of $[a, b]$ such that

$$U(f, P) - L(f, P) < \delta^2.$$

We claim that for this partition P we also have

$$U(g \circ f, P) - L(g \circ f, P) = \sum_{i=1}^{n} [M_i(g \circ f) - m_i(g \circ f)](\Delta x_i) < \varepsilon.$$

To show this, we shall separate the set of indices of the partition P into two disjoint sets. On the first set we shall make $M_i(g \circ f) - m_i(g \circ f)$ small and on the second set we shall make $\sum \Delta x_i$ small. Let

$$A = \{i\colon M_i - m_i < \delta\} \quad \text{and} \quad B = \{i\colon M_i - m_i \geqslant \delta\}.$$

Then if $i \in A$ and $x, y \in [x_{i-1}, x_i]$, we have

$$|f(x) - f(y)| \leqslant M_i - m_i < \delta,$$

so that $|g \circ f(x) - g \circ f(y)| < \varepsilon'$. But then $M_i(g \circ f) - m_i(g \circ f) \leqslant \varepsilon'$. It follows that

$$\sum_{i \in A} [M_i(g \circ f) - m_i(g \circ f)](\Delta x_i) \leqslant \varepsilon' \sum_{i \in A} \Delta x_i \leqslant \varepsilon'(b - a).$$

On the other hand, if $i \in B$, then $(M_i - m_i)/\delta \geqslant 1$, so that

$$\sum_{i \in B} \Delta x_i \leqslant \frac{1}{\delta} \sum_{i \in B} (M_i - m_i)(\Delta x_i) \leqslant \frac{1}{\delta} [U(f, P) - L(f, P)] < \delta < \varepsilon'.$$

Thus since $M_i(g \circ f) - m_i(g \circ f) \leqslant 2K$ for all i we have

$$\sum_{i \in B} [M_i(g \circ f) - m_i(g \circ f)](\Delta x_i) \leqslant 2K \sum_{i \in B} \Delta x_i < 2K\varepsilon'.$$

Now when we combine all the indices we obtain

$$\begin{aligned} U(g \circ f, P) - L(g \circ f, P) &= \sum_{i \in A} [M_i(g \circ f) - m_i(g \circ f)](\Delta x_i) \\ &\quad + \sum_{i \in B} [M_i(g \circ f) - m_i(g \circ f)](\Delta x_i) \\ &\leqslant \varepsilon'(b - a) + 2K\varepsilon' = \varepsilon'(b - a + 2K) < \varepsilon. \end{aligned}$$

Hence $g \circ f$ is integrable on $[a, b]$ by Theorem 29.9. ▮

30.8 COROLLARY Let f be integrable on $[a, b]$. Then $|f|$ is integrable on $[a, b]$ and

$$\left| \int_a^b f \right| \leqslant \int_a^b |f|.$$

Proof: Since f is integrable on $[a, b]$, there exists $B > 0$ such that $|f(x)| \leqslant B$ for all $x \in [a, b]$. Define $g: [-B, B] \to \mathbb{R}$ by $g(t) = |t|$. Then g is continuous on $[-B, B]$ and $g \circ f = |f|$. (See Exercise 21.8.) It follows from Theorem 30.7 that $|f|$ is integrable.

To establish the inequality between the integrals, we note that $-|f(x)| \leqslant f(x) \leqslant |f(x)|$ for all $x \in [a, b]$, so that Practice 30.5 implies that

$$-\int_a^b |f| \leqslant \int_a^b f \leqslant \int_a^b |f|.$$

But then $|\int_a^b f| \leqslant \int_a^b |f|$, as desired. ▮

30.9 PRACTICE Suppose that f is integrable on $[a, b]$ and let $n \in \mathbb{N}$. Use Theorem 30.7 to prove that f^n is integrable on $[a, b]$. (We define f^n by $f^n(x) = [f(x)]^n$ for all $x \in [a, b]$.)

ANSWERS TO PRACTICE PROBLEMS

30.5 Here are two different approaches: (1) Given any partition P of $[a, b]$, since $f(x) \leqslant g(x)$ for all $x \in [a, b]$, we have $M_i(f) \leqslant M_i(g)$ for all i. Thus for any partition P, we have $U(f, P) \leqslant U(g, P)$, so $\int_a^b f = U(f) \leqslant U(g) = \int_a^b g$. (2) By Theorem 30.4, the function $h = g - f$ is integrable on $[a, b]$. Since $h(x) \geqslant 0$ for all $x \in [a, b]$, Exercise 29.6 implies that $\int_a^b h = L(h) \geqslant 0$. But then $\int_a^b g - \int_a^b f \geqslant 0$, by Theorem 30.4 again.

30.9 Since f is integrable on $[a, b]$, there exists $B > 0$ such that $|f(x)| \leqslant B$ for all $x \in [a, b]$. Define $g: [-B, B] \to \mathbb{R}$ by $g(t) = t^n$. Then g is continuous on $[-B, B]$ and $g \circ f = f^n$. Thus f^n is integrable by Theorem 30.7.

EXERCISES

30.1 Let f be integrable on $[a, b]$, let P be a partition of $[a, b]$, and let $k < 0$.
(a) Show that $L(kf, P) = k \cdot U(f, P)$ and $U(kf, P) = k \cdot L(f, P)$.
(b) Show that $L(kf) = k \cdot U(f)$ and $U(kf) = k \cdot L(f)$.
(c) Complete the proof of Theorem 30.4(a).

30.2 Prove the case in Theorem 30.1 where f is decreasing.

30.3 Let f be continuous on $[a, b]$ and suppose that $f(x) \geqslant 0$ for all $x \in [a, b]$. Prove that if there exists a point $c \in [a, b]$ such that $f(c) > 0$, then $\int_a^b f > 0$.

30.4 Let f be continuous on $[a, b]$ and suppose that, for every integrable function g defined on $[a, b]$, $\int_a^b fg = 0$. Prove that $f(x) = 0$ for all $x \in [a, b]$.

30.5 Let $f\colon [0, 1] \to [0, 1]$ be the Dirichlet function of Example 30.3 and let h: $[0, 1] \to [0, 1]$ be the function of Example 29.8 (with domain restricted to $[0, 1]$). Find an integrable function g: $[0, 1] \to [0, 1]$ such that $h = g \circ f$, thereby showing that the composition of two integrable functions need not be integrable.

30.6 Suppose that f is integrable on $[a, b]$ and that there exists $k > 0$ such that $f(x) \geqslant k$ for all $x \in [a, b]$. Prove that $1/f$ is integrable on $[a, b]$.

30.7 Let f and g be integrable on $[a, b]$.
*(a) Show that their product fg is integrable on $[a, b]$.
(b) Show that $\max(f, g)$ and $\min(f, g)$ are integrable on $[a, b]$. (See Example 21.11.)

30.8 Find an example of a function $f\colon [0, 1] \to \mathbb{R}$ such that f is not integrable on $[0, 1]$ but $|f|$ is integrable on $[0, 1]$.

***30.9** Let f be integrable on $[a, b]$ and suppose that $m \leqslant f(x) \leqslant M$ for all $x \in [a, b]$. Show that $m(b - a) \leqslant \int_a^b f \leqslant M(b - a)$.

30.10 Prove the mean value theorem for integrals: If f is continuous on $[a, b]$, then there exists $c \in (a, b)$ such that

$$f(c) = \frac{1}{b - a} \int_a^b f.$$

30.11 Let f and g be continuous on $[a, b]$, and suppose that $\int_a^b f = \int_a^b g$. Prove that there exists $c \in [a, b]$ such that $f(c) = g(c)$.

30.12 Let f and g be integrable on $[a, b]$ and suppose that h is defined on $[a, b]$ such that $f(x) \leqslant h(x) \leqslant g(x)$ for all $x \in [a, b]$. Prove that if $\int_a^b f = \int_a^b g$, then h is integrable on $[a, b]$ and $\int_a^b h = \int_a^b f$.

***30.13** (a) Prove the extended mean value theorem for integrals: If f and g are continuous on $[a, b]$ and $g(x) \geqslant 0$ for all $x \in [a, b]$, then there exists $c \in [a, b]$ such that

$$\int_a^b (fg) = f(c) \int_a^b g.$$

(b) Show that Exercise 30.10 is a special case of part (a).

30.14 Let f be integrable on $[a, b]$. Use the inequality in Exercise 11.4 to show that $|f|$ is integrable on $[a, b]$ without using Theorem 30.7.

30.15 Let f be integrable on $[a, b]$ and let $S = \{s_1, s_2, \ldots, s_k\}$ be a finite subset of $[a, b]$. Suppose that g is a bounded function on $[a, b]$ such that $g(x) = f(x)$ for all $x \notin S$. Prove that g is integrable on $[a, b]$ and that $\int_a^b g = \int_a^b f$.

30.16 Let (y_n) be a sequence of members of $[a, b]$ converging to y_0 in $[a, b]$. Suppose that f is bounded on $[a, b]$ and that f is continuous on $[a, b]$ except at the points y_i, $i = 0, 1, 2, \ldots$. Prove that f is integrable on $[a, b]$.

30.17 Define $f: [0, 1] \to \mathbb{R}$ by $f(x) = \sin(1/x)$ if $x \neq 0$ and $f(0) = 0$. Show that f is integrable on $[0, 1]$. *Hint*: Given any $c \in (0, 1)$, f is integrable on $[c, 1]$ since it is continuous there.

***30.18** Generalize Exercise 30.17 to the following: Let f be bounded on $[a, b]$ and integrable on $[c, b]$ for each $c \in (a, b)$. Prove that f is integrable on $[a, b]$.

Section 31 THE FUNDAMENTAL THEOREM OF CALCULUS

The fundamental theorem of calculus is really two theorems, each expressing that differentiation and integration are inverse operations. The first result says in essence that "the derivative of the integral of a function is the original function," and the second result establishes the reverse: "The integral of the derivative of a function is again the same function." (Of course, there are certain conditions that the function involved must satisfy.)

Historically, the operations of integration and differentiation were developed to solve seemingly unrelated problems. These problems may be described geometrically as finding the area under a curve and finding the slope of a curve at a point. The proof of their inverse relationship was one of the important theoretical (and practical) contributions of Newton and Leibniz in the seventeenth century.

In our discussion of integrals we have defined $\int_a^b f$ only when $a < b$. It will be convenient now to extend this and let $\int_b^a f = -\int_a^b f$. We also set $\int_a^a f = 0$.

31.1 THEOREM (The Fundamental Theorem of Calculus I) Let f be integrable on $[a, b]$. For each $x \in [a, b]$, let

$$F(x) = \int_a^x f(t)\, dt.$$

Then F is uniformly continuous on $[a, b]$. Furthermore, if f is continuous at $c \in [a, b]$, then F is differentiable at c and

$$F'(c) = f(c).$$

Proof: Since f is integrable on $[a, b]$, it is bounded there. That is, there exists $B > 0$ such that $|f(x)| \leqslant B$ for all $x \in [a, b]$. To see that F is uniformly continuous on $[a, b]$, let $\varepsilon > 0$ be given. If $x, y \in [a, b]$ with $x < y$ and $|x - y| < \varepsilon/B$, then

$$|F(y) - F(x)| = \left| \int_a^y f(t)\,dt - \int_a^x f(t)\,dt \right| = \left| \int_x^y f(t)\,dt \right|$$
$$\leqslant \int_x^y |f(t)|\,dt \leqslant \int_x^y B\,dt = B(y - x) < \varepsilon,$$

where we have used Theorem 30.6 and Corollary 30.8. Thus F is uniformly continuous on $[a, b]$.

Now suppose that f is continuous at $c \in (a, b)$. Then given any $\varepsilon > 0$ there exists $\delta > 0$ such that $|f(t) - f(c)| < \varepsilon$ whenever $t \in (a, b)$ and $|t - c| < \delta$. Since $f(c)$ is a constant, we may write

$$f(c) = \frac{1}{x - c} \int_c^x f(c)\,dt, \qquad \text{for } x \neq c.$$

Then for any $x \in (a, b)$ with $0 < |x - c| < \delta$, we have

$$\left| \frac{F(x) - F(c)}{x - c} - f(c) \right| = \left| \frac{1}{x - c} \left[\int_a^x f(t)\,dt - \int_a^c f(t)\,dt \right] - f(c) \right|$$
$$= \left| \frac{1}{x - c} \int_c^x f(t)\,dt - \frac{1}{x - c} \int_c^x f(c)\,dt \right|$$
$$= \left| \frac{1}{x - c} \int_c^x [f(t) - f(c)]\,dt \right|$$
$$\leqslant \left| \frac{1}{x - c} \right| \left| \int_c^x |f(t) - f(c)|\,dt \right|$$
$$< \left| \frac{1}{x - c} \right| \varepsilon |x - c| = \varepsilon,$$

where we have used Theorem 30.6 and Corollary 30.8 again. [Also note that $|t - c| < \delta$ since t is between c and x, and this enabled us to use $|f(t) - f(c)| < \varepsilon$.] Since $\varepsilon > 0$ was arbitrary, we conclude that

$$F'(c) = \lim_{x \to c} \frac{F(x) - F(c)}{x - c} = f(c). \qquad \blacksquare$$

31.2 THEOREM (The Fundamental Theorem of Calculus II) If f is differentiable on $[a, b]$ and f' is integrable on $[a, b]$, then

$$\int_a^b f' = f(b) - f(a).$$

Proof: Let $P = \{x_0, x_1, \ldots, x_n\}$ be any partition of $[a, b]$. By applying the mean value theorem (26.3) to each subinterval $[x_{i-1}, x_i]$, we obtain points $t_i \in (x_{i-1}, x_i)$ such that

$$f(x_i) - f(x_{i-1}) = f'(t_i)(x_i - x_{i-1}).$$

Thus we have

$$f(b) - f(a) = \sum_{i=1}^{n} [f(x_i) - f(x_{i-1})] = \sum_{i=1}^{n} f'(t_i)(x_i - x_{i-1}).$$

Since $m_i(f') \leqslant f'(t_i) \leqslant M_i(f')$ for all i, it follows that

$$L(f', P) \leqslant f(b) - f(a) \leqslant U(f', P).$$

Since this last inequality holds for each partition P, we also have

$$L(f') \leqslant f(b) - f(a) \leqslant U(f').$$

But f' is assumed to be integrable on $[a, b]$, so $L(f') = U(f') = \int_a^b f'$. Thus $f(b) - f(a) = \int_a^b f'$. ∎

As any first-year calculus student knows, one important application of Theorem 31.2 is in the evaluation of integrals.

31.3 EXAMPLE Let $f(x) = x^3$ for $x \in \mathbb{R}$. Then $f'(x) = 3x^2$, so we have, for example,

$$\int_1^4 3x^2\, dx = f(4) - f(1) = 64 - 1 = 63.$$

There are also useful theoretical results that follow from Theorem 31.2. For example, we have the familiar formula for "integration by parts."

31.4 PRACTICE (Integration by Parts) Suppose that f and g are differentiable on $[a, b]$ and that f' and g' are integrable on $[a, b]$. Prove that

$$\int_a^b (fg') = [f(b)g(b) - f(a)g(a)] - \int_a^b f'g.$$

Hint: Let $h = fg$ so that $h' = fg' + f'g$.

Improper Integrals

Throughout our discussion of the Riemann integral we have assumed that the function being integrated is bounded and that the integral is being taken over a bounded interval. Under these conditions the integral, when it exists, is said to be a **proper** integral. There are situations, however, when we wish to relax one or both of the boundedness restrictions. By so doing we may sometimes obtain an improper integral that is given as a limit of proper integrals.

31.5 DEFINITION Let f be defined on $(a, b]$ and integrable on $[c, b]$ for every $c \in (a, b]$. If $\lim_{c \to a+} \int_c^b f$ exists, then the **improper integral** of f on $(a, b]$, denoted $\int_a^b f$, is given by

$$\int_a^b f = \lim_{c \to a+} \int_c^b f.$$

If it happens that the function f in Definition 31.5 is bounded on $[a, b]$, then it will also be integrable on $[a, b]$ as a proper integral. (See Exercise 30.18.) It follows from the continuity part of Theorem 31.1 that in this case the proper and improper integrals agree. Thus the only significant use of Definition 31.5 is when f is unbounded on $[a, b]$. We also note that $f(x)$ need not be defined at $x = a$, and if it is defined there, its value does not affect the integrability of the function or change the value of the integral.

If $L = \lim_{c \to a+} \int_c^b f$ is a finite number, the improper integral $\int_a^b f$ is said to **converge** to L. If $L = \infty$ (respectively, $-\infty$), the integral is said to **diverge** to ∞ (respectively, $-\infty$).

31.6 EXAMPLES (a) Let $f(x) = x^{-1/3}$ for $x \in (0, 1]$. Then f is not integrable on $[0, 1]$ since it is not bounded there. But for each $c \in (0, 1)$ we have

$$\int_c^1 x^{-1/3}\, dx = \frac{3}{2} x^{2/3} \Big|_c^1 = \frac{3}{2}(1 - c^{2/3}).$$

[The notation $g(x)|_a^b$ represents $g(b) - g(a)$.] If we take the limit as $c \to 0+$, we find that the improper integral of f on $(0, 1]$ exists and is equal to $\frac{3}{2}$:

$$\int_0^1 x^{-1/3}\, dx = \lim_{c \to 0+} \frac{3}{2}(1 - c^{2/3}) = \frac{3}{2}.$$

(b) Let $g(x) = 1/x$ for $x \in (0, 1]$. Then for each $c \in (0, 1)$, we have

$$\int_c^1 \frac{1}{x}\, dx = \log 1 - \log c = -\log c.$$

Since $\lim_{c \to 0+} (-\log c) = \infty$, the improper integral $\int_0^1 (1/x)\, dx$ diverges to ∞ and we write

$$\int_0^1 \frac{1}{x}\, dx = \infty.$$

The improper integral of a function that is unbounded at the right endpoint is defined in a way analogous to Definition 31.5. To handle the case when the interval itself is unbounded, we have the following definition.

31.7 DEFINITION Let f be defined on $[a, \infty)$ and integrable on $[a, c]$ for every $c > a$. If $\lim_{c \to \infty} \int_a^c f$ exists, then the **improper integral** of f on $[a, \infty)$, denoted by $\int_a^\infty f$, is given by

$$\int_a^\infty f = \lim_{c \to \infty} \int_a^c f.$$

31.8 PRACTICE Let $f(x) = x^{-2}$ for $x \geqslant 1$. Find $\int_1^\infty f$, if it exists.

31.9 EXAMPLE Let $g(x) = \cos x$ for $x \geqslant 0$. Then for each $c > 0$ we have

$$\int_0^c \cos x \, dx = \sin c - \sin 0 = \sin c.$$

Since $\lim_{c \to \infty} \sin c$ does not exist, the expression $\int_0^\infty \cos x \, dx$ is not an improper integral and it has no meaning.

ANSWERS TO PRACTICE PROBLEMS

31.4 If $h = fg$, then $h' = fg' + f'g$ by Theorem 25.7. Now f and g are differentiable, hence continuous, and therefore integrable on $[a, b]$. Thus Theorem 30.4 and Exercise 30.7 imply that h' is integrable on $[a, b]$. From the fundamental theorem (31.2), we then obtain $\int_a^b h' = h(b) - h(a)$. That is,

$$\int_a^b (fg') + \int_a^b (f'g) = f(b)g(b) - f(a)g(a).$$

31.8 For each $c > 1$ we have $\int_1^c x^{-2} \, dx = -1/x|_1^c = 1 - 1/c$. Thus

$$\int_1^\infty x^{-2} \, dx = \lim_{c \to \infty} \left(1 - \frac{1}{c}\right) = 1.$$

EXERCISES

31.1 Let f be continuous on $[a, b]$ and let g be differentiable on $[c, d]$, where $g([c, d]) \subseteq [a, b]$. Define

$$F(x) = \int_a^{g(x)} f, \qquad \text{for all } x \in [c, d].$$

Prove that F is differentiable and that $F'(x) = (f \circ g)(x)g'(x)$.

31.2 Let f be continuous on $[a, b]$. For each $x \in [a, b]$ let $F(x) = \int_x^b f$. Show that F is differentiable and that $F'(x) = -f(x)$.

31.3 Use Theorem 31.1 and the previous exercises to find a formula for the derivative of each function.

(a) $F(x) = \int_0^x \sqrt{1 + t^2}\, dt$ (b) $F(x) = \int_{-x}^x \sqrt{1 + t^2}\, dt$

(c) $F(x) = \int_0^{\sin x} \cos t^2\, dt$ (d) $F(x) = \int_x^{x^2} \sqrt{1 + t^2}\, dt$

31.4 Repeat Exercise 31.3 for the following functions.

(a) $F(x) = \int_1^{x^2} (1 + t^2)^{-2}\, dt$

(b) $F(x) = \int_0^x \sqrt{1 + t^2}\, dt + \int_x^2 \sqrt{1 + t^2}\, dt$

(c) $F(x) = \int_0^{3x} \sin t^2\, dt$

(d) $F(x) = \int_{x^2}^{x^4} \cos \sqrt{t}\, dt$

31.5 Let $F(x) = \int_0^x xe^{t^2} dt$ for $x \in [0, 1]$. Find $F''(x)$ for $x \in (0, 1)$. *Caution*: $F'(x) \neq xe^{x^2}$.

31.6 Let $f(t) = t$ for $0 \leqslant t \leqslant 2$ and $f(t) = 3$ for $2 < t \leqslant 4$.

(a) Find an explicit expression for $F(x) = \int_0^x f(t)\, dt$ as a function of x.

(b) Sketch F and determine where F is differentiable.

(c) Find a formula for $F'(x)$ whenever F is differentiable.

31.7 Use Theorem 31.1 to prove Theorem 31.2 for the case when f is differentiable on $[a, b]$ and f' is continuous on $[a, b]$.

31.8 Use Theorem 31.1 to evaluate $\lim_{x \to 0} (1/x) \int_0^x \sqrt{9 + t^2}\, dt$.

31.9 Let f be continuous on $[a, b]$. Suppose that $\int_a^x f = \int_x^b f$ for all $x \in [a, b]$. Prove that $f(x) = 0$ for all $x \in [a, b]$.

31.10 Let f be continuous on $[0, \infty)$. Suppose that $f(x) \neq 0$ for all $x > 0$ and that $[f(x)]^2 = 2 \int_0^x f$ for all $x \geqslant 0$. Prove that $f(x) = x$ for all $x \geqslant 0$. *Hint*: What is $f'(x)$?

31.11 In the statement of Theorem 31.2 we require that f' be integrable on $[a, b]$. Show that this requirement is necessary. That is, find a differentiable function f such that f' is not integrable.

31.12 Prove the following "change of variable" theorem: Let g be differentiable on $[c, d]$ and g' be integrable on $[c, d]$. Suppose that f is continuous on the range of g. If $g(c) = a$ and $g(d) = b$, then

$$\int_c^d (f \circ g)(x) g'(x)\, dx = \int_a^b f(x)\, dx.$$

Hint: Define $F(x) = \int_a^x f(t)\, dt$ for $x \in$ range g and $h(x) = (F \circ g)(x)$ for $x \in [c, d]$. Then look at $\int_c^d h'(x)\, dx$.

31.13 Use Exercise 31.12 to evaluate $\int_0^2 3x^2 \sqrt{x^3 + 1}\, dx$. Identify the functions f and g that you have used and explicitly write down $\int_a^b f(x)\, dx$.

31.14 Repeat Exercise 31.13 for $\int_0^{\pi/2} (\cos x)(1 + \sin x)^3\, dx$.

31.15 Find the value of each improper integral.

(a) $\int_0^1 x^{-2}\, dx$ (b) $\int_0^2 (2 - x)^{-1/2}\, dx$

(c) $\int_0^\infty e^{-x}\, dx$ (d) $\int_1^\infty x^{-1/3}\, dx$

31.16 Prove the following form of Taylor's theorem (28.2), in which the remainder term is expressed as an integral: Suppose that f and its first $n+1$ derivatives are continuous on a closed interval I containing the point a. Then for each $x \in I$,

$$f(x) = f(a) + f'(a)(x-a) + \frac{f''(a)}{2}(x-a)^2 + \cdots + \frac{f^{(n)}(a)}{n!}(x-a)^n + R_n,$$

where the remainder R_n is given by

$$R_n = \frac{1}{n!}\int_a^x (x-t)^n f^{(n+1)}(t)\,dt.$$

Hint: Use integration by parts and induction.

31.17 Use Exercise 31.16 and 30.13 to prove Taylor's theorem (28.2) for the case when $f^{(n+1)}$ is continuous on $[a, b]$.

8

Infinite Series

Having looked carefully in Chapter 4 at the convergence properties of sequences, we now apply these properties to infinite series of real numbers. In Section 32 we define infinite series and discuss convergence. In Section 33 we develop several useful tests for determining whether a given series is convergent or divergent. Finally, in Section 34 we consider power series.

Section 32 CONVERGENCE OF INFINITE SERIES

Let (a_n) be a sequence of real numbers. We use the notation

$$\sum_{k=m}^{n} a_k \quad \text{or} \quad \textstyle\sum_{k=m}^{n} a_k$$

to denote the sum $a_m + a_{m+1} + \cdots + a_n$, where $n \geqslant m$. Using (a_n) we can define a new sequence (s_n) of **partial sums** given by

$$s_n = \sum_{k=1}^{n} a_k = a_1 + a_2 + \cdots + a_n.$$

We also refer to the sequence (s_n) of partial sums as the **infinite series** (or simply the series)

$$\sum_{n=1}^{\infty} a_n.$$

If (s_n) converges to a real number s, we say that the series $\sum_{n=1}^{\infty} a_n$ is **convergent** and we write

$$\sum_{n=1}^{\infty} a_n = s.$$

We also refer to s as the **sum** of the series $\sum_{n=1}^{\infty} a_n$. A series that is not convergent is called **divergent**. If $\lim s_n = +\infty$, we say that the series $\sum_{n=1}^{\infty} a_n$ **diverges to** $+\infty$, and we write $\sum_{n=1}^{\infty} a_n = +\infty$.

As defined above, the symbol $\sum_{n=1}^{\infty} a_n$ is used in two ways: It is used to denote the sequence (s_n) of partial sums, and it is also used to denote the limit of the sequence of partial sums, provided that this limit exists. This dual usage should not cause confusion since the context will make the intended meaning clear.

Sometimes we want to consider infinite series whose terms begin with a_m instead of a_1. We denote such a series by $\sum_{n=m}^{\infty} a_n$ and the corresponding partial sums are given by $s_n = \sum_{k=m}^{n} a_k$, where $n \geqslant m$. If the precise beginning of the series is not of immediate concern, the notation is often abbreviated to $\sum a_n$. If the particular terms in a series are of interest to us, we may expand the symbol $\sum_{n=1}^{\infty} a_n$ and write

$$a_1 + a_2 + a_3 + \cdots + a_n + \cdots.$$

In so doing we must remember that we really mean

$$\lim_{n \to \infty} (a_1 + a_2 + \cdots + a_n).$$

32.1 EXAMPLE For the infinite series $\sum_{n=1}^{\infty} 1/[n(n+1)]$, we have the partial sums given by

$$\begin{aligned} s_n &= \frac{1}{1 \cdot 2} + \frac{1}{2 \cdot 3} + \frac{1}{3 \cdot 4} + \cdots + \frac{1}{n(n+1)} \\ &= \left(\frac{1}{1} - \frac{1}{2}\right) + \left(\frac{1}{2} - \frac{1}{3}\right) + \left(\frac{1}{3} - \frac{1}{4}\right) + \cdots + \left(\frac{1}{n} - \frac{1}{n+1}\right) \\ &= 1 - \frac{1}{n+1}. \end{aligned}$$

This is an example of a "telescoping" series, so called because of the way in which the terms in the partial sums cancel. (We could also evaluate s_n using the formula in Exercise 10.4.) Since the sequence (s_n) of partial sums converges to 1, we have

$$\sum_{n=1}^{\infty} \frac{1}{n(n+1)} = 1.$$

32.2 EXAMPLE The **harmonic series** $\sum_{n=1}^{\infty} 1/n$ has partial sums

$$s_n = 1 + \frac{1}{2} + \frac{1}{3} + \cdots + \frac{1}{n}.$$

In Example 18.13 we saw that the sequence (s_n) is divergent. Thus the harmonic series is also divergent. Since the partial sums form an increasing sequence, we must have $\lim s_n = +\infty$, so

$$\sum_{n=1}^{\infty} \frac{1}{n} = +\infty.$$

32.3 PRACTICE In Exercise 10.6 we found that

$$\frac{1}{3} + \frac{1}{15} + \frac{1}{35} + \cdots + \frac{1}{4n^2 - 1} = \frac{n}{2n+1}, \qquad \text{for all } n \in \mathbb{N}.$$

Use this to find the sum of the infinite series $\sum_{n=1}^{\infty} 1/(4n^2 - 1)$.

The following theorem is sometimes useful in determining convergence and evaluating sums of series. Its proof follows from the corresponding result for sequences, Theorem 17.1.

32.4 THEOREM Suppose that $\sum a_n = s$ and $\sum b_n = t$. Then $\sum (a_n + b_n) = s + t$ and $\sum (ka_n) = ks$, for every $k \in \mathbb{R}$.

Proof: Exercise 32.4. ▌

If a series $\sum a_n$ is to be convergent, it would seem that the terms a_n would have to get close to zero. This observation is justified in our next theorem. The converse, however, is definitely not true. In Example 32.2 we saw that the harmonic series $\sum 1/n$ was divergent even though $\lim 1/n = 0$.

32.5 THEOREM If $\sum a_n$ is a convergent series, then $\lim a_n = 0$.

Proof: If $\sum a_n$ converges, then the sequence (s_n) of partial sums must have a finite limit. But $a_n = s_n - s_{n-1}$, so $\lim a_n = \lim s_n - \lim s_{n-1} = 0$. ▌

Corresponding to the Cauchy criterion for sequences (Theorem 18.12), we have the following result for series. Its usefulness is in being able to establish the convergence of a series without having to know the limit of the partial sums. (See Theorem 33.5.)

32.6 THEOREM (Cauchy Criterion for Series) The infinite series $\sum a_n$ converges iff for each $\varepsilon > 0$ there exists a number N such that, if $n \geqslant m > N$, then

$$|a_m + a_{m+1} + \cdots + a_n| < \varepsilon.$$

Proof: Suppose that $\sum a_n$ converges. Then the sequence (s_n) of partial sums converges, and by Theorem 18.12, (s_n) is Cauchy. Thus given any $\varepsilon > 0$ there exists a number N such that $m, n > N$ implies that $|s_n - s_m| < \varepsilon$. Hence if $n \geqslant m > N + 1$, then $m - 1 > N$, so that

$$|a_m + a_{m+1} + \cdots + a_n| = |s_n - s_{m-1}| < \varepsilon.$$

Conversely, given any $\varepsilon > 0$, suppose that there exists N such that $n \geqslant m > N$ implies that $|a_m + a_{m+1} + \cdots + a_n| < \varepsilon$. Then for $n > m > N$ we have $m + 1 > N$, so that

$$|s_n - s_m| = |a_{m+1} + a_{m+2} + \cdots + a_n| < \varepsilon.$$

Thus the sequence (s_n) of partial sums is Cauchy and the series is convergent. ∎

32.7 EXAMPLE One of the most useful series is the **geometric series** $\sum_{n=0}^{\infty} r^n$. In Exercise 10.5 we saw that for $r \neq 1$ we have

$$s_n = \sum_{k=0}^{n} r^k = 1 + r + r^2 + \cdots + r^n = \frac{1 - r^{n+1}}{1 - r}$$

for all $n \in \mathbb{N}$. But Exercise 16.3(f) shows that $\lim r^{n+1} = 0$ when $|r| < 1$, so we conclude that

$$\sum_{n=0}^{\infty} r^n = \lim s_n = \frac{1}{1 - r}, \qquad \text{for } |r| < 1.$$

If $|r| \geqslant 1$, then the sequence (r^n) does not converge to zero and it follows from Theorem 32.5 that the series $\sum r^n$ diverges.

There is a clever geometric argument that illustrates the convergence of the geometric series for $0 < r < 1$ as follows: Let $A = (0, 0)$ and $B = (1, 0)$ be points in the plane. Draw a line of slope r through A and a line of slope 1 through B. (See Figure 32.1.) Since $0 < r < 1$, these lines intersect at a point, say C. Let P_1 be the point on the line segment $\overline{AC}$ that lies directly above B. Since $\overline{AC}$ has slope r and $\ell(\overline{AB}) = 1$, we have $\ell(\overline{P_1 B}) = r$. [$\ell(\overline{AB})$ denotes the length of segment $\overline{AB}$.] Let P_2 be the point on $\overline{BC}$ at the same height as P_1. Since the slope of $\overline{BC}$ is 1, $\ell(\overline{P_1 P_2}) = r$. Continue in this manner with vertical and horizontal lines. If $\overline{CD}$ is the perpendicular from C to the line through A and B, then $\ell(\overline{AD})$ is the sum $s = 1 + r + r^2 + r^3 + \cdots$. Since $\ell(\overline{BD}) = \ell(\overline{CD})$, we have

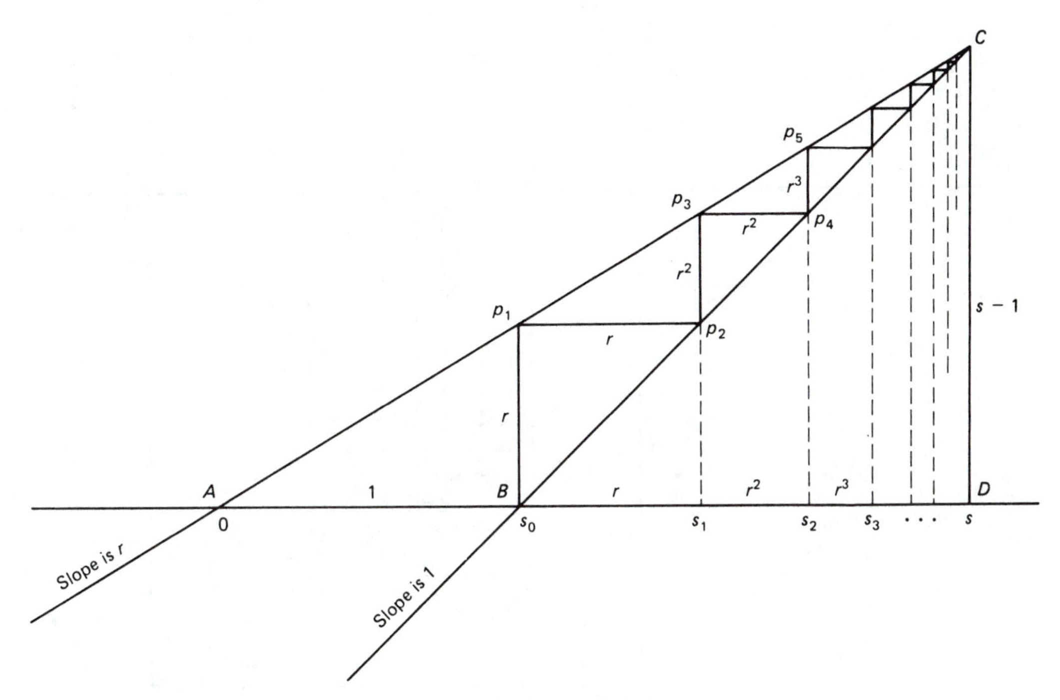

Figure 32.1 The Geometric Series

$\ell(\overline{CD}) = s - 1$. Now the triangles ΔABP_1 and ΔADC are similar, so their corresponding sides are proportional. That is,

$$\frac{\ell(\overline{AD})}{\ell(\overline{AB})} = \frac{\ell(\overline{CD})}{\ell(\overline{P_1B})} \quad \text{or} \quad \frac{s}{1} = \frac{s-1}{r}.$$

Solving for s, we obtain the expected formula $s = 1/(1 - r)$.

In Figure 32.2 we show the corresponding diagram for the alternating series

$$\sum_{n=0}^{\infty} (-1)^n r^n = 1 - r + r^2 - r^3 + r^4 - r^5 + \cdots, \qquad \text{for } 0 < r < 1.$$

This time the line through A is drawn with slope $-r$. We find $\ell(\overline{AB}) = 1$, $\ell(\overline{AD}) = s$, and $\ell(\overline{DC}) = \ell(\overline{DB}) = 1 - s$, so that

$$\frac{\ell(\overline{AD})}{\ell(\overline{AB})} = \frac{\ell(\overline{DC})}{\ell(\overline{BP_1})} \quad \text{or} \quad \frac{s}{1} = \frac{1-s}{r}.$$

Solving for s, we obtain $s = 1/(1 + r)$.

In the next section we derive several tests for determining whether or not a given series is convergent. The problem of actually finding the sum

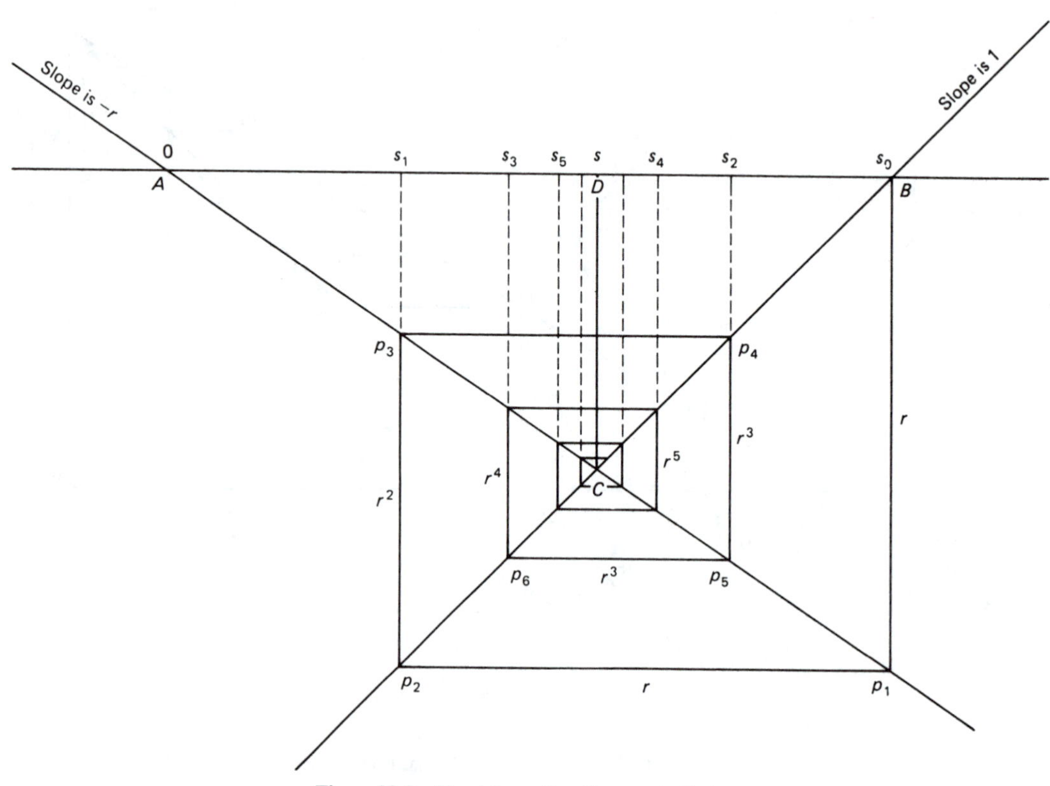

Figure 32.2 The Alternating Geometric Series

of a convergent series is often more difficult than proving convergence. One of the main techniques to use in finding the sum of a given series is to try to relate it to a geometric series by using Theorem 32.4.

32.8 PRACTICE Find the sum of the series

$$2 + \frac{4}{3} + \frac{8}{9} + \frac{16}{27} + \cdots + \frac{2^{n+1}}{3^n} + \cdots = \sum_{n=0}^{\infty} 2\left(\frac{2}{3}\right)^n.$$

ANSWERS TO PRACTICE PROBLEMS

32.3 Since $s_n = \dfrac{n}{2n+1} = \dfrac{1}{2 + 1/n}$, $\lim s_n = \dfrac{1}{2}$. Thus the series converges and its sum is $\frac{1}{2}$.

32.8 $\displaystyle\sum_{n=0}^{\infty} 2\left(\frac{2}{3}\right)^n = 2\sum_{n=0}^{\infty}\left(\frac{2}{3}\right)^n = 2\left(\frac{1}{1-\frac{2}{3}}\right) = 2(3) = 6$

EXERCISES

32.1 (a) Suppose that $\sum_{n=1}^{\infty} a_n$ is a convergent series and let $m \in \mathbb{N}$ with $m > 1$. Prove that $\sum_{n=m}^{\infty} a_n$ is convergent and that

$$\sum_{n=m}^{\infty} a_n = \left(\sum_{n=1}^{\infty} a_n\right) - (a_1 + a_2 + \cdots + a_{m-1}).$$

(b) Suppose that $m \in \mathbb{N}$ with $m > 1$ and that $\sum_{n=m}^{\infty} a_n$ is convergent. If $a_1, \ldots, a_{m-1}$ are real numbers, prove that $\sum_{n=1}^{\infty} a_n$ is convergent and that $\sum_{n=1}^{\infty} a_n = a_1 + \cdots + a_{m-1} + \sum_{n=m}^{\infty} a_n$.

32.2 Show that each series is divergent.

(a) $\sum (-1)^n$ (b) $\sum \dfrac{n}{n+1}$

(c) $\sum \dfrac{n}{\sqrt{n^2+1}}$ (d) $\sum \cos \dfrac{n\pi}{2}$

32.3 Find the sum of each series.

(a) $\sum_{n=1}^{\infty} \left(\dfrac{1}{3}\right)^n$ (b) $\sum_{n=3}^{\infty} \left(\dfrac{1}{2}\right)^n$

(c) $\sum_{n=0}^{\infty} 2\left(-\dfrac{1}{2}\right)^n$ (d) $\sum_{n=1}^{\infty} \left(-\dfrac{2}{3}\right)^n$

(e) $\sum_{n=2}^{\infty} \dfrac{1}{n(n-1)}$ (f) $\sum_{n=1}^{\infty} \dfrac{1}{(2n-1)(2n+1)}$

(g) $\sum_{n=1}^{\infty} \dfrac{1}{(3n-2)(3n+1)}$ (h) $\sum_{n=1}^{\infty} \dfrac{2}{n^2+2n}$

(i) $\sum_{n=1}^{\infty} \dfrac{1}{n(n+1)(n+2)}$ (j) $\sum_{n=1}^{\infty} \dfrac{(n-1)!}{(n+k)!}$ where $k \in \mathbb{N}$ is fixed

32.4 Prove Theorem 32.4.

***32.5** Given series $\sum a_n$ and $\sum b_n$, suppose that there exists a number N such that $a_n = b_n$ for all $n > N$. Prove that $\sum a_n$ is convergent iff $\sum b_n$ is convergent. Thus the convergence of a series is not affected by changing a finite number of terms. (Of course, the value of the sum may change.)

***32.6** Let (a_n) be a sequence of nonnegative real numbers. Prove that $\sum a_n$ converges iff the sequence of partial sums is bounded.

32.7 Determine whether or not the series $\sum_{n=1}^{\infty} 1/(\sqrt{n+1} + \sqrt{n})$ converges. Justify your answer.

32.8 Let (x_n) be a sequence of real numbers and let $y_n = x_n - x_{n+1}$ for each $n \in \mathbb{N}$.

(a) Prove that the series $\sum_{n=1}^{\infty} y_n$ converges iff the sequence (x_n) converges.
(b) If $\sum_{n=1}^{\infty} y_n$ converges, what is the sum?

32.9 Prove that if $\sum |a_n|$ converges and (b_n) is a bounded sequence, then $\sum a_n b_n$ converges.

32.10 Suppose that (a_n) is a sequence of positive numbers. For each $n \in \mathbb{N}$, let $b_n = (a_1 + a_2 + \cdots + a_n)/n$. Prove that $\sum b_n$ diverges to $+\infty$.

32.11 A series $\sum b_n$ is said to arise from a given series $\sum a_n$ by **grouping of terms** if every b_n is the sum of a finite number of consecutive terms of $\sum a_n$, and every pair of terms a_p and a_q, where $p < q$, appear as terms in a unique pair of terms b_r and b_s, respectively, where $r \leqslant s$. For example, the grouping

$$(a_1 + a_2) + (a_3 + a_4) + (a_5 + a_6) + \cdots$$

gives rise to the series $\sum b_n$, where $b_1 = a_1 + a_2$, $b_2 = a_3 + a_4, \ldots$.

*(a) Prove that any series arising from a convergent series by grouping of terms is convergent and has the same sum as the original series.

(b) Show by an example that grouping of terms may change a divergent series into a convergent series.

(c) Prove that any series arising from a divergent nonnegative series by grouping of terms is divergent.

32.12 Suppose that $r > 1$. Construct diagrams as in Figures 32.1 and 32.2 to illustrate the divergence of the geometric and alternating geometric series.

Section 33 CONVERGENCE TESTS

In this section we derive several tests that are useful in determining whether or not a given series is convergent. We begin by considering series of nonnegative terms. They are simpler to work with since their partial sums are monotone. Our first test is easy to prove and has wide application.

33.1 THEOREM (Comparison Test) Let $\sum a_n$ and $\sum b_n$ be infinite series of nonnegative terms. That is, $a_n \geqslant 0$ and $b_n \geqslant 0$ for all n. Then

(a) If $\sum a_n$ converges and $0 \leqslant b_n \leqslant a_n$ for all n, then $\sum b_n$ converges.
(b) If $\sum a_n = +\infty$ and $0 \leqslant a_n \leqslant b_n$ for all n, then $\sum b_n = +\infty$.

Proof: Since $b_n \geqslant 0$ for all n, the sequence (t_n) of partial sums of $\sum b_n$ is an increasing sequence. In part (a) this sequence is bounded above by the sum of the series $\sum a_n$, so (t_n) converges by the monotone convergence theorem (18.3). Thus $\sum b_n$ converges. In part (b) the sequence (t_n) must be unbounded, for otherwise $\sum a_n$ would have to converge. But then $\lim t_n = +\infty$ by Theorem 18.8, so that $\sum b_n = +\infty$. ∎

33.2 EXAMPLE Consider the series $\sum 1/(n+1)^2$. Now for all $n \in \mathbb{N}$ we have

$$0 < \frac{1}{(n+1)^2} < \frac{1}{n(n+1)}.$$

In Example 32.1 we saw that the series $\sum 1/[n(n+1)]$ is convergent, so Theorem 33.1 implies that $\sum 1/(n+1)^2$ is also convergent.

33.3 PRACTICE Use the comparison test to show that the series $\sum_{n=2}^{\infty} 1/(n-\sqrt{2})$ is divergent.

Most of the series we have seen as examples so far have had nonnegative terms. If a series $\sum a_n$ has some negative terms, then we often consider the related series $\sum |a_n|$. It turns out that the convergence of $\sum |a_n|$ implies the convergence of $\sum a_n$, although the converse is not true. (See Example 33.17.) We prove this following a preliminary definition.

33.4 DEFINITION If $\sum |a_n|$ converges, then the series $\sum a_n$ is said to **converge absolutely** (or to be **absolutely convergent**). If $\sum a_n$ converges but $\sum |a_n|$ diverges, then $\sum a_n$ is said to **converge conditionally** (or to be **conditionally convergent**).

33.5 THEOREM If a series converges absolutely, then it converges.

Proof: Suppose that $\sum a_n$ is absolutely convergent, so that $\sum |a_n|$ converges. By the Cauchy criterion for series (Theorem 32.6), given any $\varepsilon > 0$, there exists an N such that $n \geqslant m > N$ implies that $\big||a_m| + \cdots + |a_n|\big| < \varepsilon$. But then

$$|a_m + \cdots + a_n| \leqslant \big||a_m| + \cdots + |a_n|\big| < \varepsilon$$

by the triangle inequality (Exercise 11.5), so $\sum a_n$ also converges. ∎

When the terms of a series are nonnegative, convergence and absolute convergence are really the same thing. Thus the comparison test can be viewed as a test for absolute convergence. If we have a series $\sum b_n$ with some negative terms, the corresponding series $\sum |b_n|$ will have only nonnegative terms. If it happens that $0 \leqslant |b_n| \leqslant a_n$ for all n and if $\sum a_n$ converges, then we can conclude that $\sum |b_n|$ converges. That is, $\sum b_n$ converges absolutely.

Furthermore, changing the first few terms in a series will affect the value of the sum of the series, but it will not change whether or not the series is convergent. (See Exercise 32.5.) Thus, given a nonnegative convergent series $\sum a_n$ and a second series $\sum b_n$, to conclude that $\sum |b_n|$ is convergent it suffices to show that $0 \leqslant |b_n| \leqslant a_n$ for all n greater than some N.

33.6 EXAMPLE To show that the series $\sum (-1)^n/n^2$ converges absolutely, we shall first show that for $n > 2$ we have

$$\left|\frac{(-1)^n}{n^2}\right| = \frac{1}{n^2} \leqslant \frac{2}{(n+1)^2}.$$

Indeed, if $n > 2$, then $2/n \leqslant \frac{2}{3}$ and $1/n^2 \leqslant \frac{1}{3}$, so that

$$(n+1)^2 = n^2 + 2n + 1 = n^2\left(1 + \frac{2}{n} + \frac{1}{n^2}\right)$$

$$\leqslant n^2\left(1 + \frac{2}{3} + \frac{1}{3}\right) = 2n^2,$$

and the desired inequality follows. We know from Example 33.2 that the series $\sum 1/(n+1)^2$ is convergent, so Theorem 32.4 implies that $\sum 2/(n+1)^2$ also converges. We now apply the comparison test as above to conclude that $\sum (-1)^n/n^2$ converges absolutely.

As an alternative approach, we observe that for $n > 1$ the nth term of $\sum 1/n^2$ is the same as the $(n-1)$th term of $\sum 1/(n+1)^2$. Since the latter series converges (Example 32.2), so does $\sum 1/n^2$. Thus the series $\sum (-1)^n/n^2$ converges absolutely.

By using the comparison test and the geometric series, we can derive two more useful tests for convergence. Before studying their proofs it may be helpful for the reader to review the properties of the limit superior and the limit inferior of a sequence, as in Section 19.

The Ratio and Root Tests

33.7 THEOREM (Ratio Test) Let $\sum a_n$ be a series of nonzero terms.

(a) If $\limsup |a_{n+1}/a_n| < 1$, then the series converges absolutely.
(b) If $\liminf |a_{n+1}/a_n| > 1$, then the series diverges.
(c) Otherwise, $\liminf |a_{n+1}/a_n| \leqslant 1 \leqslant \limsup |a_{n+1}/a_n|$ and the test gives no information about convergence or divergence.

Proof: Let $\limsup |a_{n+1}/a_n| = L$. If $L < 1$, then choose r so that $L < r < 1$. By Theorem 19.11 there exists $N \in \mathbb{N}$ such that $n \geqslant N$ implies that $|a_{n+1}/a_n| \leqslant r$. That is, $|a_{n+1}| \leqslant r|a_n|$. It follows easily by induction that

$$|a_{N+k}| \leqslant r^k |a_N|, \qquad \text{for } k \in \mathbb{N}.$$

Since $0 < r < 1$, the geometric series $\sum_{k=1}^{\infty} r^k$ is convergent. Thus $\sum_{k=1}^{\infty} |a_N| r^k$ is also convergent, and $\sum a_n$ is absolutely convergent by the comparison test.

If $\liminf |a_{n+1}/a_n| > 1$, then it follows that $|a_{n+1}| > |a_n|$ for all n sufficiently large. Thus the sequence (a_n) cannot converge to zero, and by Theorem 32.5 the series $\sum a_n$ must diverge.

Part (c) follows from the observation that the series $\sum 1/n^2$ converges and the harmonic series $\sum 1/n$ diverges, as we have already seen. In both series we have $\lim |a_{n+1}/a_n| = 1$. ▌

33.8 THEOREM (Root Test) Given a series $\sum a_n$, let $\alpha = \lim\sup |a_n|^{1/n}$.

(a) If $\alpha < 1$, then the series converges absolutely.
(b) If $\alpha > 1$, then the series diverges.
(c) Otherwise, $\alpha = 1$ and the test gives no information about convergence or divergence.

Proof: If $\alpha < 1$, choose r so that $\alpha < r < 1$. Then from Theorem 19.10(a) we have $|a_n|^{1/n} \leqslant r$ for all n greater than some N. That is, $|a_n| \leqslant r^n$ for $n > N$. Since $0 < r < 1$, the geometric series $\sum r^n$ is convergent, so $\sum a_n$ is absolutely convergent by the comparison test.

If $\alpha > 1$, then $|a_n|^{1/n} \geqslant 1$ for infinitely many indices n. That is, $|a_n| \geqslant 1$ for infinitely many terms. Thus the sequence (a_n) cannot converge to zero and, by Theorem 32.5, the series $\sum a_n$ must diverge.

Again, part (c) follows from considering the convergent series $\sum 1/n^2$ and the divergent series $\sum 1/n$. In Example 16.11 we showed that $\lim n^{1/n} = 1$. Thus

$$\lim \left|\frac{1}{n^2}\right|^{1/n} = \frac{1}{\lim (n^2)^{1/n}} = \frac{1}{\lim (n^{1/n})^2} = \frac{1}{1^2} = 1.$$

Similarly, $\lim |1/n|^{1/n} = 1$, so the root test yields $\alpha = 1$ for both series. Thus when $\alpha = 1$ we can draw no conclusion about the convergence or divergence of a given series. ▌

33.9 EXAMPLE Consider the series $\sum n/2^n$. Letting $a_n = n/2^n$, we have

$$\frac{a_{n+1}}{a_n} = \frac{(n+1)(2^n)}{(2^{n+1})(n)} = \frac{n+1}{2n},$$

so $\lim |a_{n+1}/a_n| = \frac{1}{2}$. Similarly, $|a_n|^{1/n} = n^{1/n}/2$, so $\lim |a_n|^{1/n} = \frac{1}{2}$. (Recall that $\lim n^{1/n} = 1$.) Thus both the ratio test and the root test indicate that $\sum n/2^n$ converges.

It can be shown (Exercise 33.12) that whenever the ratio test determines convergence or divergence, the root test does, too. The ratio test is still important, however, since it is often easier to apply. The value of the root test is that it is more general, as we see in our next example.

33.10 EXAMPLE Consider the series

$$\frac{1}{2} + \frac{1}{3^2} + \frac{1}{2^3} + \frac{1}{3^4} + \frac{1}{2^5} + \frac{1}{3^6} + \cdots.$$

That is, we have $\sum_{n=1}^{\infty} a_n$, where $a_n = 1/2^n$ for n odd and $a_n = 1/3^n$ for n even. Now for n odd we have

$$\frac{a_{n+1}}{a_n} = \frac{2^n}{3^{n+1}} = \frac{1}{3}\left(\frac{2}{3}\right)^n \to 0$$

as $n \to \infty$. For n even we have

$$\frac{a_{n+1}}{a_n} = \frac{3^n}{2^{n+1}} = \frac{1}{2}\left(\frac{3}{2}\right)^n \to +\infty$$

as $n \to \infty$. Thus $\liminf |a_{n+1}/a_n| = 0$ and $\limsup |a_{n+1}/a_n| = +\infty$, so that the ratio test fails. But considering the root test we find $\limsup |a_n|^{1/n} = 1/2$, so that the series converges.

33.11 PRACTICE Determine whether or not the series $\sum n^2/2^n$ converges.

33.12 EXAMPLE Any series of the form $\sum 1/n^p$, where $p \in \mathbb{R}$, is called a p-series. It is easy to see that the ratio and root test both fail to determine convergence. For $p \leqslant 1$ we have $n^p \leqslant n$, so that $1/n^p \geqslant 1/n$. Since the harmonic series $\sum 1/n$ diverges, we conclude by the comparison test that the p-series diverges for $p \leqslant 1$. To handle the case when $p > 1$, we need another technique known as the integral test.

The Integral Test

33.13 THEOREM (Integral Test) Let f be a continuous function defined on $[0, \infty)$, and suppose that f is positive and decreasing. That is, if $x_1 < x_2$, then $f(x_1) \geqslant f(x_2) > 0$. Then the series $\sum (f(n))$ converges iff $\lim_{n\to\infty} (\int_1^n f(x)\, dx)$ exists as a real number.

Proof: Let $a_n = f(n)$ and $b_n = \int_n^{n+1} f(x)\, dx$. Since f is decreasing, given any $n \in \mathbb{N}$ we have $f(n+1) \leqslant f(x) \leqslant f(n)$ for all $x \in [n, n+1]$. It follows from Exercise 30.9 that

$$f(n+1) \leqslant \int_n^{n+1} f(x)\, dx \leqslant f(n).$$

Geometrically, this means that the area under the curve $y = f(x)$ from $x = n$ to $x = n + 1$ is between $f(n+1)$ and $f(n)$. (See Figure 33.1.) Thus $0 < a_{n+1} \leqslant b_n \leqslant a_n$ for each n. By the comparison test applied twice, $\sum a_n$ converges iff $\sum b_n$ converges. But the partial sums of $\sum b_n$ are the integrals $\int_1^n f(x)\, dx$, so $\sum b_n$ converges precisely when $\lim_{n\to\infty} [\int_1^n f(x)\, dx]$ exists as a real number. ∎

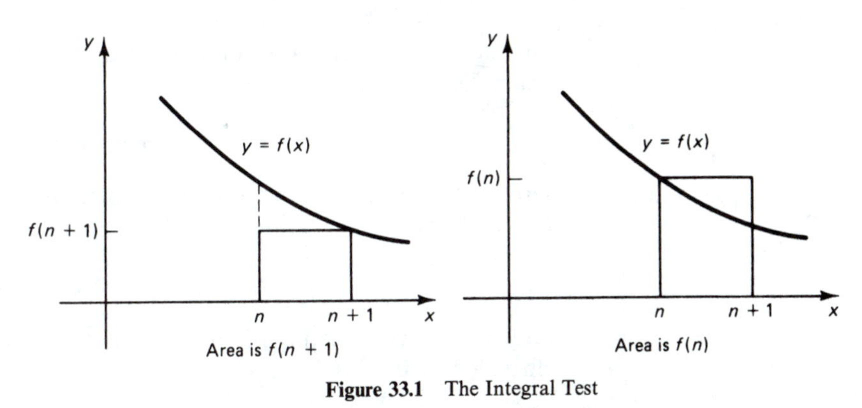

Figure 33.1 The Integral Test

33.14 EXAMPLE Let us return to the p-series, $\sum 1/n^p$. Let $f(x) = 1/x^p$ and recall that for $p \neq 1$,

$$\int_1^n \frac{1}{x^p}\,dx = \frac{1}{1-p}\,(n^{1-p} - 1).$$

The limit of this as $n \to \infty$ will be finite if $p > 1$ and infinite if $p < 1$. Thus by the integral test $\sum 1/n^p$ converges if $p > 1$ and diverges if $p < 1$. When $p = 1$ we get the harmonic series, which we already know is divergent. Thus, to summarize, the p-series $\sum 1/n^p$ converges if $p > 1$ and diverges if $p \leqslant 1$.

33.15 PRACTICE Use the integral test to show that the harmonic series $\sum 1/n$ diverges.

The Alternating Series Test

So far all of our tests for convergence have indicated whether or not a given series is absolutely convergent. As indicated earlier, some series converge but do not converge absolutely. If the terms in a series alternate between positive and negative values, the series is called an **alternating series**. Our final test gives us a simple criterion for determining the convergence of an alternating series. Using it we will be able to show, for example, that the alternating harmonic series $\sum (-1)^n/n$ is conditionally convergent.

33.16 THEOREM (Alternating Series Test) If (a_n) is a decreasing sequence of positive numbers and $\lim a_n = 0$, then the series $\sum (-1)^{n+1} a_n$ converges.

Proof: We shall show that the sequence (s_n) of partial sums converges. There are two kinds of partial sums depending on whether we are adding an odd or an even number of terms. For example,

$$s_{2n} = a_1 - a_2 + a_3 - a_4 + \cdots + a_{2n-1} - a_{2n}.$$

Now since (a_n) is decreasing we have $a_k - a_{k+1} \geqslant 0$ for all $k \in \mathbb{N}$. Thus

$$s_{2(n+1)} - s_{2n} = a_{2n+1} - a_{2n+2} \geqslant 0,$$

and the subsequence (s_{2n}) is increasing. It is also bounded above since, for each $n \in \mathbb{N}$,

$$s_{2n} = a_1 - (a_2 - a_3) - \cdots - (a_{2n-2} - a_{2n-1}) - a_{2n} \leqslant a_1.$$

Hence by the monotone convergence theorem (18.3), the subsequence (s_{2n}) converges to some real number s.

For the odd numbered partial sums we have

$$s_{2n+1} = a_1 - a_2 + a_3 - a_4 + \cdots + a_{2n-1} - a_{2n} + a_{2n+1}.$$

An argument similar to the above (Exercise 33.8) shows that the subsequence (s_{2n+1}) is decreasing and bounded below by 0. Thus it also converges. Furthermore, since $s_{2n+1} = s_{2n} + a_{2n+1}$ and $\lim a_{2n+1} = 0$, we must have $\lim s_{2n+1} = \lim s_{2n} = s$.

Combining these results, we see that given any $\varepsilon > 0$ there exist N_1 and N_2 such that $n > N_1$ implies that $|s_{2n} - s| < \varepsilon$ and $n > N_2$ implies that $|s_{2n+1} - s| < \varepsilon$. Thus for $N = \max\{2N_1, 2N_2 + 1\}$ we have $n > N$ implies that $|s_n - s| < \varepsilon$, so that the sequence (s_n) of partial sums converges. ∎

33.17 EXAMPLE Since the sequence $(1/n)$ is decreasing and $\lim (1/n) = 0$, we see that the alternating harmonic series $\sum_{n=1}^{\infty} (-1)^{n+1}/n$ converges. Since it does not converge absolutely, we conclude that it converges conditionally.

ANSWERS TO PRACTICE PROBLEMS

33.3 Since $1/(n - \sqrt{2}) > 1/n$ for all $n \geqslant 2$ and the harmonic series $\sum_{n=2}^{\infty} 1/n$ diverges, Theorem 33.1 implies that $\sum_{n=2}^{\infty} 1/(n - \sqrt{2})$ also diverges.

33.11 Either the ratio test or the root test can easily be used to show that the series converges. We have

$$\left|\frac{a_{n+1}}{a_n}\right| = \frac{1}{2}\left(\frac{n+1}{n}\right)^2 \to \frac{1}{2} \quad \text{and} \quad |a_n|^{1/n} = \frac{1}{2}(n^{1/n})^2 \to \frac{1}{2}.$$

33.15 Let $f(x) = 1/x$. Since

$$\lim_{n\to\infty}\left(\int_1^n \frac{1}{x}\,dx\right) = \lim_{n\to\infty}(\log n - \log 1) = +\infty,$$

the integral test implies that $\sum 1/n$ diverges.

EXERCISES

33.1 Determine whether each series converges or diverges. Justify your answer.

(a) $\sum \frac{n^3}{2^n}$ (b) $\sum \frac{2^n}{n!}$

(c) $\sum \frac{n}{n^2+2}$ (d) $\sum \frac{n!}{(2^n)^3}$

(e) $\sum \frac{n!}{n^n}$ (f) $\sum \frac{1}{\sqrt{n}}$

(g) $\sum \frac{1}{\sqrt{n(n+1)}}$ (h) $\sum \frac{1}{n\sqrt{n+1}}$

(i) $\sum (\sqrt{n+1} - \sqrt{n})$ (j) $\sum \frac{\sqrt{n+1}-\sqrt{n}}{n}$

(k) $\sum n^{-1-1/n}$ (l) $\sum \frac{\sin^2 n}{n^2}$

(m) $\sum 2^n e^{-n}$ (n) $\sum 3^n e^{-n}$

(o) $\sum \frac{(n!)^2}{(2n)!}$ (p) $\sum \frac{n!}{3\cdot 5\cdot 7\cdot\dots\cdot(2n+1)}$

33.2 Determine the values of p for which each series converges.

(a) $\sum_{n=2}^{\infty} \frac{1}{n(\log n)^p}$ (b) $\sum_{n=3}^{\infty} \frac{1}{n(\log n)(\log\log n)^p}$

33.3 Determine whether each series converges conditionally, converges absolutely, or diverges. Justify your answers.

(a) $\sum_{n=2}^{\infty} \frac{(-1)^n}{\log n}$ (b) $\sum \frac{(-2)^n}{n^2}$

(c) $\sum \frac{(-3)^n}{n!}$ (d) $\sum \frac{(-5)^n}{2^n}$

(e) $\sum \frac{\cos n\pi}{\sqrt{n}}$ (f) $\sum_{n=2}^{\infty} \frac{(-1)^n}{n\log n}$

(g) $\sum \frac{(-1)^n n}{n+1}$ (h) $\sum \frac{(-1)^n}{\sqrt{n^2+1}}$

(i) $\sum_{n=2}^{\infty} \frac{(-1)^n \log n}{n}$ (j) $\sum \left(\frac{1}{\sqrt{n}} - \frac{1}{n}\right)$

33.4 Find an example to show that the convergence of $\sum a_n$ and the convergence of $\sum b_n$ do not necessarily imply the convergence of $\sum (a_n b_n)$. (Compare with Exercise 32.9.)

33.5 Prove the ratio comparison test: If $a_n > 0$ and $b_n > 0$ for all n, if $\sum a_n$ converges, and if $b_{n+1}/b_n \leqslant a_{n+1}/a_n$ for all n, then $\sum b_n$ converges.

33.6 Let $\sum a_n$ and $\sum b_n$ be two series of positive terms and suppose that the sequence (a_n/b_n) converges to a nonzero number. Prove that $\sum a_n$ converges iff $\sum b_n$ converges. (This is sometimes called the limit comparison test.)

33.7 Suppose that (a_n) is a sequence of numbers such that for all n, $|a_{n+1} - a_n| \leqslant b_n$, where $\sum b_n$ is convergent. Show that (a_n) converges.

33.8 Finish the proof of the alternating series test (33.16). That is, show that the subsequence (s_{2n+1}) of partial sums is decreasing and bounded below.

33.9 Show that the series

$$1 - \frac{1}{2} + \frac{1}{3} - \frac{1}{2^2} + \frac{1}{5} - \frac{1}{2^3} + \frac{1}{7} - \frac{1}{2^4} + \cdots$$

diverges. Why doesn't this contradict the alternating series test?

***33.10** Let (a_n) be a decreasing sequence of positive numbers such that $\lim a_n = 0$. Show that the sum s of the alternating series $\sum (-1)^{n+1} a_n$ lies between any pair of successive partial sums. That is, show that $s_{2n} \leqslant s \leqslant s_{2n+1}$. Then use this to conclude that, for all n, $|s - s_n| \leqslant a_{n+1}$. Thus the error made in stopping at the nth term does not exceed the absolute value of the next term.

33.11 (a) Let (a_n) be a decreasing sequence of nonnegative terms. Prove the Cauchy condensation test: The series $\sum_{n=1}^{\infty} a_n$ converges iff the series $\sum_{k=0}^{\infty} 2^k a_{2^k}$ converges.

(b) Use part (a) to show that the p-series $\sum 1/n^p$ converges if $p > 1$ and diverges if $0 < p \leqslant 1$.

(c) Use parts (a) and (b) to show $\sum_{n=2}^{\infty} 1/[n(\log n)^p]$ converges if $p > 1$ and diverges if $0 < p \leqslant 1$.

***33.12** Let (a_n) be a sequence of positive numbers. Prove that

$$\liminf \frac{a_{n+1}}{a_n} \leqslant \liminf (a_n)^{1/n} \leqslant \limsup (a_n)^{1/n} \leqslant \limsup \frac{a_{n+1}}{a_n}.$$

Conclude that whenever the ratio test determines convergence or divergence, the root test does, too. *Hint*: To prove the inequality on the right, let $\alpha = \limsup a_{n+1}/a_n$. If $\alpha = +\infty$, the result is obvious. If α is finite, choose $\beta > \alpha$. By Theorem 19.10 there exists $N \in \mathbb{N}$ such that $a_{n+1}/a_n < \beta$ for all $n \geqslant N$. That is, for $n > N$ we have

$$a_n < \beta a_{n-1},\ a_{n-1} < \beta a_{n-2}, \ldots, a_{N+1} < \beta a_N.$$

Combine these $n - N$ inequalities to obtain $a_n < c\beta^n$ for a positive constant c. Argue from this that $\limsup (a_n)^{1/n} \leqslant \beta$. Since this holds for each $\beta > \alpha$, the desired inequality follows.

33.13 Prove that if a series is conditionally convergent then the series of negative terms is divergent.

33.14 Let $f\colon \mathbb{N} \to \mathbb{N}$ be a bijective function. Given a series $\sum a_n$, let $b_n = a_{f(n)}$ for each $n \in \mathbb{N}$. Then the series $\sum b_n$ is said to be a **rearrangement** of the series $\sum a_n$. Prove Dirichlet's theorem: If $\sum a_n$ converges absolutely, then every rearrangement of $\sum a_n$ converges absolutely, and they all converge to the same sum. *Hint*: Prove it first for nonnegative series. Suppose that $\sum a_n = A$ and let $\sum b_n$ be a rearrangement of $\sum a_n$. Show that each partial sum t_n of $\sum b_n$ satisfies $t_n \leqslant A$. Then argue that $\sum b_n$ converges to a sum B with $B \leqslant A$. Do the same thing with the partial sums of $\sum a_n$ to obtain $A \leqslant B$. Finally, to generalize to an arbitrary absolutely convergent series, consider the nonnegative and nonpositive parts separately.

33.15 Suppose that $\sum a_n$ is a conditionally convergent series and let $s \in \mathbb{R}$.

(a) Prove that there exists a rearrangement of $\sum a_n$ that converges conditionally to s.

(b) Prove that there exists a rearrangement of $\sum a_n$ that diverges.

Section 34 POWER SERIES

Up to this point we have dealt with infinite series whose terms were fixed numbers. We broaden our perspective now to consider series whose terms are variables. The simplest kind is known as a power series, and the main question will be determining the set of values of the variable for which the series is convergent.

34.1 DEFINITION Let $(a_n)_{n=0}^{\infty}$ be a sequence of real numbers. The series

$$\sum_{n=0}^{\infty} a_n x^n = a_0 + a_1 x + a_2 x^2 + a_3 x^3 + \cdots$$

is called a **power series**. The number a_n is called the nth **coefficient** of the series.

34.2 EXAMPLE Consider the power series whose coefficients are all equal to 1: $\sum x^n$. In Example 32.7 we found that this (geometric) series converged iff $|x| < 1$.

34.3 THEOREM Let $\sum a_n x^n$ be a power series and let $\alpha = \limsup |a_n|^{1/n}$. Define R by

$$R = \begin{cases} \dfrac{1}{\alpha}, & \text{if } 0 < \alpha < +\infty \\ +\infty, & \text{if } \alpha = 0 \\ 0, & \text{if } \alpha = +\infty. \end{cases}$$

Then the series converges absolutely whenever $|x| < R$ and diverges whenever $|x| > R$. (When $R = +\infty$ we take this to mean that the series converges absolutely for all real x. When $R = 0$ then the series converges only at $x = 0$.)

Proof: Let $b_n = a_n x^n$ and apply the root test (Theorem 33.8). If $\alpha \in \mathbb{R}$ we have

$$\beta = \limsup |b_n|^{1/n} = \limsup |a_n x^n|^{1/n} = |x| \limsup |a_n|^{1/n} = |x|\alpha$$

by Theorem 19.14. Thus if $\alpha = 0$, then $\beta = 0$ and the series converges for all real x. If $0 < \alpha < +\infty$, then the series converges when $|x|\alpha < 1$ and diverges when $|x|\alpha > 1$. That is, $\sum a_n x^n$ converges when $|x| < 1/\alpha = R$ and diverges when $|x| > 1/\alpha = R$.

Finally, if $\alpha = +\infty$, then for $x \neq 0$ we have $\beta = +\infty$ (Exercise 19.13), so $\sum a_n x^n$ diverges when $|x| > 0 = R$. Certainly, the series will converge when $x = 0$, for then all the terms except the first are zero. ∎

From Theorem 34.3 we see that the set of values C for which a power series converges will either be $\{0\}$, $\mathbb{R}$, or a bounded interval centered at 0. The R that is obtained in the theorem is referred to as the **radius of convergence** and the set C is called the **interval of convergence**. In doing so, we think of $\{0\}$ as an interval of zero radius and $\mathbb{R}$ as an interval of infinite radius. When $R = +\infty$, we may denote the interval of convergence $\mathbb{R}$ by $(-R, R)$.

Notice that when R is a positive real number the theorem says nothing about the convergence or divergence of the series at the endpoints of the interval of convergence. It is usually necessary to check the endpoints individually for convergence using one of the other tests in Section 33. Before illustrating this with several examples, we derive a ratio criterion for determining the radius of convergence that is often easier to apply than Theorem 34.3.

34.4 THEOREM (Ratio Criterion) The radius of convergence R of a power series $\sum a_n x^n$ is equal to $\lim |a_n/a_{n+1}|$, provided that this limit exists.

Proof: It follows from Exercise 33.12 that $\lim |a_n|^{1/n} = \lim |a_{n+1}/a_n|$, provided that the latter limit exists. Suppose that $\lim |a_n/a_{n+1}|$ exists and is equal to L. If $0 < L < +\infty$, then $a_n \neq 0$ for sufficiently large n, so Theorem 17.1(d) implies that $1/L = \lim |a_{n+1}/a_n| = \lim |a_n|^{1/n}$. But then $1/L = 1/R$ by Theorem 34.3, so $L = R$.

If $L = 0$, then $\lim |a_{n+1}/a_n| = +\infty$ by Theorem 17.13, so $\lim |a_n|^{1/n} = +\infty$ and $L = R = 0$. Similarly, if $L = +\infty$, then $0 = \lim |a_{n+1}/a_n| = \lim |a_n|^{1/n}$, so $L = R = +\infty$. ∎

Note that the ratio used in the ratio criterion for finding the radius of convergence is the reciprocal of the ratio used in the ratio convergence test (Theorem 33.7).

34.5 EXAMPLES (a) We have already seen that the interval of convergence of the series $\sum x^n$ is (1, 1). We find that $R = \lim |a_n/a_{n+1}| = 1$, as expected.

(b) For the series $\sum_{n=1}^{\infty} (1/n)x^n$ we have

$$\lim \left| \frac{a_n}{a_{n+1}} \right| = \lim \frac{n+1}{n} = 1,$$

so the radius of convergence is 1. Since the series converges (conditionally) at $x = -1$ and diverges at $x = 1$, the interval of convergence is $[-1, 1)$.

(c) For the series $\sum_{n=1}^{\infty} (1/n^2)x^n$ we have

$$\lim \left| \frac{a_n}{a_{n+1}} \right| = \lim \frac{(n+1)^2}{n^2} = 1,$$

so again the radius of convergence is 1. Since the series converges at both endpoints, the interval of convergence is $[-1, 1]$.

34.6 PRACTICE Find a series whose interval of convergence is $(-1, 1]$.

34.7 EXAMPLES (a) For the series $\sum_{n=0}^{\infty} (1/n!)x^n$ we have

$$\lim \left| \frac{a_n}{a_{n+1}} \right| = \lim \frac{(n+1)!}{n!} = \lim (n+1) = +\infty.$$

Thus the radius of convergence is $+\infty$ and the interval of convergence is $\mathbb{R}$.

(b) For the series $\sum_{n=1}^{\infty} n^n x^n$ it is easier to use the root formula:

$$\alpha = \lim |a_n|^{1/n} = \lim n = +\infty.$$

Thus $R = 0$ and the interval of convergence is $\{0\}$.

34.8 PRACTICE Find the radius of convergence and the interval of convergence for the series $\sum_{n=0}^{\infty} 2^{-n}x^n$.

34.9 EXAMPLE Consider the series

$$\frac{1}{1 \cdot 3} x^2 + \frac{1}{2 \cdot 3^2} x^4 + \frac{1}{3 \cdot 3^3} x^6 + \frac{1}{4 \cdot 3^4} x^8 + \cdots = \sum_{n=1}^{\infty} \frac{3^{-n}}{n} x^{2n}.$$

Letting $y = x^2$, we may apply the ratio criterion to the series $\sum_{n=1}^{\infty} (3^{-n}/n)y^n$ and obtain

$$\lim \left| \frac{a_n}{a_{n+1}} \right| = \lim \frac{3^{n+1}(n+1)}{3^n(n)} = 3.$$

Thus the series in y converges when $|y| < 3$. Since it also converges when $y = -3$ but diverges when $y = 3$, its interval of convergence is $[-3, 3)$. But $y = x^2$, so the original series has $(-\sqrt{3}, \sqrt{3})$ as its interval of convergence. Note that $x = -\sqrt{3}$ is not included because this corresponds to $y = +3$.

As an alternative approach, we may think of the original series in x as having zero coefficients for the odd-numbered terms:

$$0 \cdot x + \frac{1}{1 \cdot 3} x^2 + 0 \cdot x^3 + \frac{1}{2 \cdot 3^2} x^4 + 0 \cdot x^5 + \frac{1}{3 \cdot 3^3} x^6 + 0 \cdot x^7 + \cdots.$$

In this approach we cannot use the ratio criterion, but by considering the subsequence (a_{2k}) of nonzero terms, the root formula yields

$$\alpha = \limsup |a_n|^{1/n} = \lim_{k \to \infty} \left| a_{2k} \right|^{\frac{1}{2k}} = \lim_{k \to \infty} \left(\frac{3^{-k}}{k} \right)^{\frac{1}{2k}}$$

$$= \lim_{k \to \infty} \frac{3^{-1/2}}{(k^{1/k})^{1/2}} = \frac{1}{\sqrt{3}}.$$

Once again we have $R = \sqrt{3}$.

In some situations it is useful to consider more general power series of the form

$$\sum_{n=0}^{\infty} a_n(x - x_0)^n,$$

where x_0 is a fixed real number. By making the substitution $y = x - x_0$, we can apply the familiar techniques to the series $\sum_{n=0}^{\infty} a_n y^n$. If we find that the series in y converges when $|y| < R$, we conclude that the original series converges when $|x - x_0| < R$.

34.10 EXAMPLE For the series

$$1 + (x - 1) + (x - 1)^2 + (x - 1)^3 + \cdots = \sum_{n=0}^{\infty} (x - 1)^n$$

we find $R = 1$, so it converges when $|x - 1| < 1$. Since it diverges at $x = 0$ and $x = 2$, the interval of convergence is $(0, 2)$. In fact, using the formula for the geometric series we see that for all x in $(0, 2)$ the sum of the series is equal to

$$\frac{1}{1 - (x - 1)} = \frac{1}{2 - x}.$$

ANSWERS TO PRACTICE PROBLEMS

34.6 One example is $\sum_{n=1}^{\infty} (-1)^n(1/n)x^n$.

34.18 We have

$$R = \lim \left| \frac{a_n}{a_{n+1}} \right| = \lim \frac{2^{-n}}{2^{-(n+1)}} = \lim \frac{2^{n+1}}{2^n} = 2.$$

Since the series diverges when $|x| = 2$, the interval of convergence is $(-2, 2)$. Alternatively, we may use the root formula to obtain

$$\alpha = \lim |a_n|^{1/n} = \lim (2^{-n})^{1/n} = 2^{-1} = \frac{1}{2},$$

so that $R = 1/\alpha = 2$.

EXERCISES

34.1 Find the radius of convergence R and the interval of convergence C for each series.

(a) $\sum nx^n$ (b) $\sum \frac{n^2}{2^n} x^n$

(c) $\sum \frac{2^n}{n} x^n$ (d) $\sum \frac{n^3}{n!} x^n$

(e) $\sum \left(\frac{x}{n}\right)^n$ (f) $\sum \frac{(-4)^{-n}}{n} x^n$

(g) $\sum \frac{1}{\sqrt{n}} x^n$ (h) $\sum (n3^{-n})(x-2)^n$

(i) $\sum \frac{3^{2n}}{n} (x-1)^n$ (j) $\sum (2^{-n})(x-5)^{2n}$

34.2 Let R be the radius of convergence for the power series $\sum a_n x^n$. If infinitely many of the coefficients a_n are nonzero integers, prove that $R \leqslant 1$.

34.3 Find the radius of convergence for each series.

(a) $\sum \frac{(2n)!}{(n!)^2} x^n$ (b) $\sum \frac{(3n)!}{(n!)^2} x^n$

(c) $\sum \frac{n!}{n^n} x^n$ (d) $\sum \frac{n!}{(2n)^n} x^n$

34.4 Suppose that the power series $\sum a_n x^n$ converges for $x = x_0$, where $x_0 \neq 0$. Without using Theorem 34.3, prove that the series converges absolutely for all x such that $|x| < |x_0|$.

34.5 Suppose that the sequence (a_n) is bounded but that the series $\sum a_n$ diverges. Prove that the radius of convergence of the power series $\sum a_n x^n$ is equal to 1.

34.6 If $A \subseteq \mathbb{R}$, define $-A$ to be $\{-a: a \in A\}$. Let A be the interval of convergence of the series $\sum a_n x^n$. Prove that the interval of convergence of $\sum (-1)^n a_n x^n$ is $-A$.

34.7 Suppose that the series $\sum a_n x^n$ has radius of convergence 2. Find the radius of convergence of each series, where k is a fixed positive integer.

(a) $\sum a_n^k x^n$ (b) $\sum a_n x^{kn}$ (c) $\sum a_n x^{n^2}$

34.8 Prove that the series $\sum_{n=0}^{\infty} a_n x^n$ and $\sum_{n=0}^{\infty} n a_n x^n$ have the same radius of convergence (finite or infinite).

9

Sequences and Series of Functions

We concluded Chapter 8 by developing some of the properties of power series. In the present chapter we at first consider more general sequences and series of functions. In Section 35 we define two kinds of convergence of functions. Pointwise convergence is the natural extension of convergence of sequences and series of constants, but it lacks several important properties. The stronger notion of uniform convergence is seen in Section 36 to have these desired properties. Finally, in Section 37, we return to power series and apply our earlier results.

Section 35 POINTWISE AND UNIFORM CONVERGENCE

35.1 DEFINITION Let (f_n) be a sequence of functions defined on a subset S of $\mathbb{R}$. Then (f_n) **converges pointwise** on S if for each $x \in S$ the sequence of numbers $(f_n(x))$ converges. If (f_n) converges pointwise on S, then we define $f: S \to \mathbb{R}$ by

$$f(x) = \lim_{n \to \infty} f_n(x)$$

for each $x \in S$, and we say that (f_n) converges to f pointwise on S.

The main question that we wish to examine is whether or not various properties are preserved in taking the limit operation. For example, if we know that each of the functions f_n is continuous on a set S and that f_n

converges pointwise to f on S, can we conclude that f is continuous on S? Similarly, if each f_n is integrable (or differentiable), what can we say about the limit function f?

Suppose that S is an interval and that $x \in S$. By Theorem 21.2, to say that f is continuous at x means that

$$\lim_{t \to x} f(t) = f(x).$$

Similarly, if each f_n is continuous at x, then

$$f(x) = \lim_{n \to \infty} f_n(x) = \lim_{n \to \infty} \left[\lim_{t \to x} f_n(t) \right].$$

Thus the limit of a sequence of continuous functions will be continuous at x if

$$\lim_{t \to x} \left[\lim_{n \to \infty} f_n(t) \right] = \lim_{n \to \infty} \left[\lim_{t \to x} f_n(t) \right].$$

That is, we wish to know when the order of the limit processes may be reversed.

Since derivatives and integrals are also defined in terms of limits, the question of whether these properties are preserved by a limit function can also be viewed as asking whether the order of the limit processes may be interchanged. In the next three examples we see that in general the order may not be reversed without affecting the result.

35.2 EXAMPLE For $x \in [0, 1]$ and $n \in \mathbb{N}$, define $f_n(x) = x^n$. Then for $x \in [0, 1)$ we have $\lim_{n \to \infty} f_n(x) = 0$, and $\lim_{n \to \infty} f_n(1) = 1$. Thus (f_n) is pointwise convergent on $S = [0, 1]$, and the limit function f is given by

$$f(x) = \begin{cases} 0, & \text{if } 0 \leqslant x < 1 \\ 1, & \text{if } x = 1. \end{cases}$$

(See Figure 35.1.) We note that each of the functions f_n is continuous and differentiable on $[0, 1]$, but that f is neither continuous nor differentiable at $x = 1$.

35.3 EXAMPLE For $x \in [0, 1]$ and $n \geqslant 2$, define

$$f_n(x) = \begin{cases} n^2 x, & \text{if } 0 \leqslant x \leqslant \dfrac{1}{n} \\[2ex] -n^2\left(x - \dfrac{2}{n}\right), & \text{if } \dfrac{1}{n} < x \leqslant \dfrac{2}{n} \\[2ex] 0, & \text{if } \dfrac{2}{n} < x \leqslant 1. \end{cases}$$

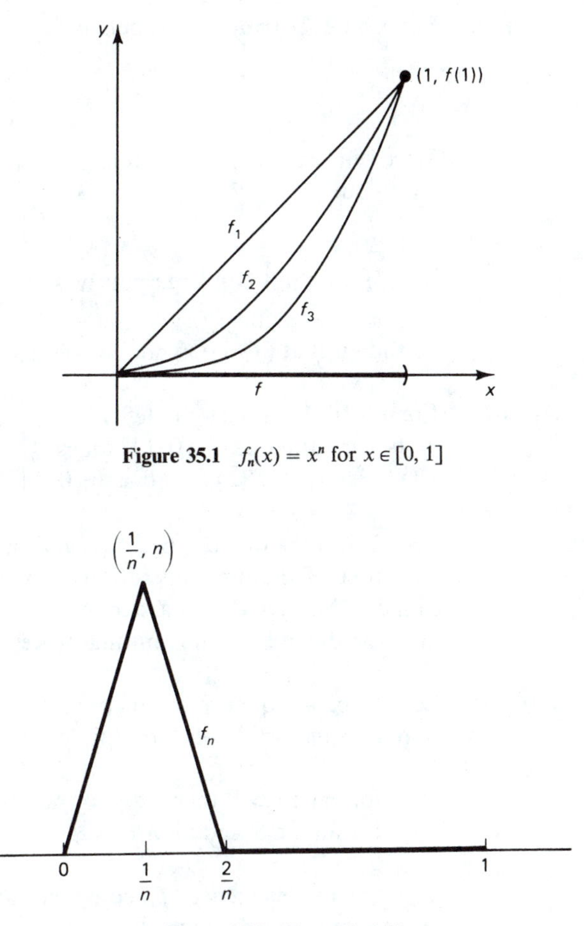

Figure 35.1 $f_n(x) = x^n$ for $x \in [0, 1]$

Figure 35.2 Example 35.3

(See Figure 35.2.) Given any $x > 0$, let $M = 2/x$. Then for $n > M$ we have $2/n < 2/M = x$, so that $f_n(x) = 0$. Since $f_n(0) = 0$ for all n, the limit function f is identically zero. That is, $f(x) = 0$ for all $x \in [0, 1]$. Now each of the functions f_n is continuous on $[0, 1]$, so it is integrable. Since f_n encloses an area bounded by a triangle with base $2/n$ and altitude n, we have

$$\int_0^1 f_n(x)\,dx = 1, \qquad \text{for } n \geqslant 2.$$

But the integral of the limit function is 0, so

$$\lim_{n \to \infty} \left[\int_0^1 f_n(x)\,dx \right] = 1 \neq 0 = \int_0^1 \left[\lim_{n \to \infty} f_n(x) \right] dx.$$

35.4 EXAMPLE For $x \in [0, 2\pi]$ and $n \in \mathbb{N}$, define

$$f_n(x) = \frac{\sin nx}{\sqrt{n}}.$$

Since $|\sin nx| \leq 1$, $f(x) = \lim_{n\to\infty} f_n(x) = 0$ for all x. Thus $f'(x) = 0$ for all x. But

$$f_n'(x) = \sqrt{n} \cos nx,$$

so (f_n') does not converge pointwise to f' on $[0, 2\pi]$. For example, $f_n'(0) = \sqrt{n} \to +\infty$ and $f_n'(\pi) = \sqrt{n}(-1)^{n+1}$ has no limit at all. [In fact, it can be shown that $(f_n'(x))$ is not convergent for any x.]

35.5 PRACTICE For $x \in [0, 1]$ and $n \in \mathbb{N}$, define $f_n(x) = 2x + x/n$. Find the limit function f. Is f continuous on $[0, 1]$? Does $\int_0^1 f(x)\,dx = \lim_{n\to\infty} \int_0^1 f_n(x)\,dx$? Does $f'(x) = \lim_{n\to\infty} f_n'(x)$ for all x in $[0, 1]$?

From the practice problem and the examples we see that sometimes the order of the limit operations can be interchanged, but in general it cannot. We now define a stronger form of convergence that will behave more predictably in the limiting process.

35.6 DEFINITION Let (f_n) be a sequence of functions defined on a subset S of $\mathbb{R}$. Then (f_n) **converges uniformly** on S to a function f defined on S if

for each $\varepsilon > 0$ there exists a number N such that $|f_n(x) - f(x)| < \varepsilon$ for all $x \in S$ and all $n > N$.

To say that a sequence (f_n) converges uniformly on S is to say that there exists a function f to which (f_n) converges uniformly on S.

The difference between pointwise and uniform convergence can be seen as follows: If (f_n) converges pointwise to f on S, then for every $\varepsilon > 0$ and each $x \in S$, there is a number N (depending on both ε and x) such that $|f_n(x) - f(x)| < \varepsilon$ whenever $n > N$; if (f_n) converges uniformly to f on S, then it is possible for each $\varepsilon > 0$ to find *one* number N that will work for *all* $x \in S$. This is the same type of quantifier reversal that produced uniform continuity from continuity.

Geometrically, uniform convergence means that when $n > N$ the entire graph of f_n must lie within the strip bounded by the graphs of $f - \varepsilon$ and $f + \varepsilon$. (See Figure 35.3.)

35.7 EXAMPLE Let $f_n(x) = x^n$ as in Example 35.2. If $\varepsilon = \frac{1}{2}$, then each f_n has values farther than ε away from the limit 0. Indeed, given any $n \in \mathbb{N}$, if $2^{-n} < x < 1$, then $f_n(x) > \frac{1}{2}$. Thus the convergence of f_n to 0 is not uniform on $[0, 1]$.

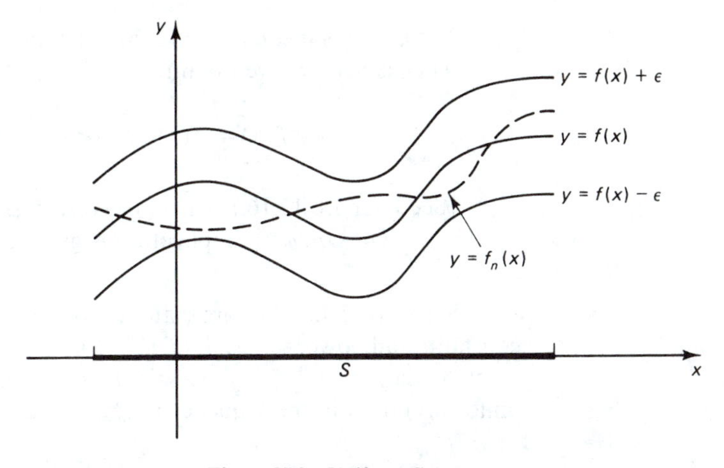

Figure 35.3 Uniform Convergence

35.8 EXAMPLE Let (f_n) be defined as in Example 35.3. Since the graph of each f_n goes above the line $y = 1$, each function assumes values that are more than $\varepsilon = 1$ away from the limit 0. Once again the convergence is not uniform on $[0, 1]$.

35.9 PRACTICE Is the convergence in Example 35.4 uniform on $[0, 2\pi]$?

As we might expect from our previous experience with sequences of numbers, there is a Cauchy criterion for uniform convergence of sequences of functions.

35.10 THEOREM Let (f_n) be a sequence of functions defined on a subset S of $\mathbb{R}$. There exists a function f such that (f_n) converges to f uniformly on S iff the following condition (called the Cauchy criterion) is satisfied:

For every $\varepsilon > 0$ there exists a number N such that $|f_n(x) - f_m(x)| < \varepsilon$ for all $x \in S$ and all $m, n > N$.

Proof: Suppose that the Cauchy criterion is satisfied. Since for each $x \in S$, $(f_n(x))$ is a Cauchy sequence, Theorem 18.12 implies that each sequence $(f_n(x))$ is convergent. Thus for each $x \in S$ we may define $f(x) = \lim_{n\to\infty} f_n(x)$, and the sequence of functions (f_n) converges pointwise on S to f. We claim that this convergence is also uniform on S.

Given any $\varepsilon > 0$, there exists a number N such that $m, n > N$ implies that

$$|f_n(x) - f_m(x)| < \frac{\varepsilon}{2}, \qquad \text{for all } x \in S.$$

Fix n in this inequality and take the limit as $m \to \infty$. Since $f_m(x) \to f(x)$ as $m \to \infty$, we obtain

$$|f_n(x) - f(x)| \leqslant \frac{\varepsilon}{2} < \varepsilon, \qquad \text{for all } x \in S.$$

(See Exercise 17.16.) Thus (f_n) converges uniformly on S to f.
The converse implication is given as Exercise 35.7. ▮

Series of functions are handled in a way analogous to series of constants and power series. If $(f_n)_{n=0}^{\infty}$ is a sequence of functions defined on a set S, the series $\sum_{n=0}^{\infty} f_n$ is said to converge pointwise (respectively, uniformly) on S iff the sequence $(s_n)_{n=0}^{\infty}$ of partial sums, given by

$$s_n(x) = \sum_{k=0}^{n} f_k(x),$$

converges pointwise (respectively, uniformly) on S. Using Theorem 35.10, we can derive a very useful test for establishing the uniform convergence of a series of functions.

35.11 THEOREM (Weierstrass M-test) Suppose that (f_n) is a sequence of functions defined on S and (M_n) is a sequence of nonnegative numbers such that

$$|f_n(x)| \leqslant M_n, \qquad \text{for all } x \in S \text{ and all } n \in \mathbb{N}.$$

If $\sum M_n$ converges, then $\sum f_n$ converges uniformly on S.

Proof: We shall show that the partial sums, $s_n(x) = \sum_{k=0}^{n} f_k(x)$, satisfy the Cauchy criterion of Theorem 35.10. Given any $\varepsilon > 0$, the convergence of $\sum M_n$ and Theorem 32.6 imply that there exists a number N such that if $n \geqslant m > N$ then

$$M_m + M_{m+1} + \cdots + M_n < \varepsilon.$$

(We have dropped the absolute-value symbols since each M_k is nonnegative.) Thus, if $n > m > N$, we have

$$\begin{aligned} |s_n(x) - s_m(x)| &= |f_{m+1}(x) + \cdots + f_n(x)| \\ &\leqslant |f_{m+1}(x)| + \cdots + |f_n(x)| \\ &\leqslant M_{m+1} + \cdots + M_n < \varepsilon, \end{aligned}$$

for all $x \in S$. It follows from Theorem 35.10 that (s_n) converges uniformly on S; hence $\sum f_n$ also converges uniformly on S. ▮

35.12 EXAMPLE Consider the series of functions $\sum_{n=0}^{\infty} f_n$ where $f_n(x) = x^n/n!$ for all $x \in \mathbb{R}$. In Example 34.7(a) we saw that this series is (pointwise) convergent on $\mathbb{R}$. Let us now show that the convergence is not uniform on $\mathbb{R}$, but given any $t \in \mathbb{R}$ the convergence is uniform on the closed interval $[-t, t]$.

To show that the convergence of the series $\sum f_n$ is not uniform on $\mathbb{R}$, we show that the sequence (s_n) of partial sums does not satisfy the Cauchy criterion of Theorem 35.10. To this end, let $\varepsilon = 1$. Then given any $n \in \mathbb{N}$, let $x_n = n$. It follows that

$$|s_n(x_n) - s_{n-1}(x_n)| = |f_n(x_n)| = f_n(n) = \frac{n^n}{n!} \geqslant 1 = \varepsilon.$$

Hence the series is not uniformly convergent on $\mathbb{R}$.

On the other hand, if $t \in \mathbb{R}$, then let $M_n = t^n/n!$. For any $x \in [-t, t]$ we have

$$|f_n(x)| = \left|\frac{x^n}{n!}\right| \leqslant \frac{t^n}{n!} = M_n.$$

Since $\sum M_n$ is convergent (by the ratio test), it follows from the Weierstrass M-test that $\sum x^n/n!$ is uniformly convergent on $[-t, t]$.

We shall return to the discussion of power series in Section 37, but first we must show that the uniform convergence of a sequence actually does enable us to interchange the order of certain limit operations. We undertake this task in the next section.

ANSWERS TO PRACTICE PROBLEMS

35.5 We find $f(x) = 2x$ for all x, and this is continuous on $[0, 1]$. Now $\int_0^1 f(x)\,dx = 1$ and $\lim_{n\to\infty} \int_0^1 f_n(x)\,dx = \lim_{n\to\infty} [1 + 1/(2n)] = 1$. Similarly, $f'(x) = 2$ for all x, and $\lim_{n\to\infty} f_n'(x) = \lim_{n\to\infty} (2 + 1/n) = 2$.

35.9 This time the convergence *is* uniform on the given set $[0, 2\pi]$. Given $\varepsilon > 0$ and any $x \in [0, 2\pi]$, let $M = 1/\varepsilon^2$. Then for $n > M$ we have

$$\left|\frac{\sin nx}{\sqrt{n}} - 0\right| \leqslant \frac{1}{\sqrt{n}} < \frac{1}{\sqrt{M}} = \varepsilon.$$

In the next section (Theorem 36.7) we obtain some additional requirements that will enable us to conclude that the derivative of the limit is the limit of the derivatives. It is apparent from this example that the uniform convergence of (f_n) is not enough.

EXERCISES

35.1 Let $f_n(x) = x^n/n$ for $x \in [-1, 1]$. Find $f(x) = \lim f_n(x)$ and determine whether or not the convergence is uniform on $[-1, 1]$. Justify your answer.

35.2 Let $f_n(x) = x/(x + n)$ for $x \geqslant 0$.
(a) Show that $f(x) = \lim f_n(x) = 0$ for all $x \geqslant 0$.
(b) Show that if $t > 0$ the convergence is uniform on $[0, t]$.
(c) Show that the convergence is not uniform on $[0, \infty)$.

35.3 Let $f_n(x) = nx/(1 + nx)$ for $x \geqslant 0$.
(a) Find $f(x) = \lim f_n(x)$.
(b) Show that if $t > 0$ the convergence is uniform on $[t, \infty)$.
(c) Show that the convergence is not uniform on $[0, \infty)$.

35.4 Repeat Exercise 35.3 with $f_n(x) = nx/(1 + n^2x^2)$ for $x \geqslant 0$.

35.5 Let $f_n(x) = 1/(1 + x^n)$ for $x \in [0, 1]$.
(a) Find $f(x) = \lim f_n(x)$.
(b) Show that if $0 < t < 1$ the convergence is uniform on $[0, t]$.
(c) Show that the convergence is not uniform on $[0, 1]$.

35.6 Let $f_n(x) = x^n/(1 + x)$ for $x \in [0, 2]$.
(a) Find the set S for which $f(x) = \lim f_n(x)$ is defined as a real-valued function.
(b) Show that if $0 < t < 1$ the convergence is uniform on $[0, t]$.
(c) Show that the convergence is not uniform on $[0, 1]$.

35.7 Prove the converse part of Theorem 35.10.

35.8 Let (f_n) be a sequence of functions defined on a set S. Prove that (f_n) converges uniformly to a function f on S iff

$$\lim_{n\to\infty} [\sup \{|f(x) - f_n(x)| : x \in S\}] = 0.$$

35.9 Let $f_n(x) = nx/e^{nx}$ for $x \in [0, 2]$.
(a) Show that $\lim f_n(x) = 0$ for $x \in [0, 2]$.
(b) Use Exercise 35.8 to show that the convergence is not uniform on $[0, 2]$.
(c) Let $0 < t < 2$. Determine on which interval, $[0, t]$ or $[t, 2]$, the convergence is uniform. Justify your answer.

35.10 Let $f_n(x) = x(1 - x)^n$ for $x \in [0, 1]$. Find $f(x) = \lim f_n(x)$ and then use Exercise 35.8 to determine whether or not the convergence is uniform on $[0, 1]$.

35.11 Let $f_n(x) = nx^n(1 - x)$ for $x \in [0, 1]$.
(a) Find $f(x) = \lim f_n(x)$.
(b) Use Exercise 35.8 to show that the convergence is not uniform on $[0, 1]$.
(c) Does $\lim_{n\to\infty} \int_0^1 f_n(x)\,dx = \int_0^1 f(x)\,dx$?

35.12 If (f_n) and (g_n) converge uniformly on a set S, prove that $(f_n + g_n)$ converges uniformly on S.

35.13 Suppose that the sequence (f_n) converges uniformly to f on a set S and that, for each $n \in \mathbb{N}$, f_n is bounded on S.

*(a) Prove that f is bounded on S.

(b) Prove that the sequence (f_n) is uniformly bounded on S. That is, there exists a number M such that $|f_n(x)| \leqslant M$ for all $x \in S$ and all $n \in \mathbb{N}$.

(c) Find a sequence of bounded functions that converges (pointwise) to a function that is not bounded.

35.14 Let $f_n(x) = x + 1/n$ and $f(x) = x$ for $x \in \mathbb{R}$.

(a) Show that (f_n) converges uniformly to f on $\mathbb{R}$.

(b) Show that (f_n^2) converges pointwise to f^2 on $\mathbb{R}$, but not uniformly.

35.15 If (f_n) and (g_n) are sequences of bounded functions that converge uniformly on S, prove that $(f_n g_n)$ converges uniformly on S.

35.16 Suppose that (f_n) converges pointwise to f on a set S. Prove that (f_n) converges uniformly to f on every finite subset of S.

35.17 Determine whether or not the given series of functions converges uniformly on the indicated set. Justify your answers.

(a) $\sum \dfrac{x^{2n}}{(n+x)^2}$ for $x \in [0, 1]$ (b) $\sum \dfrac{1}{n}\sqrt{\dfrac{\sin nx}{n}}$ for $x \in \mathbb{R}$

(c) $\sum n^{-x}$ for $x > \sqrt{2}$ (d) $\sum \dfrac{x^2}{n^2}$ for $x \in [0, 5]$

(e) $\sum \dfrac{x^2}{n^2}$ for $x \geqslant 5$ (f) $\sum \dfrac{1}{1+n^2x^2}$ for $x \in (0, 1]$

(g) $\sum \dfrac{1}{1+n^2x^2}$ for $x \geqslant 1$ (h) $\sum \dfrac{x^2}{1+n^2x^2}$ for $x \in \mathbb{R}$

Section 36 APPLICATIONS OF UNIFORM CONVERGENCE

Having seen several methods of determining whether or not a given sequence (or series) converges uniformly, we now turn our attention to showing that uniform convergence actually does behave in a more predictable way. In particular, the uniform limit of a sequence of continuous functions is continuous.

36.1 THEOREM Let (f_n) be a sequence of continuous functions defined on a set S and suppose that (f_n) converges uniformly on S to a function $f: S \to \mathbb{R}$. Then f is continuous on S.

Proof: The argument is based on the inequality

$$|f(x) - f(c)| \leqslant |f(x) - f_n(x)| + |f_n(x) - f_n(c)| + |f_n(c) - f(c)|.$$

The idea is to make the terms $|f(x) - f_n(x)|$ and $|f_n(c) - f(c)|$ small by using the uniform convergence of (f_n) and choosing n sufficiently large. Once an n is selected, the continuity of f_n will enable us to make the term $|f_n(x) - f_n(c)|$ small by requiring x to be close to c.

More precisely, let $c \in S$ and let $\varepsilon > 0$. Then there exists $N \in \mathbb{N}$ such that $n > N$ implies that $|f_n(x) - f(x)| < \varepsilon/3$ for all $x \in S$. Since $c \in S$ we also have $|f_n(c) - f(c)| < \varepsilon/3$ whenever $n > N$. In particular, these inequalities both hold when $n = N + 1$. Now since f_{N+1} is continuous at c, there exists $\delta > 0$ such that

$$|f_{N+1}(x) - f_{N+1}(c)| < \frac{\varepsilon}{3}, \qquad \text{whenever } |x - c| < \delta \text{ and } x \in S.$$

Thus for all $x \in S$ with $|x - c| < \delta$ we have

$$\begin{aligned}|f(x) - f(c)| &\leqslant |f(x) - f_{N+1}(x)| + |f_{N+1}(x) - f_{N+1}(c)| \\ &\quad + |f_{N+1}(c) - f(c)| \\ &< \frac{\varepsilon}{3} + \frac{\varepsilon}{3} + \frac{\varepsilon}{3} = \varepsilon.\end{aligned}$$

Hence f is continuous at c, and since c is any point in S, we conclude that f is continuous on S. ▌

When applied to series of functions, Theorem 36.1 yields the following corollary.

36.2 COROLLARY Let $\sum_{n=0}^{\infty} f_n$ be a series of functions defined on a set S. Suppose that each f_n is continuous on S and that the series converges uniformly to a function f on S. Then $f = \sum_{n=0}^{\infty} f_n$ is continuous on S.

Proof: Since each partial sum $s_n = \sum_{k=0}^{n} f_k$ is continuous, Theorem 36.1 implies that f, the limit of the partial sums, is also continuous. ▌

One useful application of Theorem 36.1 is in showing that a given sequence does *not* converge uniformly. For example, let $f_n(x) = x^n$ for $x \in [0, 1]$. Then each f_n is continuous on $[0, 1]$, but the limit function

$$f(x) = \begin{cases} 0, & \text{if } 0 \leqslant x < 1 \\ 1, & \text{if } x = 1 \end{cases}$$

is not continuous at $x = 1$. Thus the convergence cannot be uniform on $[0, 1]$. This can be proved directly as in Example 35.7, but using Theorem 36.1 is easier.

36.3 PRACTICE Let $f_n(x) = 1/(1 + x^n)$ for $x \in [0, 1]$. Show that the sequence (f_n) does not converge uniformly on $[0, 1]$.

The relationship between uniform convergence and integration is established in the next theorem.

36.4 THEOREM Let (f_n) be a sequence of continuous functions defined on an interval $[a, b]$ and suppose that (f_n) converges uniformly on $[a, b]$ to a function f. Then

$$\lim_{n\to\infty} \int_a^b f_n(x)\, dx = \int_a^b f(x)\, dx.$$

Proof: Theorem 36.1 implies that f is continuous on $[a, b]$, so the functions f, f_n, and $f_n - f$ are all integrable on $[a, b]$ by Theorem 30.2. Given any $\varepsilon > 0$, since (f_n) converges uniformly to f on $[a, b]$, there exists a number N such that

$$|f_n(x) - f(x)| < \frac{\varepsilon}{b-a}, \qquad \text{for all } x \in [a, b] \text{ and all } n > N.$$

(We may assume that $a \neq b$, for otherwise the integrals are all zero and the result is trivial.) Thus whenever $n > N$ we have

$$\begin{aligned}\left|\int_a^b f_n(x)\, dx - \int_a^b f(x)\, dx\right| &= \left|\int_a^b [f_n(x) - f(x)]\, dx\right| \\ &\leqslant \int_a^b |f_n(x) - f(x)|\, dx \\ &\leqslant \int_a^b \frac{\varepsilon}{b-a}\, dx = \varepsilon,\end{aligned}$$

where we have used Corollary 30.8 for the inequality. It follows that $\lim_{n\to\infty} \int_a^b f_n(x)\, dx = \int_a^b f(x)\, dx$. ∎

36.5 COROLLARY Let $\sum_{n=0}^{\infty} f_n$ be a series of functions defined on an interval $[a, b]$. Suppose that each f_n is continuous on $[a, b]$ and that the series converges uniformly to a function f on $[a, b]$. Then $\int_a^b f(x)\, dx = \sum_{n=0}^{\infty} \int_a^b f_n(x)\, dx$.

Proof: Let $s_n = \sum_{k=0}^{n} f_k$ be the nth partial sum of the series so that (s_n) converges uniformly to f on $[a, b]$. Then

$$\int_a^b s_n(x)\, dx = \int_a^b \left[\sum_{k=0}^{n} f_k(x)\right] dx = \sum_{k=0}^{n} \left[\int_a^b f_k(x)\, dx\right].$$

Thus by Theorem 36.4,

$$\begin{aligned}\int_a^b f(x)\, dx &= \lim_{n\to\infty} \int_a^b s_n(x)\, dx \\ &= \lim_{n\to\infty} \sum_{k=0}^{n} \left[\int_a^b f_k(x)\, dx\right] \\ &= \sum_{n=0}^{\infty} \int_a^b f_n(x)\, dx.\end{aligned}$$

∎

36.6 EXAMPLE Consider the geometric series $\sum_{n=0}^{\infty} (-t)^n$ and suppose that $0 < r < 1$. When $t \in [-r, r]$ then the series converges to $1/(1+t)$. In fact, since $|-t^n| \leqslant r^n$ and $\sum r^n$ converges when $0 < r < 1$, the Weierstrass M-test (35.11) shows that the convergence is uniform on any interval $[-r, r]$, where $0 < r < 1$. Thus, if $x \in (-1, 1)$, we can integrate the series term by term from 0 to x:

$$\int_0^x \frac{dt}{1+t} = \sum_{n=0}^{\infty} (-1)^n \int_0^x t^n \, dt = \sum_{n=0}^{\infty} (-1)^n \frac{x^{n+1}}{n+1}.$$

But from calculus we recall that

$$\int_0^x \frac{dt}{1+t} = \log(1+x),$$

so we conclude that for any x in $(-1, 1)$ we have

$$\log(1+x) = \sum_{n=0}^{\infty} (-1)^n \frac{x^{n+1}}{n+1} = x - \frac{x^2}{2} + \frac{x^3}{3} - \frac{x^4}{4} + \cdots.$$

We saw in Example 35.4 and Practice 35.9 that the uniform convergence of a sequence of functions (f_n) to a function f is not sufficient in itself to enable us to conclude that $\lim f_n' = f'$. It turns out that we must require the sequence of derivatives (f_n') to converge uniformly.

36.7 THEOREM Suppose that (f_n) converges to f on an interval $[a, b]$. Suppose also that each f_n' exists and is continuous on $[a, b]$, and the sequence (f_n') converges uniformly on $[a, b]$. Then $\lim_{n\to\infty} f_n'(x) = f'(x)$ for each $x \in [a, b]$.

Proof: For each $x \in [a, b]$ let $g(x) = \lim f_n'(x)$ so that the sequence (f_n') converges uniformly to g on $[a, b]$. Since each f_n' is continuous we may apply Theorem 36.4 to obtain

$$\int_a^x g(t)\, dt = \lim_{n\to\infty} \int_a^x f_n'(t)\, dt, \qquad \text{where } x \in [a, b].$$

Now $\int_a^x f_n'(t)\, dt = f_n(x) - f(a)$ by Theorem 31.2, and $\lim f_n(x) = f(x)$. Thus

$$\int_a^x g(t)\, dt = f(x) - f(a)$$

and we conclude that

$$f(x) = \int_a^x g(t)\, dt + f(a), \qquad \text{for all } x \in [a, b].$$

Since each f_n' is continuous and (f_n') converges uniformly to g on $[a, b]$, we know from Theorem 36.1 that g is continuous on $[a, b]$. Thus by Theorem 31.1 we see that f is differentiable on $[a, b]$ and that $f'(x) = g(x) = \lim f_n'(x)$ for all $x \in [a, b]$. [Since $f(a)$ is a constant, its derivative is zero.] ▮

36.8 COROLLARY Let $\sum_{n=0}^{\infty} f_n$ be a series of functions that converges to a function f on an interval $[a, b]$. Suppose that, for each n, f_n' exists and is continuous on $[a, b]$ and that the series of derivatives $\sum_{n=0}^{\infty} f_n'$ is uniformly convergent on $[a, b]$. Then $f'(x) = \sum_{n=0}^{\infty} f_n'(x)$ for all $x \in [a, b]$.

Proof: Exercise 36.10. ▮

As an important application of uniform convergence, we now show that there exists a continuous function defined on $\mathbb{R}$ that is nowhere differentiable. This surprising result was first demonstrated by Karl Weierstrass in the mid-nineteenth century and it exposed the inadequacy of the intuitive approach followed by most of his contemporaries. Certainly, a continuous curve that has no tangent at any point is unintuitive! The construction used here follows that of Rudin (1976).

36.9 THEOREM There exists a continuous function defined on $\mathbb{R}$ that is nowhere differentiable.

Proof: Define $g(x) = |x|$ for $x \in [-1, 1]$ and extend the definition of g to all of $\mathbb{R}$ by requiring that $g(x + 2) = g(x)$ for all x. (See the graph in Figure 36.1.) Now, given any $r, s \in \mathbb{R}$, we claim that

$$|g(r) - g(s)| \leqslant |r - s|.$$

Indeed, if $r, s \in [-1, 1]$, then

$$|g(r) - g(s)| = \big||r| - |s|\big| \leqslant |r - s|$$

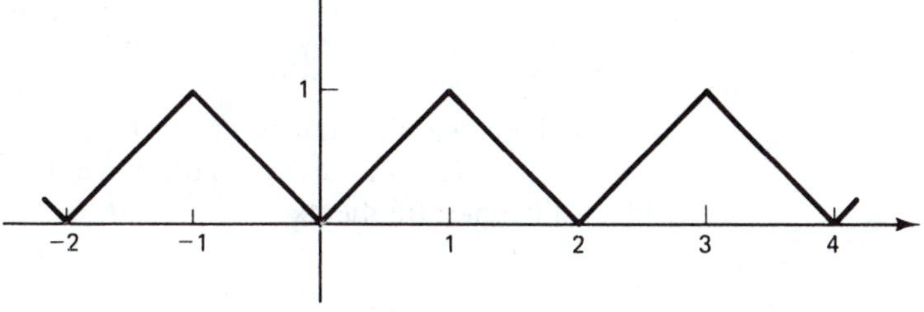

Figure 36.1 The Graph of g

by Exercise 11.4. By translation, this is seen to apply to any pair of points that are no more than two apart. On the other hand, if $|r - s| > 2$, then

$$|g(r) - g(s)| \leqslant |g(r)| + |g(s)| \leqslant 1 + 1 = 2 < |r - s|.$$

Hence $|g(r) - g(s)| \leqslant |r - s|$ for all r and s, and it follows that g is continuous on $\mathbb{R}$.

For each integer $n \geqslant 0$ let $g_n(x) = (\frac{3}{4})^n g(4^n x)$. Thus $g_0(x) = g(x)$, $g_1(x) = \frac{3}{4}g(4x)$, and so on. In general, each g_{n+1} oscillates four times as fast, but only three-fourths as high, as g_n. (See Figure 36.2.) Now define f on $\mathbb{R}$ by

$$f(x) = \sum_{n=0}^{\infty} g_n(x) = \sum_{n=0}^{\infty} \left(\frac{3}{4}\right)^n g(4^n x).$$

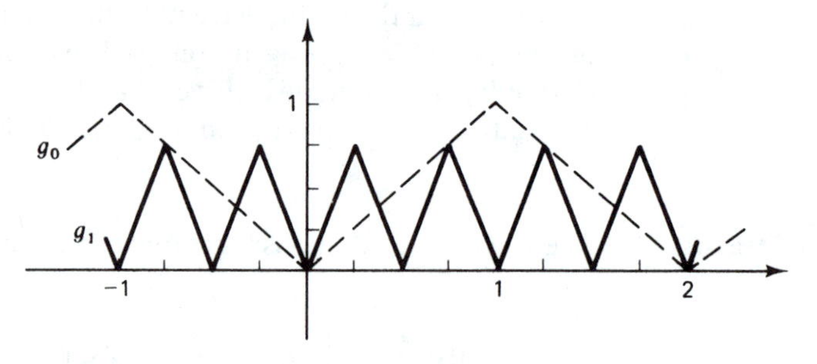

Figure 36.2 The Graphs of g_0 and g_1

For all $x \in \mathbb{R}$ we have $0 \leqslant g(x) \leqslant 1$, so that $0 \leqslant g_n(x) \leqslant (\frac{3}{4})^n$. Thus the Weierstrass M-test implies that the series converges uniformly to f on $\mathbb{R}$. Since each g_n is continuous on $\mathbb{R}$ (why?), Corollary 36.2 implies that f is continuous on $\mathbb{R}$.

To see that f is nowhere differentiable, we fix $x \in \mathbb{R}$ and $m \in \mathbb{N}$. Let

$$\delta_m = \pm \frac{4^{-m}}{2},$$

where the sign is chosen so that no integer is between $4^m x$ and $4^m(x + \delta_m)$. (This is possible since $4^m|\delta_m| = \frac{1}{2}$.) We claim that the difference quotients

$$\left| \frac{f(x + \delta_m) - f(x)}{\delta_m} \right|$$

are bounded below by a sequence that diverges to $+\infty$ as $m \to \infty$. Since $\delta_m \to 0$ as $m \to \infty$, this will imply (Theorem 25.3) that f is not differentiable at x. Now

$$\begin{aligned} f(x+\delta_m) - f(x) &= \sum_{n=0}^{\infty} g_n(x+\delta_m) - \sum_{n=0}^{\infty} g_n(x) \\ &= \sum_{n=0}^{\infty} [g_n(x+\delta_m) - g_n(x)] \\ &= \sum_{n=0}^{\infty} \left(\frac{3}{4}\right)^n [g(4^n x + 4^n \delta_m) - g(4^n x)]. \end{aligned}$$

Thus we need to look at $g(4^n x + 4^n \delta_m) - g(4^n x)$ for each n.

When $n > m$, then $4^n \delta_m = \pm 4^{n-m}/2$ is an even integer. Since $g(t) = g(t+2)$ for all t, it follows that

$$g(4^n x + 4^n \delta_m) - g(4^n x) = 0.$$

When $n < m$, then since $|g(r) - g(s)| \leqslant |r - s|$ for all $r, s \in \mathbb{R}$, we have

$$|g(4^n x + 4^n \delta_m) - g(4^n x)| \leqslant 4^n |\delta_m|.$$

Finally, suppose that $n = m$. Since there is no integer between $4^m x$ and $4^m(x + \delta_m)$, the graph of g between these points is a straight line of slope ± 1. Thus

$$|g(4^m x + 4^m \delta_m) - g(4^m x)| = 4^m |\delta_m|.$$

Combining the results, we obtain

$$\begin{aligned} \left|\frac{f(x+\delta_m) - f(x)}{\delta_m}\right| &= \left|\sum_{n=0}^{m} \left(\frac{3}{4}\right)^n \frac{g(4^n x + 4^n \delta_m) - g(4^n x)}{\delta_m}\right| \\ &\geqslant 3^m - \sum_{n=0}^{m-1} \left(\frac{3}{4}\right)^n \frac{4^n |\delta_m|}{|\delta_m|} \\ &= 3^m - \sum_{n=0}^{m-1} 3^n \\ &= 3^m - \frac{1 - 3^m}{1 - 3} \\ &= \frac{1}{2}(3^m + 1), \end{aligned}$$

where we have used the triangle inequality and Exercise 10.5. Since this last expression does indeed diverge to $+\infty$ as $m \to \infty$, we conclude that f is not differentiable at x. But x was arbitrary, so f is nowhere differentiable. ∎

ANSWERS TO PRACTICE PROBLEMS

36.3 Each f_n is continuous on $[0, 1]$, but the limit function

$$f(x) = 1 \quad \text{if } x \in [0, 1) \quad \text{and} \quad f(1) = \frac{1}{2}$$

is not continuous at $x = 1$. Thus the convergence cannot be uniform on $[0, 1]$.

EXERCISES

36.1 Let $f_n(x) = nx/(1 + nx)$ for $x \in [0, 1]$. Show that the sequence (f_n) does not converge uniformly on $[0, 1]$ by using Theorem 36.1

36.2 Repeat Exercise 36.1 for $f_n(x) = 1/(1 + x^n)$.

36.3 Let $f_n(x) = (n + \sin nx)/(3n + \sin^2 nx)$ for $x \in \mathbb{R}$.

(a) Show that (f_n) converges uniformly on $\mathbb{R}$.

(b) Use Theorem 36.4 to evaluate $\lim_{n\to\infty} \int_0^\pi f_n(x)\, dx$.

36.4 Find a sequence of functions (f_n) defined on $[0, 1]$ such that each f_n is discontinuous at each point of $[0, 1]$ and such that (f_n) converges uniformly to a function f that is continuous on $[0, 1]$.

36.5 Show that the series $\sum_{n=1}^\infty (-1)^{n+1}/(n + x^2)$ is uniformly convergent on $\mathbb{R}$ but is not absolutely convergent for any $x \in \mathbb{R}$.

36.6 Show that the series $\sum_{n=1}^\infty (-1)^{n+1}(x^2 + n)/n^2$ is uniformly convergent on every bounded set S, but is not absolutely convergent for any $x \in \mathbb{R}$.

36.7 Let $0 < k < 1$. Show that the series $\sum_{n=0}^\infty x^n$ converges uniformly on $[-k, k]$, but does not converge uniformly on $(-1, 1)$.

36.8 Let $f_n(x) = x^2/(1 + x^2)^n$ for $x \in \mathbb{R}$ and consider

$$f(x) = \sum_{n=0}^{\infty} f_n(x) = \sum_{n=0}^{\infty} \frac{x^2}{(1 + x^2)^n}.$$

(a) Use the geometric series to show that $f(x) = 1 + x^2$ for $x \neq 0$.

(b) Does the series converge uniformly on $\mathbb{R}$? Does the series converge uniformly on $[-1, 1]$? Justify your answers.

36.9 Using Corollary 36.5, integrate the geometric series

$$\frac{1}{1 - t} = 1 + t + \cdots + t^n + \cdots$$

term by term from $-x$ to x, where $x \in (-1, 1)$, and obtain a series for $\log[(1 + x)/(1 - x)]$.

36.10 Prove Corollary 36.8.

36.11 Let (f_n) be a sequence of functions that are uniformly continuous on a set S. Suppose that (f_n) converges uniformly to a function f on S. Prove that f is uniformly continuous on S.

36.12 Let (f_n) be a sequence of functions that are integrable on $[a, b]$ and that converge uniformly on $[a, b]$ to a function f. Prove that f is integrable and that

$$\int_a^b f = \lim_{n\to\infty} \int_a^b f_n.$$

36.13 Let (f_n) be a sequence of continuous functions that converges (pointwise) to a continuous function f on a compact set S. Suppose that for each $x \in S$ the sequence $(f_n(x))$ is monotone decreasing. Prove that (f_n) converges uniformly to f on S. (This is known as Dini's theorem.) *Hint*: Given $\varepsilon > 0$, for each $n \in \mathbb{N}$ let

$$B_n = \{x \in S: |f(x) - f_n(x)| \geqslant \varepsilon\}.$$

Since f and all the f_n's are continuous, each set B_n is closed (Exercise 22.10) and hence compact (Exercise 14.6). Since the sequence is monotone we have $B_{n+1} \subseteq B_n$ for all $n \in \mathbb{N}$. Show that $\bigcap_{n=1}^{\infty} B_n = \varnothing$ and then apply Theorem 14.7 to conclude that there exists $N \in \mathbb{N}$ such that $B_n = \varnothing$ for all $n \geqslant N$. Then $n \geqslant N$ and $x \in S$ imply that $|f(x) - f_n(x)| < \varepsilon$.

36.14 (a) Show that the monotone property is essential in Exercise 36.13. That is, find a sequence of continuous functions (f_n) that converges to a continuous function f on a compact set S, but such that the convergence is not uniform.

(b) Show that the compactness of S is essential in Exercise 36.13. That is, find a sequence of continuous functions (f_n) that converges to a continuous function f on a set S and such that $(f_n(x))$ is decreasing for each $x \in S$, but such that the convergence is not uniform.

36.15 A sequence of functions (f_n) defined on a set S is said to be **equicontinuous** on S if for every $\varepsilon > 0$ there exists a $\delta > 0$ such that $|f_n(x) - f_n(y)| < \varepsilon$ whenever $|x - y| < \delta$, $x \in S$, $y \in S$, and $n \in \mathbb{N}$. Prove the following.

(a) If an equicontinuous sequence of functions (f_n) converges pointwise to f on a set S, then f is uniformly continuous on S.

(b) If a sequence of continuous functions (f_n) converges uniformly on a compact set S, then the sequence is equicontinuous.

Section 37 UNIFORM CONVERGENCE OF POWER SERIES

In Section 34 we derived some of the basic properties of power series. In particular we showed (Theorem 34.3) that each power series $\sum a_n x^n$ has a radius of convergence R, where $0 \leqslant R \leqslant +\infty$. Recall that the series converges absolutely for $|x| < R$ and diverges for $|x| > R$. We now establish the uniform convergence of power series and apply some of the results of Section 36.

37.1 THEOREM Let $\sum a_n x^n$ be a power series with radius of convergence R, where $0 < R \leqslant +\infty$. If $0 < K < R$, then the power series converges uniformly on $[-K, K]$.

Proof: If $x \in [-K, K]$, then $|a_n x^n| \leqslant |a_n| K^n$. Since $0 < K < R$, the series $\sum |a_n| K^n$ is convergent. Thus by the Weierstrass M-test (35.11), $\sum a_n x^n$ converges uniformly on $[-K, K]$. ▮

Combining Theorem 37.1 with Corollary 36.5, we see that a power series may be integrated term by term on any compact interval within the interior of its interval of convergence. (See Example 36.6.) It can also be differentiated term by term, as we see in our next theorem. Note that for power series we do not have to assume that the differentiated series is uniformly convergent, as we did in Theorem 36.7.

37.2 THEOREM Suppose that a power series converges to a function f on $(-R, R)$, where $R > 0$. Then the series can be differentiated term by term, and the differentiated series converges on $(-R, R)$ to f'. That is, if $f(x) = \sum_{n=0}^{\infty} a_n x^n$, then $f'(x) = \sum_{n=1}^{\infty} n a_n x^{n-1}$, and both series have the same radius of convergence.

Proof: Since $R > 0$, Theorem 34.3 implies that the sequence $(|a_n|^{1/n})$ is bounded. Since $\lim n^{1/n} = 1$, Theorem 19.14 then implies that

$$\limsup |n a_n|^{1/n} = \limsup |a_n|^{1/n}.$$

Thus both series have the same radius of convergence. Applying Theorem 37.1, we conclude that the differentiated series converges uniformly on $[-K, K]$ for any positive $K < R$. Thus, by Corollary 36.8, $f'(x) = \sum_{n=0}^{\infty} n a_n x^{n-1} = \sum_{n=1}^{\infty} n a_n x^{n-1}$ for all x in $[-K, K]$. Since this holds for any positive $K < R$, the differentiated series converges to f' on $(-R, R)$. ▮

If we apply Theorem 37.2 repeatedly, we see that a function given by a convergent power series has derivatives of all orders, and these derivatives may be obtained by termwise differentiation of the original series within the interval of convergence.

37.3 COROLLARY Suppose that $f(x) = \sum_{n=0}^{\infty} a_n x^n$ for $x \in (-R, R)$, where $R > 0$. Then for each $k \in \mathbb{N}$, the kth derivative $f^{(k)}$ of f exists on $(-R, R)$ and

$$f^{(k)}(x) = \sum_{n=k}^{\infty} \frac{n!}{(n-k)!} a_n x^{n-k}$$

$$= k! a_k + (k+1)! a_{k+1} x + \frac{(k+2)!}{2!} a_{k+2} x^2 + \cdots.$$

Furthermore, $f^{(k)}(0) = k! a_k$.

Proof: The series for $f^{(k)}(x)$ is obtained by differentiating the terms of $\sum_{n=0}^{\infty} a_n x^n$ a total of k times. The formula for $f^{(k)}(0)$ comes from substituting $x = 0$ into the series, since only the first term is nonzero. ▮

37.4 COROLLARY If $\sum_{n=0}^{\infty} a_n x^n = \sum_{n=0}^{\infty} b_n x^n$ for all x in some interval $(-R, R)$, where $R > 0$, then $a_n = b_n$ for all $n \in \mathbb{N} \cup \{0\}$.

Proof: If the common value of the two series is denoted by $f(x)$, then by Corollary 37.3 we have

$$a_n = \frac{f^{(n)}(0)}{n!} = b_n, \qquad \text{for all } n \in \mathbb{N}.$$

If we substitute $x = 0$ into both series, we obtain $a_0 = b_0$. ▮

In our discussion so far we have begun with a power series $\sum a_n x^n$ and used it to obtain a function $f(x) = \sum a_n x^n$ throughout its interval of convergence. In practice, however, we often begin with a function f and seek a power series that converges to f over some interval. We say that a power series **represents** a function f on a set S if for each $x \in S$ the series $\sum a_n x^n$ converges to $f(x)$.

The importance of Corollary 37.4 is that it shows that a function can be represented by only one power series of the form $\sum a_n x^n$ and that this one is the Taylor series that is obtained as the limit of the Taylor polynomials. (See Section 28.)

Thus, if a function is represented by a power series, that power series must necessarily be its Taylor series. It is possible, however, that the Taylor series of a function may not represent the function on any interval. (This means, of course, that the function cannot be represented by any power series. See Exercise 37.14.) The Taylor series of a function f will represent f throughout an interval I precisely when the remainder term

$$R_n(x) = \frac{f^{(n+1)}(c)}{(n+1)!}(x - x_0)^{n+1}$$

in Taylor's theorem (28.2) has a limit of 0 for all x in I.

37.5 EXAMPLE Let $f(x) = e^x$ for $x \in \mathbb{R}$. In Example 28.3 we obtained the Taylor polynomials

$$p_n(x) = 1 + x + \frac{x^2}{2!} + \cdots + \frac{x^n}{n!}$$

and the remainder

$$R_n(x) = \frac{e^c x^{n+1}}{(n+1)!},$$

where c is some number between 0 and x. For any fixed $x \in \mathbb{R}$ it is easy to see that $\lim_{n\to\infty} R_n(x) = 0$. (See Exercise 37.11.) Thus $f(x) = e^x$ is represented by its Taylor series for all real x and we have

$$e^x = \sum_{n=0}^{\infty} \frac{x^n}{n!}.$$

37.6 PRACTICE Let $f(x) = \sin x$. Find the Taylor series for f at $x_0 = 0$ and determine for what x the series represents f.

By combining differentiation, integration, and substitution, we may obtain the series representations of various functions without actually computing the Taylor coefficients. Because of Corollary 37.4, we are assured that the series so obtained will actually be the Taylor series.

37.7 EXAMPLE We know that the geometric series

$$\frac{1}{1-x} = 1 + x + x^2 + x^3 + \cdots + x^n + \cdots$$

converges for $|x| < 1$. Differentiating, we obtain

$$\frac{1}{(1-x)^2} = 1 + 2x + 3x^2 + 4x^3 + \cdots + (n+1)x^n + \cdots,$$

which again converges for $|x| < 1$. Replacing x by x^2, we have

$$\frac{1}{(1-x^2)^2} = 1 + 2x^2 + 3x^4 + 4x^6 + \cdots + (n+1)x^{2n} + \cdots$$

and the series converges for $|x^2| < 1$, that is, $|x| < 1$. If the series for $(1-x)^{-2}$ is multiplied by x^2, we find that

$$\frac{x^2}{(1-x)^2} = x^2 + 2x^3 + 3x^4 + 4x^5 + \cdots + (n+1)x^{n+2} + \cdots$$

with convergence for $|x| < 1$. If we integrate this last series on the interval $[0, x]$, where $x < 1$, we have

$$\int_0^x \frac{t^2}{(1-t)^2}\,dt = \frac{x^3}{3} + \frac{2x^4}{4} + \frac{3x^5}{5} + \frac{4x^6}{6} + \cdots + \frac{n+1}{n+3}x^{n+3} + \cdots.$$

This also holds for x such that $-1 < x < 0$, since in this case

$$\int_0^x \frac{t^2}{(1-t)^2}\,dt = -\int_x^0 \frac{t^2}{(1-t)^2}\,dt = -\left[\sum_{n=0}^{\infty}\int_x^0 (n+1)t^{n+2}\,dt\right]$$

$$= -\left[\sum_{n=0}^{\infty}(-1)\left(\frac{n+1}{n+3}\right)x^{n+3}\right] = \sum_{n=0}^{\infty}\frac{n+1}{n+3}x^{n+3}.$$

37.8 PRACTICE Find the function given by the power series $\sum_{n=1}^{\infty} nx^n$.

Throughout our discussion of convergence in this section we have said nothing about what happens at the endpoints of a finite interval of convergence. While differentiating a convergent power series does not change the radius of convergence, it may affect the convergence at the endpoints.

37.9 EXAMPLE The series

$$f(x) = \sum_{n=2}^{\infty}\frac{x^n}{n(n-1)}$$

has radius of convergence $R = 1$. In fact, it converges for all x in $[-1, 1]$. (See Example 32.1.) When we differentiate this, we have

$$f'(x) = \sum_{n=2}^{\infty}\frac{x^{n-1}}{n-1} = \sum_{n=1}^{\infty}\frac{x^n}{n}$$

which converges only for x in $[-1, 1)$. Differentiating again we obtain

$$f''(x) = \sum_{n=1}^{\infty} x^{n-1} = \sum_{n=0}^{\infty} x^n$$

which now converges only for x in $(-1, 1)$.

Theorem 37.1 shows that a power series is uniformly convergent on any compact interval within the interior of its interval of convergence. If the interval of convergence includes an endpoint, we might hope that it would also converge uniformly on a compact interval that included this endpoint. Our hope is justified by the following result due to Niels Abel (1802–1829).

37.10 THEOREM Let $\sum_{n=0}^{\infty} a_n x^n$ be a power series with a finite positive radius of convergence R. If the series converges at $x = R$, then it converges uniformly on the interval $[0, R]$. Similarly, if the series converges at $x = -R$, then it converges uniformly on $[-R, 0]$.

Proof: We shall prove the case when $R = 1$ and the series converges at $x = 1$. The general case follows readily from this and is left to the reader (Exercise 37.13). Our strategy is to show that the Cauchy criterion of Theorem 35.10 is satisfied for the partial sums. That is, given $\varepsilon > 0$ we must find a number N such that

$$|a_m x^m + a_{m+1} x^{m+1} + \cdots + a_{m+k} x^{m+k}| < \varepsilon$$

for all $x \in [0, 1]$ and all $m > N$ and $k \geqslant 0$. Since the series is convergent at $x = 1$, Theorem 32.6 implies that there exists a number N such that

$$|a_m + a_{m+1} + \cdots + a_{m+k}| < \varepsilon$$

whenever $m > N$ and $k \geqslant 0$. Fix $m > N$ and for each integer $j \geqslant 0$ let

$$s_j = a_m + a_{m+1} + \cdots + a_{m+j}.$$

Then $|s_j| < \varepsilon$ for all $j \geqslant 0$. Now

$$\begin{aligned}
a_m x^m &+ a_{m+1} x^{m+1} + \cdots + a_{m+k} x^{m+k} \\
&= s_0 x^m + (s_1 - s_0) x^{m+1} + \cdots + (s_k - s_{k-1}) x^{m+k} \\
&= s_0(x^m - x^{m+1}) + s_1(x^{m+1} - x^{m+2}) + \cdots \\
&\quad + s_{k-1}(x^{m+k-1} - x^{m+k}) + s_k x^{m+k} \\
&= x^m(1 - x)(s_0 + s_1 x + \cdots + s_{k-1} x^{k-1}) + s_k x^{m+k}.
\end{aligned}$$

Thus for all $x \in [0, 1)$ we have

$$\begin{aligned}
|a_m x^m &+ a_{m+1} x^{m+1} + \cdots + a_{m+k} x^{m+k}| \\
&\leqslant x^m(1 - x)(\varepsilon + \varepsilon x + \cdots + \varepsilon x^{k-1}) + \varepsilon x^{m+k} \\
&= \varepsilon x^m(1 - x)(1 + x + \cdots + x^{k-1}) + \varepsilon x^{m+k} \\
&= \varepsilon x^m(1 - x^k) + \varepsilon x^{m+k} = \varepsilon x^m < \varepsilon.
\end{aligned}$$

This also holds when $x = 1$, since $|s_k| < \varepsilon$. Thus the series converges uniformly on $[0, 1]$. ∎

One useful application of Abel's theorem (37.10) is that it implies (using Theorem 36.1) that, when a power series converges at an endpoint of its interval of convergence, the function represented by the series is continuous at the endpoint.

37.11 COROLLARY Let $f(x) = \sum_{n=0}^{\infty} a_n x^n$ have a finite positive radius of convergence R. If the series converges at $x = R$, then f is continuous at $x = R$. If the series converges at $x = -R$, then f is continuous at $x = -R$.

Proof: This follows directly from Theorem 37.10 and 36.1. ∎

37.12 EXAMPLE In Example 36.6 we saw that the series

$$f(x) = \sum_{n=0}^{\infty} (-1)^n \frac{x^{n+1}}{n+1} = x - \frac{x^2}{2} + \frac{x^3}{3} - \frac{x^4}{4} + \cdots$$

is equal to $\log(1+x)$ for $x \in (-1, 1)$. Now at $x = 1$ we obtain the alternating harmonic series, which is convergent. Applying Corollary 37.11, we conclude that

$$\begin{aligned} \log 2 &= \lim_{x \to 1} \log(1+x) \\ &= \lim_{x \to 1} f(x) \\ &= f(1) = 1 - \frac{1}{2} + \frac{1}{3} - \frac{1}{4} + \cdots. \end{aligned}$$

ANSWERS TO PRACTICE PROBLEMS

37.6

$$\begin{aligned} f(x) &= \sin x & f(0) &= 0 \\ f'(x) &= \cos x & f'(0) &= 1 \\ f''(x) &= -\sin x & f''(0) &= 0 \\ f^{(3)}(x) &= -\cos x & f^{(3)}(0) &= -1 \\ &\vdots \end{aligned}$$

In general, $f^{(2n)}(0) = 0$ and $f^{(2n+1)}(0) = (-1)^n$. Since $|\sin x| \leqslant 1$ and $|\cos x| \leqslant 1$ for all real x, it follows that

$$|R_n(x)| \leqslant \frac{|x|^{n+1}}{(n+1)!} \qquad \text{for } n \in \mathbb{N} \text{ and } x \in \mathbb{R}.$$

Thus $\lim_{n \to \infty} R_n(x) = 0$ for all real x, and we have

$$\sin x = \sum_{n=0}^{\infty} \frac{(-1)^n}{(2n+1)!} x^{2n+1} \qquad \text{for all } x \in \mathbb{R}.$$

37.8 Since $(1-x)^{-2} = \sum_{n=0}^{\infty} (n+1)x^n$ as in Example 30.5, we have

$$x(1-x)^{-2} = \sum_{n=0}^{\infty} (n+1)x^{n+1} = \sum_{n=1}^{\infty} nx^n.$$

EXERCISES

37.1 (a) Find the function given by the series $\sum_{n=1}^{\infty} n^2 x^n$ for $|x| < 1$.
(b) Evaluate $\sum_{n=1}^{\infty} n^2/2^n$ and $\sum_{n=2}^{\infty} n^2/2^n$.

37.2 (a) Find the function given by the series $\sum_{n=0}^{\infty} x^{2n+1}$ for $|x| < 1$.
(b) Find the exact sum of the series $2^{-1} + 2^{-3} + 2^{-5} + 2^{-7} + \cdots$.

37.3 Find the function given by the series $\sum_{n=0}^{\infty} (-1)^n x^{2n+2}/(2n+2)$ for $|x| < 1$.

37.4 Show that $\log(1 + x^4) = \sum_{n=1}^{\infty} (-1)^{n-1}x^{4n}/n$ for $|x| < 1$.

37.5 (a) Show that $1/(1 + x^2) = \sum_{n=0}^{\infty} (-1)^n x^{2n}$ for $|x| < 1$.
(b) Show that $\arctan x = \sum_{n=0}^{\infty} (-1)^n x^{2n+1}/(2n + 1)$ for $|x| < 1$.
(c) Show that the series for $\arctan x$ in part (b) also holds when $x = 1$.
(d) Use part (c) to find a series whose sum is π.

37.6 (a) Show that

$$\int_0^x \arctan t \, dt = \sum_{n=1}^{\infty} \frac{(-1)^{n-1}x^{2n}}{2n(2n-1)}, \qquad \text{for } |x| < 1.$$

(b) Show that the formula in part (a) also holds when $x = 1$.
(c) Assuming that the series $1 - \frac{1}{2} - \frac{1}{3} + \frac{1}{4} + \frac{1}{5} - \frac{1}{6} - \frac{1}{7} + + - - \cdots$ is convergent, use part (b) to find its value.

37.7 (a) Show that

$$\int_0^x \log(1 + t) \, dt = \sum_{n=1}^{\infty} \frac{(-1)^{n-1}x^{n+1}}{n(n+1)}, \qquad \text{for } |x| < 1.$$

(b) Show that the formula in part (a) also holds when $x = 1$.
(c) Use part (b) to show that

$$\frac{1}{1\cdot 2} - \frac{1}{2\cdot 3} + \frac{1}{3\cdot 4} - \cdots = -1 + \log 4.$$

37.8 Let $f(x) = (e^x - 1)/x$ for $x \neq 0$.
(a) Find a power series $\sum a_n x^n$ that represents f for $x \neq 0$.
(b) Find a power series that represents f' for $x \neq 0$.
(c) Show that $\sum_{n=1}^{\infty} n/(n + 1)! = 1$.

37.9 Find the Taylor series for each function at $x_0 = 0$.
(a) e^{2x} (b) e^{x^2}
(c) e^{-x^2} (d) $\int_0^x e^{-t^2}\, dt$

37.10 Find the Taylor series for each function at $x_0 = 0$.
(a) $\sin 2x$ (b) $\cos x$
(c) $\sin x^2$ (d) $x \sin 3x^2$

37.11 Show that, for each $x \in \mathbb{R}$, $\lim_{n\to\infty} R_n(x) = 0$ in Example 37.5.

37.12 Let $f(x) = x^2$ for $x < 0$ and $f(x) = x^3$ for $x \geqslant 0$. Does f have a Taylor series at $x_0 = 0$? Justify your answer.

37.13 Complete the proof of Theorem 37.10 by establishing the general case.

37.14 Let $f(x) = e^{-1/x^2}$ for $x \neq 0$ and $f(0) = 0$. Show that f has derivatives of all orders at 0 and that $f^{(n)}(0) = 0$ for all $n \in \mathbb{N}$. Thus the Taylor series for f at $x_0 = 0$ does not represent f for any $x \neq 0$.

37.15 The Bessel function of order zero of the first kind may be defined by

$$J_0(x) = \sum_{n=0}^{\infty} \frac{(-1)^n x^{2n}}{4^n (n!)^2}.$$

(a) Find the radius of convergence of $J_0(x)$.
(b) Show that $y = J_0(x)$ is a solution of the differential equation $xy'' + y' + xy = 0$.

References

BLUMENTHAL, L. M. 1940. A Paradox, a Paradox, a Most Ingenious Paradox, *Am. Math. Monthly*, **47**(6), 346–353.

COX, R. H. 1968. A Proof of the Schroeder–Bernstein Theorem, *Am. Math. Monthly*, **75**(5), 508.

HAMILTON, A. G. 1982. *Numbers, Sets and Axioms*, Cambridge University Press, New York.

HENKIN, LEON, W. NORMAN SMITH, VERNE J. VARINEAU, and MICHAEL J. WALSH. 1962. *Retracing Elementary Mathematics*, Macmillan, New York.

OLMSTED, JOHN M. H. 1962. *The Real Number System*, Prentice-Hall, Englewood Cliffs, N.J.

RUDIN, WALTER. 1976. *Principles of Mathematical Analysis*, 3rd ed., McGraw-Hill, New York.

STEWART, IAN, and DAVID TALL. 1977. *The Foundations of Mathematics*, Oxford University Press, New York.

Hints for Selected Exercises

Section 1

1.1 (a) H is not a normal subgroup.
(b) The set of real numbers is not finite.
(c) Either Bob is not over 6 feet tall or Bill is not over 6 feet tall.
(d) Seven is not prime and 5 is odd.
(e) Today is not Monday and it is not hot.
(f) K is closed and bounded and not compact.

1.3 (a) Antecedent: I get there first; consequent: I raise the flag.
(b) Antecedent: normality; consequent; regularity.
(c) Antecedent: You can climb the mountain; consequent: You have the nerve.
(d) Antecedent: $x = 5$: consequent: $f(x) = 2$.

1.5 (a)

p	q	$p \Rightarrow$	$\sim q$
T	T	F	F
T	F	T	T
F	T	T	F
F	F	T	T

(c) p	q	$[p \Rightarrow$	$(q$	$\wedge$	$\sim q)]$		$\Leftrightarrow$	$\sim p$
T	T	T	F	T	F	F	T	F
T	F	T	F	F	F	T	T	F
F	T	F	T	T	F	F	T	T
F	F	F	T	F	F	T	T	T

1.7 (a) True; (b) true; (c) false; (d) true; (e) true; (f) false; (g) false; (h) true.

Section 2

2.1 (a) All pencils are not blue.
(b) Some chairs do not have four legs.
(c) $\forall x > 1, f(x) \neq 3$.
(d) $\exists x$ in $A \ni \forall y$ in B, $y \leqslant x$ or $y \geqslant 1$.
(e) $\exists x \ni \forall y \exists z \ni x + y + z > xyz$.

2.3 (a) True; (b) false; (c) false; (d) true; (e) true.

2.5 (a) $\forall x, f(-x) = f(x)$; (b) $\exists x \ni f(-x) \neq f(x)$.

2.7 (a) $\forall x$ and $\forall y$, $x \leqslant y \Rightarrow f(x) \leqslant f(y)$.
(b) $\exists x$ and $\exists y \ni x \leqslant y$ and $f(x) > f(y)$.

2.9 (a) $\forall x$ and $\forall y$, $f(x) = f(y) \Rightarrow x = y$.
(b) $\exists x$ and $\exists y \ni f(x) = f(y)$ and $x \neq y$.

2.11 (a) $\forall \varepsilon > 0, \exists \delta > 0 \ni \forall x \in D, |x - c| < \delta \Rightarrow |f(x) - f(c)| < \varepsilon$.
(b) $\exists \varepsilon > 0 \ni \forall \delta > 0, \exists x \in D \ni |x - c| < \delta$ and $|f(x) - f(c)| \geqslant \varepsilon$.

2.13 (a) $\forall \varepsilon > 0 \exists \delta > 0 \ni \forall x \in D, 0 < |x - c| < \delta \Rightarrow |f(x) - L| < \varepsilon$.
(b) $\exists \varepsilon > 0 \ni \forall \delta > 0, \exists x \in D \ni 0 < |x - c| < \delta$ and $|f(x) - L| \geqslant \varepsilon$.

Section 3

3.1 (a) If some violets are not blue, then some roses are not red.
(b) If H is not normal, then H is regular.
(c) If K is not compact, then either K is not closed or K is not bounded.

3.7

(a)	$r \Rightarrow \sim s$	hypothesis
	$\sim s \Rightarrow \sim t$	contrapositive of hypothesis: 3.12(c)
	$r \Rightarrow \sim t$	by 3.12(1)
(b)	$\sim t \Rightarrow (\sim r \vee \sim s)$	contrapositive of hypothesis: 3.12(c)
	$\sim r \vee \sim s$	by 3.12(h)
	$\sim s$	by 3.12(j)
(c)	$r \vee \sim r$	by 3.12(d)
	$\sim s \vee \sim t$	by 3.12(o)
	$\sim s \vee u$	by 3.12(o)
	$\sim s \Rightarrow \sim v$	contrapositive of hypothesis: 3.12(c)
	$\sim v \vee u$	by 3.12(o)

Section 4

4.1 Let $n = -2$. Then

$$n^2 + \frac{3}{2}n = 4 + \left(\frac{3}{2}\right)(-2) = 4 - 3 = 1,$$

as required. The integer *is* unique.

4.3 Let x be a real number greater than 3 and let $y = 2x/(3 - x)$. Then $3 - x < 0$ and $2x > 0$, so $y < 0$. Furthermore,

$$\frac{3y}{2 + y} = \frac{3\left(\dfrac{2x}{3 - x}\right)}{2 + \left(\dfrac{2x}{3 - x}\right)} = \frac{6x}{2(3 - x) + 2x} = \frac{6x}{6} = x,$$

as required.

4.5 Use a proof by contradiction.

4.7 Suppose that $x^2 + x - 6 \geqslant 0$ and $x > -3$. It follows that $(x - 2)(x + 3) \geqslant 0$ and, since $x + 3 > 0$, it must be that $x - 2 \geqslant 0$. That is, $x \geqslant 2$.

4.9 Find a counterexample.

4.11 Consider two cases depending on whether n is odd or even.

Section 5

5.1 (a) True; (b) false; (c) true; (d) true; (e) false; (f) true; (g) true; (h) true.

5.7 Suppose that $U = A \cup B$ and $A \cap B = \varnothing$. If $x \in A$, then since $A \cap B = \varnothing$, $x \notin B$. But $A \subseteq U$, so $x \in U$. Thus $x \in U \backslash B$. Therefore, $A \subseteq U \backslash B$.

On the other hand, if $x \in U \backslash B$, then $x \in U$ and $x \notin B$. Since $x \in U$, $x \in A \cup B$. But $x \notin B$, so we must have $x \in A$. Thus $U \backslash B \subseteq A$.

Since $A \subseteq U \backslash B$ and $U \backslash B \subseteq A$, we conclude that $A = U \backslash B$.

5.11

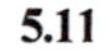

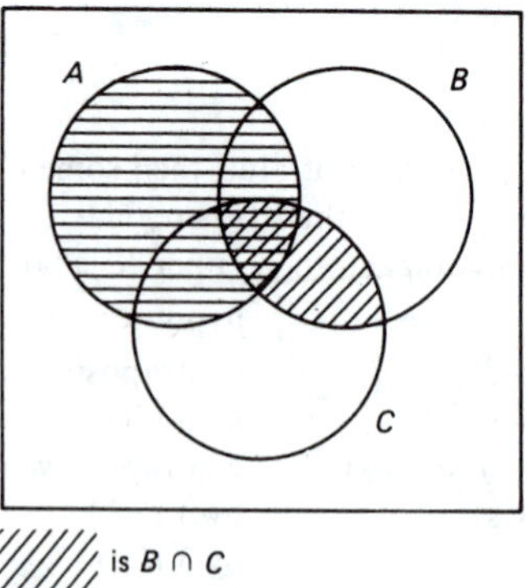

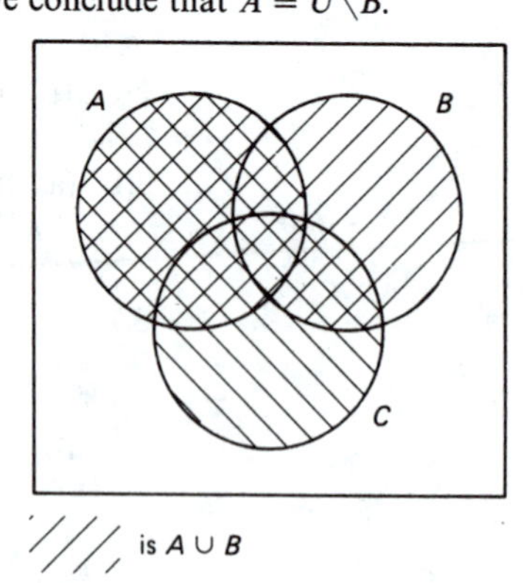

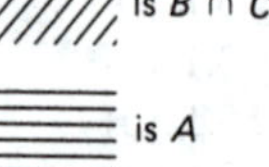

is $B \cap C$

is A

Total shaded area is $A \cup (B \cap C)$

is $A \cup B$

is $A \cup C$

is $(A \cup B) \cap (A \cup C)$

5.15 (a) $\bigcup_{B \in \mathscr{B}} B = [1, 2]$, $\bigcap_{B \in \mathscr{B}} B = \{1\}$.

Section 6

6.3 Proof: Let $(x, y) \in (A \cap B) \times C$. Then $x \in \underline{A \cap B}$ and $y \in \underline{C}$. Since $x \in A \cap B$, $x \in \underline{A}$ and $x \in \underline{B}$. Thus $(x, y) \in \underline{A \times C}$ and $(x, y) \in \underline{B \times C}$. Hence $(x, y) \in (A \times C) \cap (B \times C)$, so $\underline{(A \cap B) \times C \subseteq (A \times C) \cap (B \times C)}$.

On the other hand, suppose that $(x, y) \in \underline{(A \times C) \cap (B \times C)}$. Then $(x, y) \in \underline{A \times C}$ and $(x, y) \in \underline{B \times C}$. Since $(x, y) \in A \times C$, $x \in \underline{A}$ and $y \in \underline{C}$. Since $(x, y) \in B \times C$, $\underline{x \in B}$ and $\underline{y \in C}$. Thus $x \in A \cap B$, so $\underline{(x, y) \in (A \cap B) \times C}$ and $\underline{(A \times C) \cap (B \times C) \subseteq (A \cap B) \times C}$.

6.5 (a) Reflexive, transitive; (d) all three.

6.7 $E_{(a, b)}$ is a vertical line through the point (a, b).

Section 7

7.1 (a) $[1, \infty)$.
(c) $f(x) = (x + 2)^2 - 3$, so the range is $[-3, \infty)$.

7.3 Suppose that domain $f =$ domain g and $f(x) = g(x)$ $\forall\, x \in$ domain f. We must show that f and g (as sets of ordered pairs) are equal. If $(x, y) \in f$, then $y = f(x)$. But then $y = g(x)$, so $(x, y) \in g$. Thus $f \subseteq g$. Similarly, if $(x, y) \in g$, then $y = g(x) = f(x)$, so $(x, y) \in f$. Hence $g \subseteq f$, and we conclude that $f = g$.

The converse is trivial.

7.5 It is clear that $g \circ f$ is a relation between A and C. To show it is a function, suppose that $(x, y) \in g \circ f$ and $(x, y') \in g \circ f$. Then prove that $y = y'$.

7.7 (a) There are nine different functions: six are injective and none are surjective.

7.9 (d) Theorem: $f(C_1 \cup C_2) = f(C_1) \cup f(C_2)$.

Proof: Let $y \in f(C_1 \cup C_2)$. Then $\exists\, x \in C_1 \cup C_2 \ni f(x) = y$. Since $x \in C_1 \cup C_2$, $x \in C_1$ or $x \in C_2$. If $x \in C_1$, then $f(x) \in f(C_1)$. If $x \in C_2$, then $f(x) \in C_2$. In either case, $y = f(x) \in f(C_1) \cup f(C_2)$.

Conversely, suppose that $y \in f(C_1) \cup f(C_2)$. Then $y \in f(C_1)$ or $y \in f(C_2)$. If $y \in f(C_1)$, then $\exists\, x_1 \in C_1 \ni f(x_1) = y$. But then $x_1 \in C_1 \cup C_2$, so $y = f(x) \in f(C_1 \cup C_2)$. On the other hand, if $y \in f(C_2)$, then $\exists\, x_2 \in C_2 \ni f(x_2) = y$. Then $x_2 \in C_1 \cup C_2$, so $y = f(x_2) \in f(C_1 \cup C_2)$. Thus in both cases we have $y \in f(C_1 \cup C_2)$. ∎

7.13 (c) If f is injective, then equality holds.

7.17 (b) Use Exercise 7.3.

7.19 Suppose that S has at least two elements, say s_1 and s_2. Define two functions $f: S \to S$ and $g: S \to S$ in such a way that $f \circ g \neq g \circ f$.

Section 8

8.1 (b) Let $f(x) = 1/(n+1)$ if $\exists\, n \in \mathbb{N} \ni x = 1/n$, and $f(x) = x$ otherwise.

8.3 Given a bijection $f\colon S\backslash T \to T\backslash S$, define $g(x) = f(x)$ if $x \in S\backslash T$ and $g(x) = x$ if $x \in S \cap T$.

8.9 Suppose that S is an infinite set. Then S is not empty, so $\exists\, x_1 \in S$. Since S is not finite, $\exists\, x_2 \in S\backslash\{x_1\}$. Proceeding in this manner, given distinct points $x_1, \ldots, x_n$, since S is not finite the set $S\backslash\{x_1, \ldots, x_n\}$ is not empty and we can choose $x_{n+1} \in S\backslash\{x_1, \ldots, x_n\}$. Now define $f\colon \mathbb{N} \to S$ by $f(n) = x_n$. Clearly, f is injective, so $f(\mathbb{N})$ is a denumerable subset of S. (This proof uses the axiom of choice in a subtle way. See Section 9.)

8.11 Outline of proof: Suppose first that T is a proper subset of S and that there exists an injection $f\colon S \to T$. Obtain a bijection $g\colon S \to T$ as follows: Let f^0 be the identity function on S and $\forall\, k \in \mathbb{N}$ and $\forall\, x \in S$, define $f^k(x) = f[f^{k-1}(x)]$. Then let $B = \bigcup_{n=0}^{\infty} f^n(S\backslash T)$. Define $g\colon S \to T$ by $g(x) = f(x)$ if $x \in B$ and $g(x) = x$ if $x \in S\backslash B$. Observe that $S\backslash T \subseteq B$, $f(B) \subseteq B$, and if $m \neq n$, then $f^m(S\backslash T) \cap f^n(S\backslash T) = \varnothing$. From this conclude that $g(S) = (S\backslash B) \cup f(B) = S\backslash(S\backslash T) = T$. Since f is injective, g is bijective. Finally, generalize to the case when T is not a subset of S. [This proof comes from Cox (1968).]

8.17 (e) Let $S = \mathbb{N} \cup (0, 1)$. Since $\mathbb{N} \cap (0, 1) = \varnothing$, $|S| = \aleph_0 + c$. On the other hand, $\mathbb{R} \sim (0, 1) \subseteq S$ and $S \sim S \subseteq \mathbb{R}$, so $S \sim \mathbb{R}$ and $|S| = c$.

Section 9

9.1 Use Exercise 7.18.

9.3 Let $S = \{y\colon x \subseteq y\}$ and let $T = \bigcup\{y\colon y \in S\}$. For any set W show that $\exists$ a set $z \ni w \in z$ and $x \subseteq z$. From this conclude that T is the "set of all sets."

9.5 Apply the axiom of regularity to the set $\{x, y\}$.

9.7 The mathematician's name was Sam—short for Samantha. Do you see why?

9.9 (b) Let S be the set of points with polar coordinates $(1, n)$, where $n = 0, 1, 2, \ldots$. [Recall that when (r, θ) is the polar coordinate of a point then r is the distance from the origin and θ is the radian measure of the angle between the positive x-axis and the ray from the origin through the point (r, θ).] Then S is a subset of the unit circle centered at the origin. The set $T = S\backslash\{(1, 0)\}$ is congruent to S since a counterclockwise rotation of S through 1 radian will make S coincide with T.

(c) See the article by Blumenthal (1940).

Section 10

10.3 Use Exercise 10.2 and Example 10.3.

10.5 $1 + r + r^2 + \cdots + r^k + r^{k+1} = \dfrac{1 - r^{k+1}}{1 - r} + r^{k+1} = \dfrac{1 - r^{k+2}}{1 - r}.$

10.7 $5^{2k+2} - 1 = 5^{2k+2} - 5^2 + 5^2 - 1 = 5^2(5^{2k} - 1) + 5^2 - 1.$

10.9 (a) True for $n = 1$ and $n \geqslant 4$. For $n = 1$ we have $1^2 \leqslant 1!$, which is true. For $n = 4$ we have $4^2 = 16 \leqslant 24 = 4!$, which is also true. Now suppose that $k^2 \leqslant k!$ for some $k \geqslant 4$. Then $k + 1 = k(1 + 1/k) \leqslant k^2$ since $1 + 1/k \leqslant k$. Thus $(k+1)^2 = (k+1)(k+1) \leqslant (k+1)(k^2) \leqslant (k+1)(k!) = (k+1)!$. It follows from Theorem 10.6 that $n^2 \leqslant n!$ for all $n \geqslant 4$.

10.11 Let $P(n)$ satisfy the hypotheses of the principle of strong induction, and let $Q(n)$ be the statement "$P(j)$ is true for $1 \leqslant j \leqslant n$." Then use Theorem 10.2 to show that $Q(n)$ is true for all $n \in \mathbb{N}$.

10.13 As a check on your answer, $t_{12} = 364$.

10.15 Adapt the proof of Theorem 10.2.

10.17 (a) Let $n \in P$ and let $M = \{m \in P\colon n + m \text{ is defined}\}$. Then $1 \in M$ by D1. Now suppose that $k \in M$ so that $n + k$ is defined. Then by D2, $n + k'$ is defined to be $(n + k)'$, so $k' \in M$. Thus by axiom P4, $M = P$ and $n + m$ is defined for all $n, m \in P$.

(b) Let $M = \{n \in P\colon n + 1 = 1 + n\}$. Then certainly $1 \in M$. Now suppose that $k \in M$, so that $k + 1 = 1 + k$. Then $1 + k' = (1 + k)' = (k + 1)' = (k')' = k' + 1$ by D2 and the induction hypothesis. Thus $k' \in M$, and by axiom P4, $M = P$. Hence $n + 1 = 1 + n$ for all $n \in P$.

(c) Let $M = \{n \in P\colon m' + n = (m + n)' \ \forall\, m \in P\}$.

(d) Let $M = \{m \in P\colon n + m = m + n, \forall\, n \in P\}$. *Caution*: You cannot use the associative property since it has not been proved yet.

(e) Let $M = \{p \in P\colon (m + n) + p = m + (n + p) \ \forall\, m, n \in P\}$.

Section 11

11.1 (a) From A5 we have $x + (-x) = 0$. Then $(-x) + x = 0$ by A2. Hence $x = -(-x)$ by the uniqueness of $-(-x)$ in A5.

(e) Consider the two cases $x > 0$ and $x < 0$.

(f) Use part (e).

(j) Use induction.

(k) Use $(\sqrt{y} - \sqrt{x}) = (y - x)/(\sqrt{y} + \sqrt{x})$.

11.3 Consider the four cases where x and y are both positive, both negative, or have opposite signs.

11.5 Use induction.

11.9 (a) Let $P = \{f \in F\colon f > 0\}$. Show that P satisfies the properties of Exercise 11.6 and then apply Exercise 11.7.

(b) $-x^3 < -x + 3 < 5 < x + 2 < x^2$.

11.11 Define $(a, b) \sim (c, d)$ iff $a + d = b + c$.

Section 12

12.1 (a) 3; (c) 4; (e) 1; (g) 1; (i) no sup; (k) 1; (m) 5.

12.7 (a) Apply the well-ordering property to the set $\{m \in \mathbb{N}\colon m > y\}$.

12.9 Use the binomial theorem (Exercise 10.14) to generalize the proof of Theorem 12.11.

12.11 Let $S = \{q \in \mathbb{Q}: q < x\}$. Then S is bounded above x and we can let $y = \sup S$. Prove that $y = x$ by showing that $y < x$ and $y > x$ both lead to contradictions.

12.13 Let $S = \{m \in \mathbb{N}: ma > b\}$. Then S is nonempty by the Archimedean property and thus has a least element, call it n, by the well-ordering property (10.1). Now let $q = n - 1$. Then q is either 0 or $q \in \mathbb{N}$. Consider both cases.

Section 13

13.1 (a) $\varnothing$; (b) $(0, 5)$; (c) $\varnothing$; (d) $\varnothing$; (e) $\varnothing$.

13.5 $A \backslash B = A \cap (\mathbb{R} \backslash B)$.

13.7 Look at neighborhoods.

13.9 Suppose that some neighborhood of x contains only a finite number of points of S and then find a smaller deleted neighborhood that misses S completely.

13.11 Construct an argument similar to the proof of Theorem 13.17(b).

Section 14

14.1 (c) Let $A_n = \left(\frac{1}{n}, 2\right) \forall n \in \mathbb{N}$.

14.5 Use Theorem 13.13.

Section 15

15.1 (a) For the triangle inequality, use Theorem 12.7.

15.3 (a) For the triangle inequality, consider two cases: (1) $d(x, z) < 1$ and $d(z, y) < 1$, and (2) $d(x, z) \geqslant 1$ or $d(z, y) \geqslant 1$.

15.7 (a) Show that $S \backslash \{x\}$ is an open set.

15.9 (c) It is possible to find an example in $\mathbb{R}$ with the absolute value metric.

15.11 First show that for all $\mathbf{x}, \mathbf{y} \in \mathbb{R}$ we have $d_2(\mathbf{x}, \mathbf{y}) \leqslant d(\mathbf{x}, \mathbf{y}) \leqslant d_1(\mathbf{x}, \mathbf{y})$ and $d_1(\mathbf{x}, \mathbf{y}) \leqslant \sqrt{2} d(\mathbf{x}, \mathbf{y}) \leqslant 2d_2(\mathbf{x}, \mathbf{y})$.

Section 16

16.1 (a) 1, 4, 9, 16, 25, 36, 49; (c) $\frac{1}{2}, -\frac{1}{2}, -1, -\frac{1}{2}, \frac{1}{2}, 1, \frac{1}{2}$.

16.3 (d) $\sqrt{n}/(n+1) < \sqrt{n}/n = 1/\sqrt{n}$.

(f) Suppose first that $x \neq 0$. Since $|x| < 1$, $|x| = 1/(1 + y)$ for some $y > 0$. Use Bernoulli's inequality (Exercise 10.10) to obtain the bound $|x^n| = 1/(1 + y)^n < 1/(ny)$. The case when $x = 0$ is trivial.

16.5 (a) True, the proof follows from the inequality in Exercise 11.4; (b) false; (c) true.

16.9 Given $\varepsilon > 0$, $\exists N_1 \ni n > N_1$ implies that $|a_n - b| < \varepsilon$. In particular, $b - \varepsilon < a_n$ whenever $n > N_1$. Similarly, $\exists N_2 \ni n > N_2$ implies that $|c_n - b| < \varepsilon$, so that $c_n < b + \varepsilon$ whenever $n > N_2$. Now let $N = \max\{N_1, N_2\}$. Then for $n > N$ we have $b - \varepsilon < a_n \leqslant b_n \leqslant c_n < b + \varepsilon$, so that $|b_n - b| < \varepsilon$. Hence $b_n \to b$.

16.11 (a) Suppose that (s_n) is a sequence in $S \backslash \{x\} \ni s_n \to x$. Given any deleted neighborhood $N^*(x; \varepsilon) \exists N \ni n > N$ implies that $|s_n - x| < \varepsilon$. In particular, if k is the smallest integer $> N$, then $|s_k - x| < \varepsilon$. Since $s_k \neq x$, $s_k \in N^*(x; \varepsilon)$. Hence x is an accumulation point of S.

Conversely, suppose that x is an accumulation point of S. Given any $\varepsilon > 0 \, \exists$ a point in $S \cap N^*(x; \varepsilon)$. Thus for each $n \in \mathbb{N}$ there exists a point, say s_n, in $S \cap N^*(x; 1/n)$. Clearly, $s_n \in S \backslash \{x\} \, \forall n$, and we claim $s_n \to x$. Given any $\varepsilon > 0$, let $N = 1/\varepsilon$. Then for any $n > N$ we have $s_n \in N^*(x; 1/n)$ so that $|s_n - x| < 1/n < 1/N = \varepsilon$. Note that this half of the argument uses the axiom of choice (see Section 9).

Section 17

17.1 (a) 3/7; (b) 0.

17.3 (a) Converges to -2; (c) diverges with no limit; (e) diverges to $+\infty$; (g) converges to 0; (i) diverges to $+\infty$; (k) converges to 0.

17.13 (a) Multiply and divide by $\sqrt{n+1} + \sqrt{n}$.

17.15 (a) For $k > 0$, use the ratio test.

17.17 (c) Suppose that $\lim s_n = +\infty$. Given $M \in \mathbb{R}$, $\exists N \ni n > N$ implies that $s_n > -M$. Then for $n > N$ we have $-s_n < M$, so that $\lim(-s_n) = -\infty$. The converse is similar.

Section 18

18.1 The limits are (a) 5/3; (b) 5/3; (c) 5/2; (d) $1 + \sqrt{2}$; (e) $5 + \sqrt{8}$.

18.3 Show that $(s_{n+1})^2 - x = (s_n^2 - x)^2/(45n^2) \geqslant 0$, so that $s_n \geqslant \sqrt{x}$ for $n \geqslant 2$. To prove that (s_n) is decreasing for $n \geqslant 2$, show that $s_n - s_{n+1} \geqslant 0$.

18.9 Show that $|s_{n+2} - s_{n+1}| \leqslant k^n |x_2 - x_1|$. Note that for $m > n$, $s_m - s_n = \sum_{i=n}^{m-1} (s_{i+1} - s_i)$. Then use the triangle inequality and Exercise 10.5 to get a bound on $|s_m - s_n|$.

Section 19

19.1 (a) $S = \{-1, 1\}$, $\lim\sup s_n = 1$, $\lim\inf s_n = -1$.
(b) $S = \{0\}$, $\lim\sup t_n = \lim\inf t_n = 0$.
(c) $S = \{-\infty, 0\}$, $\lim\sup u_n = 0$, $\lim\inf u_n = -\infty$.
(d) $S = \{-\infty, 0, +\infty\}$, $\lim\sup v_n = +\infty$, $\lim\inf v_n = -\infty$.

19.3 (a) e; (b) e^2; (c) e; (d) $1/e$; (e) $\sqrt{e}$; (f) e.

19.5 Suppose that (s_n) does not converge to S. Then find a convergent subsequence of (s_n) having a limit other than s.

19.7 Use Exercise 16.11(b).

19.9 (a) Use Theorem 19.10.

Section 20

20.1 (a) 2; (b) 2; (c) 1/2; (d) 2; (e) 0; (f) 1/4; (g) -4; (h) 2.

20.3 (c) Let (s_n) be a positive sequence converging to c and use Example 17.6.

20.5 (a) and (b) do not exist; in (c), the limit is 0.

20.7 To prove that (a) $\Rightarrow$ (b), suppose that (b) is false. Let (s_n) be a sequence in D with $s_n \neq c \; \forall \, n$ such that $s_n \to c$. Since (b) is false, $(f(s_n))$ converges to some value, say L. Before we can use Theorem 20.8, we must show that given *any* sequence (t_n) in D with $t_n \neq c \; \forall \, n$ such that $t_n \to c$, we have $\lim f(t_n) = L$. We only know from the negation of (b) that $(f(t_n))$ is convergent. To see that $\lim f(t_n) = L$, consider the sequence $(u_n) = (s_1, t_1, s_2, t_2, \ldots)$ and note that $(f(s_n))$ and $(f(t_n))$ are both subsequences of $(f(u_n))$.

20.9 Use Exercise 16.9.

20.11 Use Exercises 16.8 and 16.12.

20.13 To show that $\lim_{x \to c} f(x)$ does not exist for $c \neq 0$, look at a sequence of rational numbers converging to c. Then look at a sequence of irrational numbers converging to c.

20.15 First show that $f(nx) = nf(x) \; \forall \, n \in \mathbb{N}$ and $\forall \, x \in \mathbb{R}$, so that $f(y/n) = f(y)/n \; \forall \, n \in \mathbb{N}$ and $\forall \, y \in \mathbb{R}$. Then show that if $\lim_{x \to 0} f(x)$ exists, it is equal to 0. Also show that $f(x) = f(x - c) + f(c)$, so that $f(x) - f(c) = f(x - c)$, $\forall \, x, c \in \mathbb{R}$.

Section 21

21.1 Define $f(5) = 6$.

21.3 Adapt Example 21.8 so that the two parts of the function "come together" at one point.

21.5 True.

21.7 f is also continuous at 3.

21.9 Show that $\forall \, m, n \in \mathbb{R}$, $\max\{m, n\} = \frac{1}{2}(m + n) + \frac{1}{2}|m - n|$. Do this by considering the two cases $m \geqslant n$ and $m < n$.

21.11 Let $\alpha = \frac{1}{2}f(c)$.

21.13 Use Exercise 16.11(b).

21.15 See the hint for Exercise 20.15. Show that for any rational number m/n, $f(m/n) = (m/n)f(1)$.

Section 22

22.1 Only (g) and (i) are true. Note that (g) is true for all functions. To prove (i), first show that a nonempty subset S of $\mathbb{R}$ is an interval iff $[a, b] \subseteq S$ whenever $a, b \in S$ and $a < b$. Do this by considering cases depending on whether or not S is bounded above or bounded below.

22.5 Consider the function $g(x) = f(x) - x$.

22.9 Use Theorem 21.2(c).

22.13 (a) $\forall\, x \in D\, \exists$ a neighborhood U_x of x and a number $M_x \ni f$ is bounded by M_x on U_x. Now $\{U_x\colon x \in D\}$ is an open cover for D. Since D is compact $\exists\, x_1, \dots, x_n$ in $D \ni D \subseteq U_{x_1} \cup \cdots \cup U_{x_n}$. Let $M = \max \{M_{x_1}, \dots, M_{x_n}\}$. Then f is bounded by M on D.

22.15 Use Exercise 22.9.

Section 23

23.1 (a), (c), (d), and (g) are uniformly continuous.

23.3 Note that $f(x) = \sqrt{x}$ is uniformly continuous on $[0, 2]$ by Theorem 23.5. Then prove f is uniformly continuous on $[1, \infty)$ using Definition 23.1. Given $\varepsilon > 0$, let δ_1 be the δ that works on $[0, 2]$ and let δ_2 be the δ that works on $[1, \infty)$. Then let $\delta_3 = \min\{1, \delta_1, \delta_2\}$. Show that δ_3 satisfies Definition 23.1 on $[0, \infty)$.

23.7 True.

23.9 Suppose that $f(D)$ is not bounded. Then $\exists$ a sequence (s_n) in $D \ni |f(s_n)| \geqslant n$ for each n. Now use Theorem 19.6 and 23.7.

Section 24

24.1 (a) $s_n \to (0, 5)$ (c) $s_n \to (0, 2)$

24.5 Use Theorems 24.7 and 24.9.

24.9 (c) Follow the same strategy as in Example 24.4 to show that $|d(x, D) - d(y, D)| \leqslant d(x, y)\ \forall\, x, y \in S$.

24.11 Prove the contrapositive.

Section 25

25.3 (a) Note that $x - c = (x^{1/3} - c^{1/3})(x^{2/3} + c^{1/3}x^{1/3} + c^{2/3})$.

25.5 (b) $f'(x) = 2x$ if $x < -1$ or $x > 1$, and $f'(x) = -2x$ if $-1 < x < 1$.
(d) $f'(x) = 2|x|$ for all real x.

25.7 (a) Since $f(x)$ is defined differently for positive and negative x, you have to use Definition 25.1.
(c) f' is continuous on $\mathbb{R}$, but is not differentiable at $x = 0$.

25.11 $(fgh)' = (fg)h' + (fh)g' + (gh)f'$.

Section 26

26.1 (a) Strictly increasing on $[\frac{3}{2}, 2]$ and strictly decreasing on $[0, \frac{3}{2}]$.
(b) Maximum is $f(0) = 4$ and minimum is $f(\frac{3}{2}) = \frac{7}{4}$.

26.7 Use Exercise 26.4.

26.9 (a) and (b) Use the mean value theorem.
(c) Use parts (a) and (b) and the intermediate value theorem.

26.13 Use the mean value theorem with the function $g - f$.

26.19 (c) Consider $f(x) = |x|$.

Section 27

27.1 (a) 1; (b) 0; (c) $-\frac{1}{6}$; (d) 0; (e) 1; (f) e^3; (g) e; (h) 0.

27.5 Adapt the proof of Theorem 20.8.

27.7 Use the sequential criterion in Exercise 27.5.

27.11 (a) $\lim_{x \to c} f(x) = \infty$ iff for every $M \in \mathbb{R} \ \exists \ \delta > 0 \ni f(x) > M$ whenever $x \in D$ and $0 < |x - c| < \delta$.

Section 28

28.1 (a) $1.5 < \sqrt{3} < 2.0$; (c) $1.0950 < \sqrt{1.2} < 1.0955$.

28.3 (a) $p_6(x) = x - x^3/6 + x^5/120$; (b) within 1/5040.

28.5 $\cos 1 \approx 0.54167$ with error less than 0.0014.

28.7 (a) Use L'Hospital's rule and Exercise 25.14(b).
(b) Consider $f(x) = x|x|$.

28.9 Taylor's theorem implies that for $x \in I$ we have $f(x) = f(x_0) + f^{(n)}(c)(x - x_0)^n/n!$ for some point c between x_0 and x. Since $f^{(n)}(x_0) \neq 0$ and $f^{(n)}$ is continuous, $\exists$ a neighborhood U of $x_0 \ni f^{(n)}(x)$ and $f^{(n)}(x_0)$ have the same sign $\forall \, x \in U$. Now consider cases.

Section 29

29.1 For each $n \in \mathbb{N}$ consider the partition $P_n = \{0, b/n, 2b/n, \ldots, (n-1)b/n, b\}$. Then use Example 10.3.

29.3 (a) The formula in Exercise 10.2 will be helpful. We obtain $L(f, P_n) = (1 - 1/n)^2/4$.

29.7 Use Exercise 21.11 to prove the contrapositive.

Section 30

30.1 Use Exercise 12.5.

30.3 Use Exercise 29.7.

30.7 (a) Note that $fg = [(f + g)^2 - f^2 - g^2]/2$.

30.11 Use Exercise 30.10 with the function $h = f - g$.

30.13 (a) If $g(x) = 0$ for all x, then the result is trivial. If $g(x) > 0$ for some $x \in [a, b]$, apply Exercise 30.4 to obtain $\int_a^b g > 0$. If $m \leqslant f(x) \leqslant M$ for all $x \in [a, b]$, then $m \int_a^b g \leqslant \int_a^b (fg) \leqslant M \int_a^b g$. Divide by $\int_a^b g$ and use the intermediate value theorem (22.6).

30.15 Use Exercise 29.10.

Section 31

31.1 Let $G(x) = \int_a^x f$ for $x \in [a, b]$.

31.3 (a) $F'(x) = \sqrt{1 + x^2}$.
(b) $F'(x) = 2\sqrt{1 + x^2}$.
(c) $F'(x) = [\cos(\sin^2 x)] \cos x$.
(d) $F'(x) = 2x\sqrt{1 + x^4} - \sqrt{1 + x^2}$.

31.5 $F''(x) = 2x^2 e^{x^2} + 2e^{x^2}$.

31.7 Let $F(x) = \int_a^x f'$ for all $x \in [a, b]$.

31.11 Look for a function whose derivative is not bounded on the interval.

31.13 $\int_a^b f(x)\,dx = \int_0^8 \sqrt{x + 1}\,dx = \frac{52}{3}$.

31.15 (a) Diverges to $-\infty$; (c) 1.

Section 32

32.1 (a) This may seem trivial, but there is really something to prove. Recall that the value of $\sum_{n=m}^{\infty} a_n$ is not computed as a "sum" but as a limit of a sequence. You cannot just add $(a_1 + a_2 + \cdots + a_{m-1})$ to both sides of the equation and change the lower limit of the left summation. You have to look at the sequence of partial sums. The point of the exercise is to justify future use of what appears to be just a simple algebraic manipulation.

32.3 (a) $\frac{1}{2}$; (c) $\frac{4}{3}$; (e) 1; (g) $\frac{1}{3}$; (i) $\frac{1}{4}$. In (i), look at the partial sums and note that $2/[n(n + 1)(n + 2)] = 1/[n(n + 1)] - 1/[(n + 1)(n + 2)]$. In (j), look at the partial sums and try to break up the terms as in (i).

32.5 Use Theorem 32.6.

32.7 Rationalize the denominator and look at the partial sums.

32.9 Use Theorem 32.6.

Section 33

33.1 (a), (e), (m), and (o) converge; (c), (g), (i), and (k) diverge.

33.3 (a), (e), and (i) converge conditionally; (c) converges absolutely; (g) diverges.

33.5 Look at the sequence (b_n/a_n).

33.7 Consider the series $\sum (a_{n+1} - a_n)$.

33.9 If the original series is $\sum (-1)^{n+1} a_n$, consider the series $\sum a_{2n-1}$ of positive terms and the series $\sum a_{2n}$ of negative terms. Use them to show that the sequence of partial sums of the original series is unbounded.

33.13 Given a conditionally convergent series $\sum a_n$, consider the two series $\sum p_n$ and $\sum q_n$, where $p_n = (|a_n| + a_n)/2$ and $q_n = (|a_n| - a_n)/2$.

33.15 If a series is conditionally convergent, its positive terms by themselves form a divergent series, as do the negative terms by themselves. (See Exercise 33.13.) Use this observation together with Theorem 32.5 to obtain the desired rearrangements.

Section 34

34.1 (a) $C = (-1, 1)$; (c) $C = [-\frac{1}{2}, \frac{1}{2})$; (e) $C = \mathbb{R}$; (g) $C = [-1, 1)$; (i) $C = [\frac{8}{9}, \frac{10}{9})$.

34.3 (a) $\frac{1}{4}$; (c) e.

34.7 (a) 2^k; (b) $2^{1/k}$; (c) 1.

Section 35

35.1 Uniform.

35.3 (a) $f(0) = 0$ and $f(x) = 1$ for $x > 0$.
(b) If $x \geqslant t$, then

$$\left| \frac{nx}{1+nx} - 1 \right| = \frac{1}{1+nx} \leqslant \frac{1}{1+nt} \to 0 \qquad \text{as } n \to \infty.$$

(c) Given $n \in \mathbb{N}$, if $0 < x < 1/n$, then

$$\left| \frac{nx}{1+nx} - 1 \right| = \frac{1}{1+nx} > \frac{1}{1+1} = \frac{1}{2}.$$

35.5 (a) $f(x) = 1$ for $x \in [0, 1)$, $f(1) = \frac{1}{2}$.
(c) Given $n \in \mathbb{N}$, if $(\frac{1}{2})^{1/n} < x < 1$, then $\frac{1}{2} < x^n < 1$, so

$$\left|\frac{1}{1+x^n} - 1\right| = \frac{x^n}{1+x^n} > \frac{\frac{1}{2}}{1+1} = \frac{1}{4}.$$

35.9 (a) Use Practice 17.8.
(b) Take the derivative of each f_n in order to find its maximum in [0, 2].

35.11 (b) $f_n'(x) = nx^{n-1}[n - (n+1)x] = 0$ when $x = 0$ or $x = n/(n+1)$. Since $f_n(n/(n+1)) = 1$ for all n, the convergence is not uniform.

35.13 (a) Since (f_n) converges to f uniformly on S, $\exists\, N \ni n > N$ implies that $|f_n(x) - f(x)| < 1\ \forall\, x \in S$. Fix $n_0 > N$. Since f_{n0} is bounded on S, $\exists\, K \ni |f_{n0}(x)| < K\ \forall\, x \in S$. Thus $\forall\, x \in S$,

$$|f(x)| \leqslant |f(x) - f_{n0}(x)| + |f_{n0}(x)| \leqslant 1 + K.$$

Hence f is bounded on S.

35.15 Use Exercise 35.13.

35.17 (a), (c), and (g) converge uniformly; (e) does not.

Section 36

36.3 (b) $\int_0^{\pi}(\frac{1}{3})\, dx = \pi/3$.

36.5 To show uniform convergence use Exercise 33.10.

36.7 Use Theorem 35.11 for the first part and Exercise 35.13 for the second part.

36.9 $2(x + x^3/3 + x^5/5 + \cdots)$.

36.11 Follow the approach used in the proof of Theorem 36.1.

Section 37

37.1 (a) $(x + x^2)(1 - x)^{-3}$; (b) 6 and $\frac{11}{2}$

37.3 $\frac{1}{2}\log(1 + x^2)$.

37.5 (d) $\pi = 4(1 - \frac{1}{3} + \frac{1}{5} - \frac{1}{7} + \cdots)$.

37.9 (a) $\sum_{n=0}^{\infty} \frac{2^n x^n}{n!}$; (c) $\sum_{n=0}^{\infty} \frac{(-1)^n x^{2n}}{n!}$.

37.11 Use Theorem 17.7.

Index